# Dynamic Analysis of Enzyme Systems

# Dynamic Analysis of Enzyme Systems

## An Introduction

Katsuya Hayashi
Naoto Sakamoto

With 318 Figures and 68 Tables

Springer Japan KK

Katsuya Hayashi : Faculty of Agriculture, Kyushu University, Fukuoka, 812, Japan
Naoto Sakamoto : Institute of Information Sciences and Electronics, University of Tsukuba, Ibaraki, 305, Japan

© Springer Japan 1986

Originally published by Springer-Verlag Berlin Heidelberg New York Tokyo in 1986
Softcover reprint of the hardcover 1st edition 1986

ISBN 978-3-662-11584-8          ISBN 978-3-662-11582-4 (eBook)
DOI 10.1007/978-3-662-11582-4

Sole distribution rights outside Japan granted to Springer Japan KK

# Preface

This book is concerned with a quantitative analysis of dynamic behavior of various enzymatic reaction systems by computer simulation. The authors and coworkers have been engaged in cooperative research since 1975, seeking to clarify the catalytic and regulatory characteristics of enzymatic reactions *in vivo* and control mechanisms suitable for enzyme technology. Rather than "enzyme kinetics" generally known in enzymology, this research has employed an approach called "enzyme dynamics" which concentrates on the exact schematic representation of an actual reaction mechanism, derivation of rate equation on the basis of the scheme, and computer simulation of its dynamic behavior (numerical solution of the rate equation and explanation of kinetic and regulatory properties of the enzymatic reaction).

A rate equation representing the behavior of enzymatic reactions is generally expressed by a set of nonlinear differential equations. The analytic solution of rate equations is therefore impossible in general, making it necessary to introduce some approximations in order to analyze the experimental data in enzyme kinetics. For example, under an assumption of excess substrate against enzyme in a closed system, we commonly use the linear approximation for the early period of reaction, the quasi-steady-state approximation based on putative maintenance of steady state in enzyme species, and the rapid-equilibrium approximation assuming instantaneous equilibration in complex formation and between complexes. The kinetic characteristics obtained by these approximations do not always reflect the dynamic behavior of actual enzymatic reactions. Furthermore, enzymatic reactions in *in vivo* systems and reactors in enzyme technology operate in open systems, often violating the underlying conditions for the approximations. This may also be true for some cases of multi-enzyme systems in a closed system.

A new approach to the analysis of dynamic behavior of actual enzyme systems is desired and has been developed as enzyme dynamics in conjunction with the progress in molecular description of biochemical processes and simulation techniques using computers. Quantitative analysis of dynamic behavior of biochemical reactions in the cellular environments leads to an elucidation of the relationship between the structure and function of biological systems on the molecular basis, which is one of the most important goals in biology. Dynamic analysis by computer simulation will now furnish the biochemical research in this direction with a potent and quantitative means for understanding the *in vivo* behavior of enzymatic reaction and its regulation at the molecular level. The dynamic simulation can readily be extended to the systems exploited in biochemical and biomedical engineering.

Our cooperative research has resulted in the development of a computer program of numerical integration suitable for stiff nonlinear differential equations, dynamic analysis of basic enzymatic reactions and complex enzyme systems, and formulation of optimization methods applicable to estimation of reaction schemes and determination of reaction control processes. The research results were published by Japan Scientific Societies Press in 1981 in the form of a book in Japanese titled "Dynamics of Enzymatic Reactions" with the support of a grant-in-aid for publication of scientific research results from the Ministry of Education, Science and Culture of Japan. The English translation of that edition serves as the basis for the present volume with the results of our more recent research added to revise and expand the content. Our publication objective with this edition is to broadly disseminate the methods and applications of the approach of enzyme dynamics and also to encourage research with this approach which will lead to further understanding of the dynamic characteristics of *in vivo* systems of enzymatic reactions and detailed mechanisms of metabolic processes.

The many illustrations of time courses obtained from simulations of important enzymatic reactions in the book provide a comprehensive introduction to the methods and applications of enzyme dynamics for students and researchers in biochemistry, physiology and bioengineering. Accordingly, the book is written and arranged in the form of a textbook rather than as an edited compilation of chapters by individual contributors, although the contents of each chapter are based on the results of contributors' own research. We acknowledge the cooperation of Yoichi Aso and Satoru Kuhara of Kyushu University, Masahiro Okamoto of the University of Michigan, and Yukihiro Eguchi and Kiyokazu Nemoto of the Mitsui Knowledge Industry. In writing we attempt as clear a description as possible and use the same notations throughout the book. Explanations emphasize the analysis of dynamic behavior rather than the mathematical treatment of models and rate equations. As the material reflects the results of research by the contributors, the corresponding results of other re-searchers are not necessarily reviewed or referred to exhaustively.

The introductory chapters deal with the derivation and approximation methods for rate equations of enzymatic reactions. The major portion is devoted to the dynamic simulation and analysis of some fundamental and important systems such as reactions of the Michaelis-Menten-type and allosteric enzymes, linear chain systems with feedback loops, branched re-action system, and complex systems with particular characteristics. The dynamic simulation is performed by numerical integration of the rate equa-tions with the computer procedures and their principle and applicability are described in detail. Analysis emphasizes the dynamic aspects of the sys-tems functioning under various conditions in closed and open systems as well as reaction-diffusion system. The simulation is further applied to the determination of reaction schemes and parameters for some enzyme systems.

Our cooperative research began with a research project of enzyme technology supported by the Office for Life Science Promotion of the Institute of Physical and Chemical Research. We wish to thank Prof. Akiyoshi Wada of the University of Tokyo, the project leader, and Dr. Shotaro Kohtsuki of the Mitsui Knowledge Industry for their encourage-ment and management of the research at that stage. Comments and remarks received from many scientists are appreciated regarding our research results and the Japanese edition.

The publication of this book is supported by a grant-in-aid for publication of scientific research results from the Ministry of Education, Science and Culture of Japan.

Katsuya Hayashi
Naoto Sakamoto

# Contents

## INTRODUCTION

# Role of Dynamic Analysis of Enzyme Systems in Biochemical Research

Intensive studies in biochemistry and molecular biology have sought an elucidation at the molecular level of the relationship between the structures and functions of living systems such as cell organelles, cells, tissues, organs and individual organisms. Progress in the field has revealed that the biological functions all stem from biochemical reactions within the systems. Furthermore, it is now established that such biochemical reactions are integrated into a biochemical network which circulates the flows of matter-energy and information in order to operate and regulate the biological processes responsible for certain functions of the system [1]. In fact, every network of biochemical reactions in a living system is decomposed into their constituent enzyme systems like metabolic pathways, and the sequence of individual reactions in each enzyme system is identified by the specific enzymes and metabolites involved. We have now accumulated a wealth of knowledge on the molecular mechanisms of enzymatic reactions as well as the molecular properties of enzymes and metabolites.

### Objectives of dynamic analysis of enzyme systems

The next stage of the research is concerned with the molecular dynamics of enzyme systems which describes the mechanism for generation of dynamic behavior in terms of molecular structure. That is, we attempt to examine in detail how the molecules of enzymes and metabolites participating in enzymatic reactions are integrated to play their roles in the function and structure of an enzyme system. This investigation will eventually lead to understanding of the mechanisms by which the biochemical networks generate and regulate the biological processes.

This chapter was written by Naoto Sakamoto.

   Thus, it is essential for determination of the relationship between structure and function in a biochemical network that we make a dynamic analysis of the enzyme systems working as network constituents. By dynamic analysis we mean analysis of the dynamic behavior of a system to explain input-output relations in that system. The dynamic behavior of enzyme systems can, in turn, be related to the functions of an entire network by integrating these systems into the network.

   An enzyme system is basically formed of sequential enzymatic reactions and enzymes certainly play most important roles in it. The physico-chemical properties of enzyme molecules have been extensively studied, for example, for their molecular weight and subunit structure, sedimentation and diffusion coefficients, and extinction coefficient. The catalytic activity is characterized by the Michaelis constant and maximum velocity, and further determination of the rate constants themselves is steadily progressing. The catalytic mechanisms are now investigated even at the level of quantum chemistry.

   The analysis of these data on structural and functional properties of enzymes provides fundamental knowledge for dynamic analysis in temporal and spatial dimensions of every individual enzyme in an enzyme system. Hence, we can say that we have reached the stage to start the dynamic analysis of enzyme systems. Furthermore, it is hoped that such analysis will lead to the molecular dynamics of enzyme systems.

**System model and computer simulation**
A powerful procedure for the dynamic analysis of enzyme systems is furnished by a system model and computer simulation. The model represents a postulated molecular mechanism for input-output relations of the enzyme system. Computer simulation of such a model can yield a quantitative description of the relationships between system's dynamic behavior and molecular structure.

   The approach using modeling and simulation can be related to the understanding of structure and function of a biochemical network as shown in Fig.1. "Biochemical network" is a real system to be studied, which is comprised of enzyme systems as structural components of the system. Input-output relations for the functions of the system correspond to the flows of matter-energy and information through the biochemical network. Experiments and observations on a biochemical network are performed to obtain data regarding its structure and dynamic behavior. These data are analyzed to construct a system model for the network which describes the input-output relations of the system. "Modeling" thus is the process of constructing a system model based on an analysis of the data of a real system.

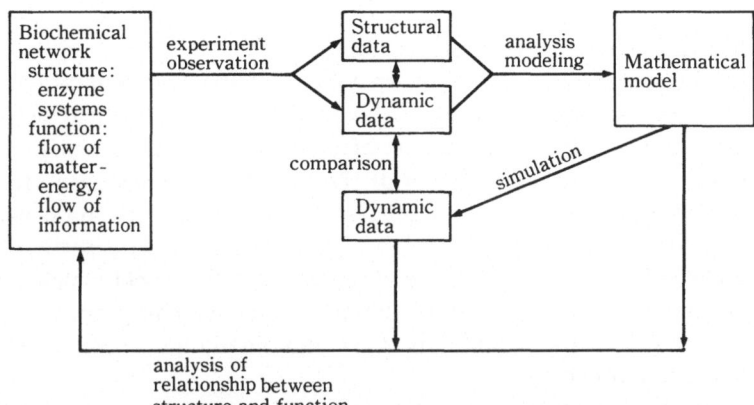

**Fig.1**   Dynamic analysis of biochemical network.

By a system model we mean a mathematical model in general, since it is basically a set of instructions for generating output (*i.e.*, behavioral data of the model) from input through "simulation." The simulation is a computational process which is now carried out commonly by a digital computer according to the model instructions suitably encoded as a program. A mathematical model representing the dynamic behavior of a biochemical network is often expressed by nonlinear differential equations. Its simulation is performed by numerical solution of these equations using computer procedures. Thus, computer simulation or, more specifically, numerical solution is often interchangeable with simulation in our terminology. With respect to the flow of data in Fig. 1, we can say essentially that a real system is a source of behavioral data, supplying "real system data" through experiment and observation, and that a system model is a source of generating other behavioral data, "model-generated data" through simulation [2].

In this approach the authenticity of the system model is of greatest concern, that is, how accurately a model represents the real system of interest. In order to understand how a real system works, we use models to embody hypotheses about the underlying and often inaccessible structure of reality. The criterion for accuracy or validity of a model is the degree of agreement between real system data and model-generated data. A model is "replicatively valid" if it matches data already acquired from the real system, and "predictively valid" if it can provide data before they are acquired from the real system. A model is "structurally valid" if it not only reproduces the observed behavior of the real system, but truly reflects the way in which the real system operates to produce this behavior [2].

### Enzyme dynamics and enzyme kinetics

In the dynamic analysis of enzyme systems, we aim at the complete construction of structurally valid models for biochemical networks. As indicated in Fig. 1, comparison of model-generated data with real system data can lead to a prediction of the unknown structure and function of a biochemical network and a suggestion of experiments and observations to be performed for their detection. The real system data newly obtained are analyzed to construct a more structurally valid model which can generate yet more refined data. With this procedure we can gradually explain the complex structures and functions and finally their relationships in the biochemical networks.

Hence, emphasizing its objectives and procedures, we call the dynamic analysis of enzyme systems "enzyme dynamics." In contrast, so-called "enzyme kinetics" is mainly concerned with the estimation of kinetic parameters and reaction schemes for enzymatic reactions. These static data from the kinetics can be related to the dynamic behavior of systems through the dynamics. An effective accomplishment with the dynamics is naturally dependent on productive experiments and good models.

### Open and closed systems

Most kinetic experiments with individual enzymatic reactions have been performed in a closed system (*i.e.*, in a test tube, or a common laboratory system), and much data are available on the kinetic properties of many enzymes. The biochemical networks of living systems, on the other hand, operate in open system in almost all situations. Hence, the relevant data of dynamic behavior in open system are essential for the dynamic analysis of enzyme systems.

Originally, open and closed systems arise from the concept defined in thermodynamics. The thermodynamic system is defined as a geometrical space in the universe, which is the objective of our interest for experiment or observation. The system is classified in three types with respect to the property of its boundary which exists between the system and its surroundings :

isolated (adiabatic) system : Neither energy nor matter can penetrate the boundary.

closed system : Energy can go through the boundary, but matter cannot.

open system : Both energy and matter are exchangeable through the boundary.

In the isolated and closed systems, equilibrium is the most stable state in which the entropy does not change according to the laws of equilibrium thermodynamics.

An open system, however, has to be treated with nonequilibrium thermodynamics, which describes the state of a system using time and boundary conditions in addition to state variables. In an open system a nonequilibrium state, called a steady state, can be generated when no change in entropy takes place because of the balance between its production in processes of the system and the influx from the exterior through the exchange of energy and matter. At a steady state in an open system neither entropy nor concentrations of chemical species change temporally. Hence, the steady state is also called "dynamic equilibrium." Although the concentrations are constant through time, a molecular exchange of chemical species continuously occurs between the system and the surroundings, because the steady state cannot be maintained without the flow of energy and matter through the boundary.

As mentioned above, the elucidation of relationships between the structures and functions of biochemical networks in living systems requires the accumulation and analysis of data concerning the dynamic behavior of enzyme systems in open system. Such experiments and observations, however, have not been seriously performed, not only because they are not fully recognized as essential, but also because an appropriate design of experimental systems is difficult. At present, dynamic analysis through modeling and simulation would be a rather efficient means of accumulating the behavioral data of enzyme systems in open system. Moreover, the analysis could suggest the experimental systems which produce useful real system data of the behavior in open system.

**Methods for dynamic analysis**
Dynamic analysis is performed for various types of enzyme systems from a single enzymatic reaction to an entire cellular metabolism consisting of thousands of reactions in linear, branched or cyclic pathways. Enzyme systems also display diversified functions such as autocatalysis, feedback and feedforward loops, a coupled reaction system subject to interactive inhibition and activation, and a two-factor system with an output of threshold characteristics. In studying the dynamic characteristics of enzyme systems, their environments such as closed or open, and homogeneous or distributed systems have to be taken into consideration in addition to their variety of structure and function. It is naturally expected that a system behaves differently in various environments and plays different roles.

Finding of a method and/or principle for treating this diversity in a unified manner would lead to the discovery of a fundamental law relating the functions to the structures in living systems. For the time being, however, the diversified enzyme systems are classified into similar groups with respect to structure or function so that each group is analyzed by its most effective method specifically devised. This would currently be the best approach to dynamic analysis, although an ultimate unification should continue to be sought. Indeed, many groups of enzyme systems can be efficiently analyzed employing the method of enzyme dynamics, which is the primary subject of this book.

As mentioned, the construction of mathematical models for enzyme systems plays an essential role in enzyme dynamics. For the process of modeling, it is of basic importance to clearly specify the conditions under which dynamic behavior is to be characterized by the model. First, we have a problem as to whether the reaction process in an enzyme system is to be described as a deterministic or stochastic process in the model. In many cases the deterministic description has been widely applied to the time course of concentration change of chemical species in a system based on the mass-action law.

On the other hand, from the microscopic viewpoint of molecular dynamics, chemical reaction evolves from an efficient collision of reactant molecules, which is a stochastic process at the molecular level. The stochastic approach, however, introduces considerably more variables and parameters in the model than the deterministic approach, resulting in practical difficulty in computation even for the behavior in steady states. If an averaging operation with respect to the number of molecules is done to reduce the computational load, the results can only prove the findings obtained from the deterministic model. Neither the steady-state assumption nor averaging operation are desirable for enzyme dynamics. It is necessary, in any event, that we first choose either the deterministic or stochastic approach to describe the individual enzymatic reactions in the model.

Secondly, the standpoint in modeling is dependent on whether the system is autonomous or nonautonomous. In enzyme systems the activity is regulated to meet one of two types of requirements. One type keeps a constant production of necessary metabolites in order to maintain the homeostasis in living systems. The other leads the time course of the system in a definite direction for the homeoresis such as development, differentiation and morphogenesis. The enzyme systems for the former type are autonomous since the structure and kinetic parameters are constant temporally and the regulatory mechanisms in the system are closed. The enzyme systems for the latter type are nonautonomous; the time term appears explicitly in the model and the structure and kinetic parameters of the system vary with respect to time. The regulatory mechanism governing the nonautonomous system, however, remains unknown as to the manner of change of the structure and kinetic parameters. Kinetic study of these systems is possible solely by imbedding the system in a more extended and complex system or by introducing a macroscopic approximation in the model.

Consequently, this book concentrates on the analytical procedure employing the deterministic and autonomous treatments, which is most commonly applicable to enzyme systems. These treatments have the following characteristics and shortcomings. The first is a problem of nonlinearity. Enzymatic reaction is intrinsically nonlinear owing to the formation mechanism of the enzyme-substrate complex and application of the mass-action law in the deterministic approach. Allosteric enzymes in particular acquire important functions in metabolic regulation from the higher-order nonlinearity due to polymeric structure and cooperativity. Hence, the enzyme system as a whole becomes very highly nonlinear and complex.

The nonlinear differential equations of rate equations represent the time course of the system, and in general have no analytic solutions. Some approximations should be employed to examine the dynamic behavior of the system. Methods for this purpose have been developed over many years and are classified into two kinds. One includes the quasi-steady-state, rapid-equilibrium and linear approximations, which essentially remove the nonlinearity from the rate equation and transform it into an appropriate form for analytical treatment. The other type of approximation is performed by the method for numerical solution of a rate equation in using digital computers.

Since the pioneering research of Michaelis and Menten on enzyme kinetics, approximation by assuming the rapid equilibrium or steady state among enzyme species has been commonly used to transform a rate equation (a system of ordinary differential equations) into a system of algebraic equations, from which the relationship between reaction rate and substrate concentration ($S$-$v$ relationship) is derived. In fact, the Michaelis constant ubiquitously employed for characterizing the kinetic property of enzyme is determined from treatment of the experimental data with a relationship based on the quasi-steady-state approximation. Enzyme kinetics thus aims mainly at an estimation of kinetic parameters and reaction scheme for every enzymatic reaction from the experimental data applying the $S$-$v$ relationship in the form of rational function of substrate concentration.

In reaction kinetics further evaluation is done for rate constants which are correlated more directly to reaction rate than the Michaelis constant. For this the relaxation method has been devised in parallel with the development of experimental instruments for rapid reaction. In this method the rate equation is treated by the linear approximation to derive the relationship between the rate constants and relaxation time of reaction. The rate constants are evaluated from the relaxation time experimentally obtained, and the reaction mechanism is possibly inferred by the existence of reaction intermediates.

The dynamic analysis of enzyme systems by numerical solution of rate equations with a computer has also advanced recently since the development of methods suitable for the stiff differential equations which are peculiar to rate equations. Analysis with numerical solutions (*i.e.*, computer simulation) now becomes a powerful method for enzyme dynamics, which treats the rate equations more precisely than the linear and quasi-steady-state approximations and best reflects the dynamic behavior of enzyme systems. Thus, computer simulation is the main analytical procedure also in this book.

The only difficulty with the simulation is that the derivation of general characteristics from the numerical solutions is not so straightforward as from the analytic solution. The accumulation of numerical solutions under various conditions is required for integration of the data into characterization of the dynamic behavior of the objective system. Therefore, this method demands the most effort and time of those now used for dynamic analysis. The efficient execution of analysis could at present be accomplished by computer simulation coupled with other approximation methods. A dynamic property of an enzyme system which is first revealed by computer simulation, can lead to evaluation of the effective range of linear and quasi-steady-state approximations, so that the analytical treatment in that range can predict the general feature of the property. The validity of generalization can then be examined by computer simulation.

The methods of systems analysis are also important in enzyme dynamics. These methods deal with the rate equation without approximations to derive the dynamic characteristics of a system, instead of analyzing the dynamic behavior from the time courses of a system obtained as the solutions of the rate equation. In contrast to enzyme kinetics, which is most concerned with the estimation of kinetic parameters and reaction schemes, systems analysis is fundamentally based on system dynamics which can examine the stability and oscillatory phenomenon of enzyme systems. The rate equation is thus desirably treated with nonlinearity retained. In fact, analysis with the relationships derived from the linear and quasi-steady-state approximations would lead to some insights into the dynamic characteristics, but it might ignore the nonlinearity inherent in enzyme systems and hence the biological processes. In this respect, a method for nonlinear system dynamics should be developed which suits for dynamic analysis of enzyme systems.

In addition, systems analysis ought to make great use of the analytical procedures based on control theories. From a functional viewpoint, the enzyme system is none other than an optimal or adaptive control system. The analysis of regulatory properties of the system is essential to its dynamic analysis. Techniques have already been developed for optimal and adaptive control problems in engineering systems. These control theories should be applied to the dynamic analysis of enzyme systems. At the same time, it is necessary to develop the optimal and adaptive control theories for nonlinear systems appropriate for the analysis of enzyme systems and biochemical networks.

The procedure employed in this book wherein the dynamic behavior of a system is examined by an analysis of numerical solutions of the rate equation representing the system dynamics eventually becomes very difficult with large-scale systems because of the capacity of current computers. For analysis of large-scale systems, many methods have been proposed from procedures such as graph methods by steady-state approximations, automaton and logical analysis by discrete treatment, and compartmentation methods. However, the actual situation at present is that a procedure exactly fitted to enzyme dynamics is not yet available. A serious future problem may arise concerning the proper treatment of large-scale systems in the analysis of biochemical networks. It is also important to extend the applicability of enzyme dynamics by further developing approaches based on the stochastic processes and reaction-diffusion systems.

**Scope of this book and its position in the research of enzyme dynamics**
This book describes the methods for dynamic analysis of enzyme systems, especially of the kinetic behavior or time courses of reactions, and presents the primary results of analysis of the most important and interesting systems of enzymatic reactions. The derivation of rate equations is first explained with examples of some enzymatic reactions. Analytical methods introduced include procedures of both approximation and numerical solution of rate equations. The main subjects of the book are concerned with the results of microscopic and macroscopic analyses of fundamental systems of enzymatic reactions by numerical methods. Estimation of reaction schemes and kinetic parameters is then mentioned, and the related topics in dynamic analysis are also discussed. The subjects of each chapter are summarized in the following.

The first three chapters are devoted to explanation of the methods for dynamic analysis. In Chapter 1 the basic procedure is introduced for derivation of rate equations by a deterministic approach, that is, according to the mass-action law. The rate equation for a homogeneous system is expressed by a set of nonlinear ordinary differential equations. Behavior in an inhomogeneous system such as reaction-diffusion system is represented by a rate equation in the form of nonlinear partial differential equations. The procedure is applied to some reaction systems to give examples of rate equations.

In Chapter 2 approximation methods are described which have been widely applied to the rate equations of enzymatic reactions. The relationships employed for practical purposes are derived in order to explain the basic concepts in the respective methods. The quasi-steady-state approximation is applied to the Michaelis-Menten-type reactions to derive the relationship between reaction rate and substrate concentration. The corresponding relationship for reactions of allosteric enzymes is obtained from the rapid-equilibrium approximation. Application of the linear approximation to a rate equation yields the relationship used for the relaxation method. Rate equations for reaction-diffusion systems are treated with the steady-state and linear approximations.

In Chapter 3 the methods for numerical solution of rate equations are explained to introduce the main procedure for computer simulation in this book. The principle in the numerical solution of ordinary differential equations is first delineated with reference to the Euler method. The linear multi-step method is then described mainly in association with the Gear method, which is suitable for stiff differential equations and employed for computer simulation in this volume. Some problems are pointed out in application of the numerical methods to solution of the rate equations. Finally, several methods are introduced for numerical solution of partial differential equations representing reaction-diffusion systems.

From Chapter 4 to Chapter 7 the results of computer simulation performed for various enzyme systems are given to reveal their dynamic behavior in closed and open, and homogeneous and inhomogeneous systems. In Chapter 4 the dynamic behavior of individual enzymatic reactions in closed and homogeneous system is analyzed for Michaelis-Menten-type and allosteric reactions. This analysis demonstrates the kinetic behavior of the enzymatic reactions at nonsteady states in the *in vitro* or experimental system. The behavior is especially compared with that obtained from the quasi-steady-state and rapid-equilibrium approximations in order to examine the validity of application of the approximations to the experimental system.

Chapter 5 deals mainly with the allosteric enzymes which play important roles in the functions of enzyme systems. The dynamic behavior is analyzed microscopically for enzyme systems in open system, attempting to examine their actual operation in the *in vivo* cellular environments. Microscopic analysis refers to the procedure in which the rate equation is derived directly from the reaction mechanism and analyzed by computer simulation. The one-step systems (reaction system of a single enzyme) for simulation are the reactions of Michaelis-Menten-type and allosteric enzymes. The two-step system comprised of a Michaelis-Menten-type enzyme and an allosteric enzyme is analyzed under the conditions of negative and positive feedback regulations. The dynamic behavior of more complex systems is also described for branched biosynthetic pathways and coupled reaction systems.

In Chapter 6 macroscopic analysis is performed for enzyme systems of a cyclic reaction system, reaction system with delay, oscillatory reaction system, feedback system with constant output, two-factor system, and system with threshold. This analysis leads to characterization of the systems as a whole by representing the kinetic properties peculiar to the constituent enzymatic reactions in a system with simplified mechanisms.

In Chapter 7 the behavior of reaction-diffusion systems is discussed and the characteristics of enzyme systems in inhomogeneous and open system delineated. The numerical solution of a rate equation (partial differential equations) is obtained and analyzed for a reaction-diffusion system with feedback loop and a partition-chromatographic column of a reacting system.

Another aspect of application of the computer simulation is demonstrated in Chapter 8. The procedure for estimation of a reaction scheme is described and applied to determination of the activation mechanism in pepsinogen and the reaction scheme in $p$-hydroxybenzoate hydroxylase, and to selection of the basic schemes for spike-type oscillation. The combination of optimization and numerical solution of a rate equation leads to the procedure for evaluation of kinetic parameters. The gradient and Powell methods of optimization are explained in conjunction with treatment of the experimental data for evaluation. The rate constants in a model scheme are evaluated for comparison of these two methods. The procedure is also applied to evaluation of the binding free energy of substrates with six subsites of lysozyme.

Finally, some topics related to dynamic analysis of enzyme systems are discussed briefly in Chapter 9: reaction as a stochastic process, control processes in enzyme systems, analysis of large-scale systems, and thermodynamics in biological processes. All have an important association with enzyme dynamics, but their detailed discussion would be beyond the scope and space of this publication.

This book thus intends to inform researchers of the current state and capacity in mathematical analysis of enzyme systems available and effective for experimental research. The analytical procedures for enzyme dynamics  are explained with a special emphasis on the computer simulation and the results of the dynamic analysis of fundamental enzyme systems are described.  If this presentation can provide an understanding of the analytical procedures for experimental data and a suggestion for new experiments in enzyme dynamics, its objectives will have been attained. Moreover, an accomplishment at a higher level is expected from the book if the experimental data on the dynamic behavior of enzyme systems are compared with the time courses obtained from computer simulation to reveal the kinetic characteristics of systems.

As mentioned repeatedly, enzyme dynamics aims at the elucidation of relationships between the structures and functions of enzyme systems.  The

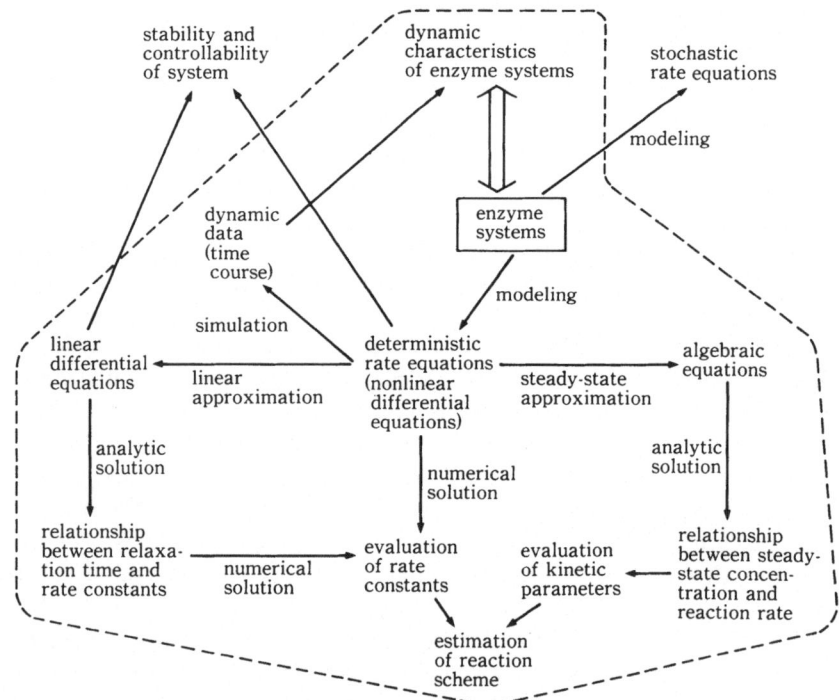

**Fig. 2**   Methods for dynamic analysis.

subjects of enzyme dynamics can be depicted by the relation in the methods for dynamic analysis, as shown in Fig.2. The area described in this book is indicated in the portion bounded by the broken line. The enzyme systems treated in this book are only a few of the most basic systems ; in fact, there exist numerous systems with more complexity and diversity in the actual biochemical networks and living systems. Hence, this book is considered as a starting point in the research of enzyme dynamics.

**Prospects for future research**

We must continue the accumulation of real system data from experiments and model-generated data from computer simulation regarding the dynamic behavior of enzyme systems. Even the dynamic analysis of individual enzymatic reactions is not yet satisfactory, and data on behavior in open system are very few so far. Further accumulation of data is desirable from the simulation with more improved procedures and appropriately designed experiments. The dynamic analysis of enzyme systems will thus advance to gradually characterize the dynamic properties. Consequently, we need to devise a method of integrating these results into approaches to the elucidation of relatioships between the structures and functions in biochemical networks.

Enzyme dynamics will have to resolve the very important problem in the future of how knowledge procured by analyses at the various levels is correlated to the structures and functions in the biochemical network. The stochastic approach to molecular dynamics of enzymatic reactions, the microscopic analysis of deterministic rate equations for enzyme systems, and the analysis of biochemical networks as large-scale systems all generate their own specific results, which are not presently complementary to each other. Therefore, dynamic analysis by computer simulation should be done more extensively with more advanced procedures. In addition, the related approaches in dynamic analysis mentioned in Chapter 9 need to be developed and, most of all, systems analysis should be made available for enzyme dynamics.

Finally, we should point out that development of application techniques of enzymatic reactions has recently been attempted with some fruitful results in the biochemical, pharmaceutical and medical industries. Progress in these fields requires essentially the clarification of kinetic and regulatory characteristics in enzyme systems and biochemical networks. The development of real application techniques cannot be expected without adequate understanding of the dynamic behavior *in vivo* of enzyme systems. Research in enzyme dynamics will advance in the future and yield valuable knowledge for the development in these fields.

# References

[ 1 ]  Miller, J.G. (1978). "Living Systems," McGraw-Hill, New York.
[ 2 ]  Zeigler, B.P. (1976). "Theory of Modelling and Simulation," John Wiley & Sons, New York.

GENERAL

1.  Segel, L.A. ed. (1980). "Mathematical Models in Molecular and Cellular Biology," Cambridge University Press, Cambridge.
2.  Reich, J.G. and E.E. Sel'kov (1981). "Energy Metabolism of the Cell. A Theoretical Treatise," Academic Press, London.
3.  Albert, B., D. Bray, J.Lewis, M.Raff, K.Roberts and J.D. Watson (1983). "Molecular Biology of the Cell," Garland Publishing, New York.

# CHAPTER 1

# Derivation of Rate Equations for Enzymatic Reactions

We can describe the dynamic behavior of a system of enzymatic reactions by a rate equation, where an enzymatic reaction *in vivo* or *in vitro* can be assumed to proceed within the traditional (deterministic) theory of chemical kinetics. This holds true for our analysis throughout this book unless mentioned otherwise. A rate equation is formulated as a set of ordinary differential equations for a homogeneous system in which we need not consider spatial distribution of chemical species, and as a set of partial differential equations for an inhomogeneous system in which chemical species are spatially distributed.

In this chapter we discuss the basic procedure for derivation of rate equations and its application to some reaction systems. Rate equations in the form of ordinary differential equations are treated in Section 1.1. Partial differential equations are derived to represent some inhomogeneous systems in Section 1.2.

Chemical kinetics deals with the rate of chemical change of chemical species in a reaction system. More exactly, the rate is an instantaneous difference in concentration due to chemical reaction. When $X$ denotes the concentration of a certain chemical species X, the reaction rate is expressed mathematically by $dX/dt$, *i.e.,* derivative of concentration with respect to time. Some authors use (X), [X], or other symbols to signify the concentration of X. Throughout this book, however, Roman capital letters identify chemical species and italic capital letters represent their concentrations.

---

This chapter was written by Yoichi Aso and Katsuya Hayashi.

## 1.1 Representation in Ordinary Differential Equations

### 1. Basic Reactions

Let us start with the simplest example of chemical reaction of Scheme 1.1:

$$A \xrightarrow{k_1}$$

Scheme 1.1. Unimolecular first-order reaction

where $k_1$ is a rate constant. This is known as a unimolecular first-order reaction and so-called self-degradation reaction. The changing rate of concentration of A is expressed by

$$\frac{dA(t)}{dt} = -k_1 A(t). \tag{1}$$

Equation (1) is a differential equation with respect to $A(t)$, indicating that a decreasing rate of $A$ is linearly proportional to $A$ at that time. In other words, compound A spontaneously changes into another compound at a rate of $k_1 A(t)$. What we want to know next is the time course of $A$ during the progress of reaction. For this purpose, we have to solve equation (1). Namely, the time course is obtained as a solution of the differential equation. Equation (1) is in a class which has the general form,

$$\frac{dX}{dt} = f(X)g(t). \tag{2}$$

Since equation (2) is expressed formally as

$$\frac{dX}{f(X)} = g(t)\,dt, \tag{3}$$

in which variables $X$ and $t$ are separated on the left- and right-hand sides, respectively, an equation in this class is called a differential equation with separable variables. Integrating the left- and right-hand sides of equation (3) from $X_0$ to $X$ and from $t_0$ to $t$, respectively, we obtain

$$\int_{X_0}^{X} \left( \frac{1}{f(\xi)} \right) d\xi = \int_{t}^{t} g(\tau)\,d\tau + C, \tag{4}$$

where $X_0$ and $t_0$ are the corresponding initial values, and $C$ is an integral constant. In the case of equation (1), we have $g(t) = 1$ and $f(X) = -k_1 X$ so that the solution in integral form is given as

$$\int_{A_0}^{A} \left( \frac{1}{A} \right) dA = -k_1 \int_{t_0}^{t} d\tau + C. \tag{5}$$

By carrying out both integrations, we obtain

$$\ln\left(\frac{A}{A_0}\right) = -k_1(t - t_0) + C,\qquad(6)$$

where $A_0$ is a value of $A$ at $t = t_0$. This is called a general solution, because it still contains an arbitrary constant $C$. Substitution into equation (6) of the relationship of $A = A_0$ at $t = t_0$, which is called an initial condition, yields $C = 0$, and then we can get a special solution,

$$A(t) = A_0\exp\{-k_1(t - t_0)\}.\qquad(7)$$

Taking $t_0 = 0$, we can get a well-known solution,

$$A(t) = A_0\exp(-k_1 t).\qquad(8)$$

Let us now consider a reaction of Scheme 1.2:

$$A + B \xrightarrow{k_1}$$

Scheme 1.2.  Bimolecular second-order reaction

where compounds A and B interactively change into other compounds. This is called a bimolecular second-order reaction. A system of ordinary differential equations,

$$\begin{aligned}\frac{dA}{dt} &= -k_1 AB \\ \frac{dB}{dt} &= -k_1 AB,\end{aligned}\qquad(9)$$

represents the rate equation for the reaction of Scheme 1.2. Each equation of (9) is rewritten in the same form as

$$\frac{dX}{dt} = k_1(A_0 - X)(B_0 - X),\qquad(10)$$

since we can express $A = A_0 - \delta A$, $B = B_0 - \delta B$, and $\delta A = \delta B = X$, where $\delta A$, $\delta B$, and $X$ represent deviations from initial concentrations. Rearranging the variables in the form of equation (3) and integrating both sides, we obtain

$$\int_0^X \frac{dX}{(A_0 - X)(B_0 - X)} = k_1 \int_0^t dt,\qquad(11)$$

where we suppose $X = 0$ at $t = 0$. When $A_0 \neq B_0$, we can transform equation (11) by partial fraction into

$$\int_0^X \frac{dX}{B_0 - X} - \int_0^X \frac{dX}{A_0 - X} = k_1(A_0 - B_0)\int_0^t dt,\qquad(12)$$

to get the relationship:

$$\ln\left(\frac{A(t)}{B(t)}\right) = k_1 (A_0 - B_0) t + \ln\left(\frac{A_0}{B_0}\right). \tag{13}$$

When $A_0 = B_0$, the solution of equation (9) becomes

$$A(t) = B(t) = \frac{A_0}{k_1 A_0 t + 1}. \tag{14}$$

Thus, we can obtain a solution only for the special case of $A_0 = B_0$. When $A_0 \neq B_0$, however, only the relationship between A and B can be obtained.

For general treatment, expanding the right-hand side of equation (10), we have

$$\frac{dX}{dt} = k_1 X^2 - k_1 (A_0 + B_0) X + k_1 A_0 B_0. \tag{15}$$

This is called a nonlinear differential equation, since it contains a power term of $X$ on the right-hand side. An ordinary differential equation is expressed by a general form,

$$F(X) = F\left(t, X, \frac{dX}{dt}, \frac{d^2 X}{dt^2}, \cdots, \frac{d^n X}{dt^n}\right) = 0, \tag{16}$$

and is called a differential equation of order $n$, because the order of the highest derivative in the equation is $n$. For example, equations (1) and (15) are formally expressed as $F(t, A, dA/dt) = 0$ and $F(t, X, dX/dt) = 0$, respectively. Both equations have the same form, but the terms involved are completely different.

Equation (16) is linear if linear relations,

$$\begin{aligned} F(y_1 + y_2) &= F(y_1) + F(y_2) \\ F(cy) &= cF(y), \end{aligned} \tag{17}$$

hold for arbitrary variable $y$ and constant $c$. Equation (1) is obviously linear, while equation (15) is nonlinear. In fact, for equation (1)

$$F\left(A_1 + A_2, \frac{d(A_1 + A_2)}{dt}\right) = F\left(A_1, \frac{dA_1}{dt}\right) + F\left(A_2, \frac{dA_2}{dt}\right) \tag{18}$$

and for equation (15)

$$F\left(x_1 + x_2, \frac{d(x_1 + x_2)}{dt}\right) \neq F\left(x_1, \frac{dx_1}{dt}\right) + F\left(x_2, \frac{x_2}{dt}\right). \tag{19}$$

We need to remember from the theory of differential equations that it is impossible in principle to get an analytic solution of a nonlinear ordinary differential equation.

For a generalized scheme like Scheme 1.3, the law of mass action allows

$$n_1A+n_2B+n_3C+\cdots\cdots\xrightarrow{k}$$

Scheme 1. 3. General representation of reaction

us to derive a general rate equation :

$$\frac{dA}{dt}=\frac{dB}{dt}=\frac{dC}{dt}=\cdots=-kA^{n_1}B^{n_2}C^{n_3}\cdots . \tag{20}$$

When $n=\sum_i n_i$, the reaction of Scheme 1.3 is called the $n$th order reaction. We can now realize that this is a nonlinear differential equation of the order $n$.

In relation to the direction of reactions, we are concerned with whether a system is closed or open (see Introduction and Chapter 9). For example, in Scheme 1.4 the system within the rectangle bounded by the broken line is

$$\longrightarrow A \longrightarrow B \longrightarrow$$

Scheme 1. 4. Reaction in an open system

closed, but it is practically open. The difference between these two systems remarkably affects the results of kinetic analysis, especially if the systems operate *in vivo*. Reactions with only an one-way path like those described above are called irreversible reactions ; on the other hand, a reaction with a two-way path as shown in Scheme 1.5 is called a reversible reaction.

$$A \underset{k_2}{\overset{k_1}{\rightleftarrows}} B$$

Scheme 1. 5. Reversible reaction

When the reaction path from A to B is taken as a forward reaction, the reaction from B to A then becomes a reverse (backward) reaction. The rate equation for Scheme 1.5 is expressed by the differential equations,

$$\frac{dA}{dt}=\left(\frac{dA}{dt}\right)_f+\left(\frac{dA}{dt}\right)_r=-k_1A+k_2B$$
$$\frac{dB}{dt}=\left(\frac{dB}{dt}\right)_f+\left(\frac{dB}{dt}\right)_r=k_1A-k_2B , \tag{21}$$

where the terms with subscripts f and r indicate the rates for forward and reverse reactions, respectively. This set of equations is called simultaneous ordinary differential equations.

## 2. Relatively Complex Reactions
### Consecutive reaction

We consider the consecutive reaction of Scheme 1.6 operating in open

$$A \xrightarrow{k_1} B \xrightarrow{k_2} C \xrightarrow{k_3}$$

Scheme 1.6. Consecutive reaction

system. The rate equation is given by the following simultaneous first-order differential equations,

$$
\begin{aligned}
\frac{dA}{dt} &= -k_1 A \\
\frac{dB}{dt} &= k_1 A - k_2 B \\
\frac{dC}{dt} &= k_2 B - k_3 C .
\end{aligned}
\tag{22}
$$

The relationship,

$$\frac{dA}{dt} + \frac{dB}{dt} + \frac{dC}{dt} = \frac{d(A+B+C)}{dt} = -k_3 C , \tag{23}$$

derived from equation (22) indicates that the total concentration of chemical species in the system decreases at a rate of $k_3 C$.

If C does not leak out of the system, we use Scheme 1.7 instead of Scheme

$$A \xrightarrow{k_1} B \xrightarrow{k_2} C$$

Scheme 1.7. Consecutive reaction in a closed system

1.6. Equations (22) and (23) are changed into

$$
\begin{aligned}
\frac{dA}{dt} &= -k_1 A \\
\frac{dB}{dt} &= k_1 A - k_2 B \\
\frac{dC}{dt} &= k_2 B
\end{aligned}
\tag{24}
$$

and

$$\frac{dA}{dt} + \frac{dB}{dt} + \frac{dC}{dt} = 0 , \tag{25}$$

respectively. Equation (25) reveals that the time course of one of the three species in Scheme 1.7 is completely determined by the other two. If $X$ denotes the total concentration of the three species, which is sometimes given by summation of initial concentrations at the start of reaction, simultaneous differential equations (24) lead to a complex system composed of differential and algebraic equations,

$$\frac{dA}{dt} = -k_1 A$$

$$\frac{dB}{dt} = k_1 A - k_2 B \tag{26}$$

$$C = X - (A + B).$$

This suggests to us the difference in characteristics between open and closed systems. It should be noted that a slight change of Scheme 1.6 significantly affects the form of rate equation.

The discussion can be easily extended to a reaction including $n$ chemical species. For open system we have Scheme 1.8. The rate constant $k_{ij}$ is for

$$\xrightarrow{k_{01}} C_1 \xrightarrow{k_{12}} C_2 \xrightarrow{k_{23}} C_3 \xrightarrow{k_{34}} \cdots \xrightarrow{k_{(n-1)n}} C_n \xrightarrow{k_{n(n+1)}}$$

Scheme 1. 8.   Irreversible consecutive reaction

a reaction path from $C_i$ to $C_j$. $k_{10}$ is the rate of influx into the system, and $k_{n(n+1)}$ is the rate constant of efflux from the system. The rate equation is given by

$$\frac{dC_i}{dt} = k_{ji} C_j - k_{im} C_i, \tag{27}$$

where $i = 1, 2, \cdots, n$, $j = i-1$, $m = i+1$, and $C_0 = 1$. For a reversible reaction

$$\underset{k_{10}}{\overset{k_{01}}{\rightleftarrows}} C_1 \underset{k_{21}}{\overset{k_{12}}{\rightleftarrows}} C_2 \underset{k_{32}}{\overset{k_{23}}{\rightleftarrows}} \cdots \underset{k_{n(n-1)}}{\overset{k_{(n-1)n}}{\rightleftarrows}} C_n \underset{k_{(n+1)n}}{\overset{k_{n(n+1)}}{\rightleftarrows}}$$

Scheme 1. 9.   Reversible consecutive reaction

system of Scheme 1.9, we can derive the rate equation as

$$\frac{dC_i}{dt} = k_{ji} C_j + k_{mi} C_m - (k_{ij} + k_{im}) C_i, \tag{28}$$

where $i = 1, 2, \cdots, n$, $j = i-1$, $m = i+1$, and $C_0 = 1$. When we need a general equation for closed system, which must be composed of $(n-1)$ differential and one algebraic equations, $k_{01}, , k_{10}, k_{n(n+1)}$ and $k_{(n+1)n}$ should be set at zero in equation (28).

We now deal with the solution of simultaneous linear differential equations. As solution methods are easily found in any textbook on differential equations, we here demonstrate just how to reformulate rate equations in order to use such methods. Let us take rate equation (22) as a simple example. Equation (22) is rewritten in matrix form as

$$\frac{d}{dt}\begin{bmatrix} A \\ B \\ C \end{bmatrix} = \begin{bmatrix} -k_1 & 0 & 0 \\ k_1 & -k_2 & 0 \\ 0 & k_2 & -k_3 \end{bmatrix}\begin{bmatrix} A \\ B \\ C \end{bmatrix},$$ (29)

or

$$\frac{dX}{dt} = KX,$$ (30)

where $X$ is a column vector with components of concentrations of chemical species, and $K$ is a matrix with elements of rate constants. Obtaining a general solution to equation (29) or (30) is reduced to the solving of the characteristic equation,

$$\begin{vmatrix} -k_1-\lambda & 0 & 0 \\ k_1 & -k_2-\lambda & 0 \\ 0 & k_2 & -k_3-\lambda \end{vmatrix} = 0,$$ (31)

or

$$(\lambda + k_1)(\lambda + k_2)(\lambda + k_3) = 0,$$ (32)

to determine eigenvalues $\lambda$ and the corresponding eigenvectors. Analysis of the mathematical structure of matrix $K$, especially its eigenvalues, would lead to the kinetic characteristics of reaction represented by linear differential equations. Linear algebra thus is a powerful modern technique used in solving simultaneous linear differential equations.

**Nonlinear consecutive reaction**

For the two-step consecutive reaction with bimolecular and unimolecular

$$A + B \underset{k_2}{\overset{k_1}{\rightleftharpoons}} C \underset{k_4}{\overset{k_3}{\rightleftharpoons}} D$$

Scheme 1.10.   Two-step consecutive reaction

steps of Scheme 1.10, the rate equation is derived as

$$\frac{dA}{dt} = -k_1 AB + k_2 C$$

$$\frac{dB}{dt} = -k_1 AB + k_2 C$$ (33)

$$\frac{dC}{dt} = -(k_2 + k_3)C + k_1 AB + k_4 D$$

$$\frac{dD}{dt} = -k_4 D + k_3 C.$$

Considering that $dA/dt + dC/dt + dD/dt = 0$, $dB/dt + dC/dt + dD/dt = 0$, and $dA/dt = dB/dt$, we can reduce equation (33) to

$$\frac{dA}{dt} = -k_1 AB + k_2 C$$

$$\frac{dC}{dt} = k_1 AB + k_4 D - (k_2 + k_3) C \tag{34}$$

$$A = X_a - (C + D)$$

$$B = X_b - (C + D),$$

where $X_a$ and $X_b$ are constants. Equation (34) is nonlinear, because it contains a product term $AB$ of unknown variables. Thus, we can say that the reaction in Scheme 1.10 is nonlinear.

**Concurrent reaction**

Reaction systems can include concurrent paths as shown in Scheme 1.11.

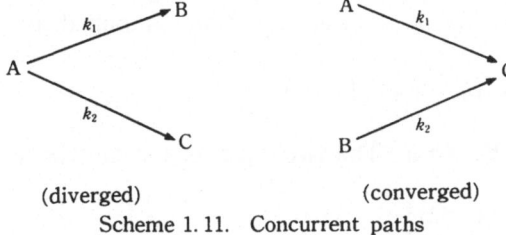

(diverged)            (converged)

Scheme 1.11.  Concurrent paths

Rate equations for the diverged and converged paths are given by

$$\frac{dA}{dt} = -(k_1 + k_2) A$$

$$\frac{dB}{dt} = k_1 A \tag{35}$$

$$\frac{dC}{dt} = k_2 A,$$

and

$$\frac{dA}{dt} = -k_1 A$$

$$\frac{dB}{dt} = -k_2 B \tag{36}$$

$$\frac{dC}{dt} = k_1 A + k_2 B,$$

respectively.  We can find many examples of concurrent paths in bio-chemical reaction networks like metabolic pathways.  In order to derive

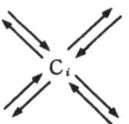

Scheme 1. 12.   A node in a reaction network

rate equation in a generalized form, let us suppose that each species coresponds to one of the nodes in a network, as graphically illustrated in Scheme 1.12. In this case, a general equation is written as

$$\frac{dC_i}{dt} = \sum_{j=1}^{n} (-k_{ij}C_i + k_{ji}C_j) , \qquad (37)$$

where $i=1,2,\cdots,n$, $j \neq i$ and $k_{ij}$ is the rate constant for the path from $C_i$ to $C_j$. The constraint of $j \neq i$ is to prohibit a meaningless path from $C_i$ to itself. If there exists no direct path, the corresponding rate constant is assumed to vanish.  Equation (37) is reduced to a more simplified form,

$$\frac{dC_i}{dt} = \sum_{j=1}^{n} k_{ji}C_j , \quad i=1,2,\cdots,n, \qquad (38)$$

where $k_{ii} = -\sum_{j=1}^{n} k_{ij}$, $j \neq i$.  The two equations in matrix form become

$$\frac{d}{dt}\begin{bmatrix} C_1 \\ C_2 \\ \vdots \\ C_n \end{bmatrix} = \begin{bmatrix} -\sum k_{1j} & k_{21} & \cdots & k_{n1} \\ k_{12} & -\sum k_{2j} \cdots & k_{n2} \\ \vdots & & \vdots \\ k_{1n} & k_{2n} & \cdots -\sum k_{nj} \end{bmatrix}\begin{bmatrix} C_1 \\ C_2 \\ \vdots \\ C_n \end{bmatrix} \qquad (39)$$

and

$$\frac{d}{dt}\begin{bmatrix} C_1 \\ C_2 \\ \vdots \\ C_n \end{bmatrix} = \begin{bmatrix} k_{11} & k_{21}\cdots k_{n1} \\ k_{12} & k_{22}\cdots k_{n2} \\ \vdots & \vdots \quad \vdots \\ k_{1n} & k_{2n}\cdots k_{nn} \end{bmatrix}\begin{bmatrix} C_1 \\ C_2 \\ \vdots \\ C_n \end{bmatrix}, \qquad (40)$$

respectively.   It should be remembered that the matrix form is not attainable if nonlinear reactions are included in a network.

**Cyclic reaction**

Cyclic reaction can also be a component of a reaction network. The simplest example of cyclic reaction with three chemical species is given in Scheme 1.13. We have the rate equation in matrix form as

$$\frac{d}{dt}\begin{bmatrix} A \\ B \\ C \end{bmatrix} = \begin{bmatrix} -(k_1+k_6) & k_2 & k_5 \\ k_1 & -(k_2+k_3) & k_4 \\ k_6 & k_3 & -(k_4+k_5) \end{bmatrix}\begin{bmatrix} A \\ B \\ C \end{bmatrix}. \qquad (41)$$

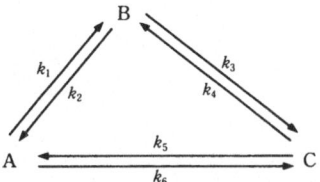

Scheme 1.13. Cyclic reaction

This is the same form as equation (39). Let us now suppose that this reaction finally reaches an equilibrium, at which we have $dA/dt = dB/dt = dC/dt = 0$. That is, we can get the following relationship:

$$
\begin{bmatrix}
-(k_1+k_6) & k_2 & k_5 \\
k_1 & -(k_2+k_3) & k_4 \\
k_6 & k_3 & -(k_4+k_5)
\end{bmatrix}
\begin{bmatrix}
\bar{A} \\
\bar{B} \\
\bar{C}
\end{bmatrix}
=
\begin{bmatrix}
0 \\
0 \\
0
\end{bmatrix},
\tag{42}
$$

where $\bar{A}$, $\bar{B}$ and $\bar{C}$ are the equilibrium concentrations of A, B and C, respectively. Since only two expressions in equation (42) are independent in a mathematical sense, two relationships,

$$
\frac{\bar{A}}{\bar{B}} = \frac{k_2k_4+k_2k_5+k_3k_5}{k_1k_4+k_1k_5+k_4k_6}
$$

$$
\frac{\bar{C}}{\bar{A}} = \frac{k_1k_3+k_2k_6+k_3k_6}{k_2k_4+k_2k_5+k_3k_5},
\tag{43}
$$

are derived. The theory of chemical equilibrium indicates that, when an entire reaction system is at equilibrium, each elementary reaction should also be at equilibrium. This allows us to derive another set of expressions:

$$
\frac{\bar{B}}{\bar{A}} = \frac{k_1}{k_2}
$$

$$
\frac{\bar{C}}{\bar{A}} = \frac{k_6}{k_5}
\tag{44}
$$

$$
\frac{\bar{B}}{\bar{C}} = \frac{k_4}{k_3}.
$$

Substituting these into equation (43), we obtain

$$
k_1k_3k_5 = k_2k_4k_6,
\tag{45}
$$

where we practically use only equation (44) to have

$$
\frac{\bar{A}}{\bar{B}}\frac{\bar{B}}{\bar{C}}\frac{\bar{C}}{\bar{A}} = \frac{k_2}{k_1}\frac{k_4}{k_3}\frac{k_6}{k_5} = 1
\tag{46}
$$

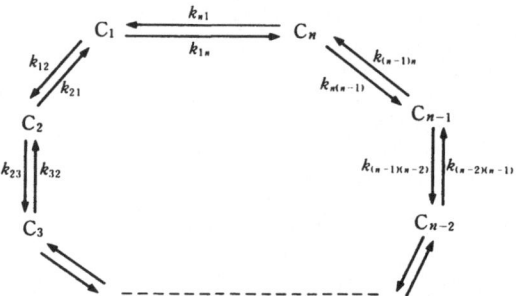

Scheme 1.14. Cyclic reaction formed of $n$ chemical species

for derivation of equation (45). This infers that we can independently change the values of five rate constants at most, for the remaining one should be determined from equation (45). For a cyclic reaction with only one loop formed of $n$ chemical species of Scheme 1.14, the relationship is given by

$$K_1 K_2 \cdots K_{n-1} = K_n, \cdot \tag{47}$$

where equilibrium constants are used instead of rate constants. This is well known as Wegscheider's condition. Using the same notation as above, we have the relationship between rate constants:

$$\frac{k_{12}}{k_{21}} \frac{k_{23}}{k_{32}} \cdot \cdots \cdot \frac{k_{(n-1)n}}{k_{n(n-1)}} = \frac{k_{1n}}{k_{n1}}. \tag{48}$$

**Autocatalytic reaction**

If we represent a catalytic reaction with catalyzer S by Scheme 1.15, the corresponding autocatalytic reaction may have a scheme like Scheme 1.16.

$$S + A \longrightarrow B \longrightarrow C + S$$

Scheme 1.15. Catalytic reaction

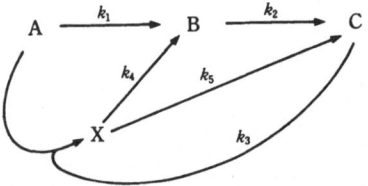

Scheme 1.16. Autocatalytic reaction

In this scheme, C is a product of the reaction as well as a catalyzer to degrade the substrate A. The rate equation becomes, of course, a nonlinear differential equations:

$$\frac{dA}{dt} = -k_1 A - k_3 AC$$
$$\frac{dB}{dt} = k_1 A + k_4 X - k_2 B$$
$$\frac{dC}{dt} = k_2 B + k_5 X - k_3 AC$$
$$\frac{dX}{dt} = k_3 AC - (k_4 + k_5) X .$$

(49)

Supposing that B is degraded at the moment of generation from A, we obtain a simpler scheme (Scheme 1.17). In this case, the rate equation is

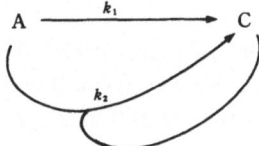

Scheme 1.17. Autocatalytic reaction without complex formation

given by

$$\frac{dA}{dt} = -(k_1 + k_2 C) A$$
$$\frac{dC}{dt} = (k_1 + k_2 C) A .$$

(50)

It follows that, with $A = A_0$ and $C = 0$ at $t = 0$, the generating rate of $C$ is accelerated from $k_1 A_0$ at $t = 0$ after the start of reaction, and hence the concentration of C increases explosively.

## 1.2 Representation in Partial Differential Equations

### 1. Reaction-Diffusion System

In inhomogeneous cellular conditions, enzymes are thought to be localized at the definite positions, and metabolites enter the inside through the cell membrane and move to the reacting sites mainly by free diffusion. Hence, the state variable (concentration) of an enzymatic reaction system in an inhomogeneous medium should be expressed as a function of time and spatial coordinates. In such a reaction–diffusion system, the changing rate of concentration in a metabolite is governed by the velocity of translation due to diffusion and the rate of reaction catalyzed by an enzyme. The reaction term in the rate equation in an inhomogeneous medium has the same form as that in a homogeneous medium. Therefore, it is only necessary to consider the translation of a metabolite by diffusion and the concentration change of the metabolite by reaction at a fixed enzyme position.

Let $C(x, t)$ and $D$ be the concentration and diffusion coefficient of the reactant (metabolite), respectively and $x$ be the spatial coordinate. We assume here that the translation of the reactant occurs in the one-dimensional space. The velocity of translation of the material through a unit area of cross-section is proportional to the concentration gradient at the cross-section. This relation is called Fick's law, and may be represented by

$$F = -D \frac{\partial C(x, t)}{\partial x},$$
(51)

where $F$ denotes the velocity of translation. Application of Fick's law to the fluxes of material into and from the small volume element at $x$ yields the partial differential equation for the diffusion process:

$$\frac{\partial C}{\partial t} = D \frac{\partial^2 C}{\partial x^2}.$$
(52)

When the diffusing material reacts during the translational process, the rate equation may be expressed by

$$\frac{\partial C}{\partial t} = D \frac{\partial^2 C}{\partial x^2} + f(C, t),$$
(53)

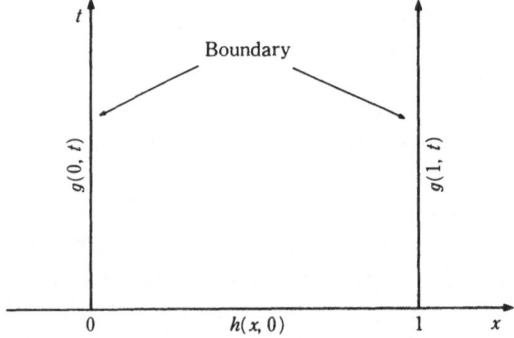

Fig. 1. 1 Boundary and initial conditions.

where $f(C, t)$ represents the rate of reaction in the same form as in a homogeneous medium. If $C(t)$ for a homogeneous medium replaces $C(x, t)$ for an inhomogeneous medium, equation (53) is reduced to

$$\frac{dC}{dt} = f(C, t).$$  (54)

For rate equation (53) we must specify the boundary conditions, $g(0, t)$ and $g(1, t)$, as well as the initial condition $h(x, 0)$ (see Fig.1.1).

*Initial condition*: We take $t \in [0, \infty]$, and $x \in [0, 1]$. If the concentration of reactant is zero in the whole space at $t = 0$ except for the boundaries, the initial condition is represented by

$$h(x, 0) = C(x, 0) = 0, \quad 0 < x < 1.$$  (55)

If the reactant is distributed in the space at $t = 0$, an appropriate function or value is given for $h(x, 0)$.

*Boundary conditions*: At the boundaries two quantities should be specified. One is the concentrations, $C(0, t)$ and $C(1, t)$, and the other is the permeation rate of reactant through the boundary. If the concentration of reactant at the boundary is held constant and the permeability of product is proportional to its concentration, the boundary conditions are expressed by $C(0, t) = C(1, t) = C_0$ (constant) for the reactant, and with a constant $a$

$$\left. \frac{\partial C}{\partial x} \right|_{x=0,1} = aC(x, t) \Big|_{x=0,1}$$  (56)

for the product.

If the product cannot permeate through the boundary, the boundary condition is represented by

$$\left. \frac{\partial C}{\partial x} \right|_{x=0,1} = 0, \quad \text{(Neumann condition)}$$  (57)

and if the concentration of product is held at a certain value $\alpha$ or zero,

$$\begin{aligned} C(0, t) = C(1, t) = \alpha \\ C(0, t) = C(1, t) = 0 \quad \text{(Dirichlet condition)} \end{aligned}$$  (58)

are used, respectively. In this case, the product permeates through the boundary so that the concentration is held at $\alpha$ or 0.

## 2. Bimolecular Reaction

We assume that the half infinite space is full of the solution of diffusible substance B at $t = 0$ and that substance A enters the space through the boundary and reacts with B in bimolecular reaction,

$$A + B \xrightarrow{\ k\ } P,$$

where $k$ is a rate constant. The rate equation and the initial and boundary conditions are represented as follows ;
Rate equation :

$$\frac{\partial A(x,t)}{\partial t} = D_A \frac{\partial^2 A(x,t)}{\partial x^2} - kA(x,t)B(x,t)$$

$$\frac{\partial B(x,t)}{\partial t} = D_B \frac{\partial^2 B(x,t)}{\partial x^2} - kA(x,t)B(x,t) \qquad (59)$$

$$\frac{\partial P(x,t)}{\partial t} = D_P \frac{\partial^2 P(x,t)}{\partial x^2} + kA(x,t)B(x,t)$$

$$t \in [0,\infty], \quad x \in [0,\infty]$$

Initial conditions :

$$\begin{aligned} A(x,0) &= 0, \quad x > 0 \\ B(x,0) &= B_0 \ (\text{constant}), \quad x \geq 0 \\ P(x,0) &= 0, \quad x \geq 0 \end{aligned} \qquad (60)$$

Boundary conditions :

$$\begin{aligned} A(0,t) &= A^* \ (\text{constant}) \\ \frac{\partial B(0,t)}{\partial x} &= 0 \\ \frac{\partial P(0,t)}{\partial x} &= 0 \\ A(\infty,t) &= P(\infty,t) = 0 \\ B(\infty,t) &= B_0. \end{aligned} \qquad (61)$$

## 3.  Enzyme System

We consider an enzyme system in the one-dimensional space of Fig. 1.2, in which enzymes $E_1$ and $E_2$ are fixed at the definite positions, $L_1$ and $L_2$, respectively and substrate S enters the region through boundary (0). The enzymatic reactions proceed according to Scheme 1.18. $C_1$ and $C_2$ are the enzyme-substrate complexes, and $P_1$ and $P_2$ are the products. Product $P_1$ serves as the substrate of $E_2$. The substrate and products are assumed to move in the region by free diffusion. It is further postulated that the substrate at boundary (1) and the products at boundaries (0) and (1) can permeate at rates proportional to their concentrations.

The rate equation and conservation law are represented by

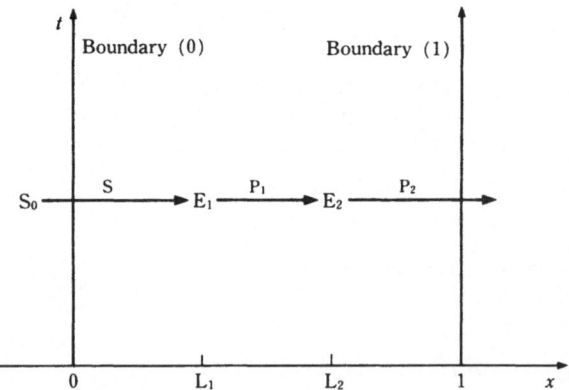

**Fig.1.2** Structure of two-enzyme system.

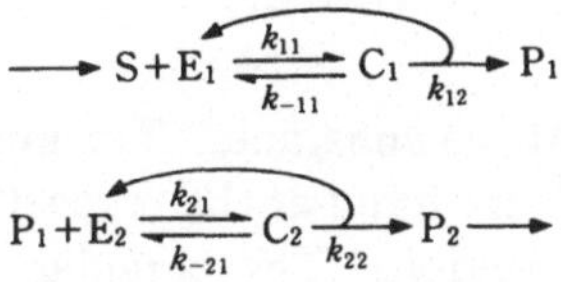

$$\longrightarrow S+E_1 \underset{k_{-11}}{\overset{k_{11}}{\rightleftharpoons}} C_1 \xrightarrow{k_{12}} P_1$$

$$P_1+E_2 \underset{k_{-21}}{\overset{k_{21}}{\rightleftharpoons}} C_2 \xrightarrow{k_{22}} P_2 \longrightarrow$$

Scheme 1.18. Two-enzyme system

$$\frac{\partial S(x,t)}{\partial t} = D_S \frac{\partial^2 S(x,t)}{\partial x^2} - k_{11}S(x,t)E_1(x,t) + k_{-11}C_1(x,t)$$

$$\frac{\partial P_1(x,t)}{\partial t} = D_{P_1} \frac{\partial^2 P_1(x,t)}{\partial x^2} + k_{12}C_1(x,t) - k_{21}P_1(x,t)E_2(x,t)$$

$$+ k_{-21}C_2(x,t) \tag{62}$$

$$\frac{\partial P_2(x,t)}{\partial t} = D_{P_2} \frac{\partial^2 P_2(x,t)}{\partial x^2} + k_{22}C_2(x,t)$$

$$\frac{\partial E_1(x,t)}{\partial t} = -k_{11}S(x,t)E_1(x,t) + k_{-11}C_1(x,t) + k_{12}C_1(x,t)$$

$$\frac{\partial E_2(x,t)}{\partial t} = -k_{21}P_1(x,t)E_2(x,t) + k_{-21}C_2(x,t) + k_{22}C_2(x,t) \tag{63}$$

$$\frac{\partial C_1(x,t)}{\partial t} = k_{11}S(x,t)E_1(x,t) - k_{-11}C_1(x,t) - k_{12}C_1(x,t)$$

$$\frac{\partial C_2(x,t)}{\partial t} = k_{21}P_1(x,t)E_2(x,t) - k_{-21}C_2(x,t) - k_{22}C_2(x,t) \tag{64}$$

$$E_1(x,t) + C_1(x,t) = E_{10}\delta_{x,L_1}$$

$$E_2(x,t) + C_2(x,t) = E_{20}\delta_{x,L_2} \tag{65}$$

where $\delta_{x,L_i}$ is Kronecker's delta and $E_{i0}$ represents the total concentration of enzyme $E_i$. The initial conditions are written as

$$
\begin{aligned}
&S(0,0) = S_0, \quad S(x,0) = 0, \quad 0 < x \leq 1 \\
&P_i(x,0) = 0, \quad i = 1,2, \quad 0 \leq x \leq 1 \\
&C_i(x,0) = 0, \quad i = 1,2, \quad 0 \leq x \leq 1 \\
&E_i(L_i,0) = E_{i0}, \quad i = 1,2 \\
&E_i(x,0) = 0, \quad x \neq L_i, \quad i = 1,2,
\end{aligned}
\tag{66}
$$

and the boundary conditions as

$$
\begin{aligned}
&S(0,t) = S_0 \\
&\frac{\partial P_i(0,t)}{\partial x} = aP_i(0,t), \quad i = 1,2 \\
&\frac{\partial S(1,t)}{\partial x} = bS(1,t) \\
&\frac{\partial P_i(1,t)}{\partial x} = dP_i(1,t), \quad i = 1,2,
\end{aligned}
\tag{67}
$$

where $a$, $b$, and $d$ are positive constants. The initial conditions indicate that at $t=0$ only the enzymes are fixed at the respective positions and that there are no reactants in the region. The boundary conditions mean that the substrate enters the region at a constant rate through boundary (0) and that the substrate at boundary (1) and the products at boundaries (0) and (1) can permeate through the boundaries at rates proportional to their concentrations.

## 4. Flow System

We consider the system described in the preceding section 3 with the assumption that a medium flows along the $x$-axis. This system is a model for an immobilized-enzyme column [1,2]. The translation of substance in the region by the flow may be represented by

$$
\frac{\partial C}{\partial x} \frac{dx}{dt} = v \frac{\partial C}{\partial x},
\tag{68}
$$

where $v$ is the flow rate. In the use of equation (68), the rate equation for the reactive species can be written in the general form of

$$
\frac{\partial C}{\partial t} = D \frac{\partial^2 C}{\partial x^2} - v \frac{\partial C}{\partial x} + f(C,x,t),
\tag{69}
$$

where $f(C,x,t)$ represents the reaction term. If the translation of the reactive species is predominantly governed by the flow and the diffusion term is negligible, then the rate equation is reduced to

$$\frac{\partial C}{\partial t} = -v \frac{\partial C}{\partial x} + f(C, x, t).$$ (70)

# References

[ 1 ] Kuhara, S., S.Iwamoto, Y.Yanase, T.Fukamizo and K.Hayashi (1982). Note on the separation of reacting system by chromatography. *J. Fac. Agr. Kyushu Univ.,* **27**,33-45.

[ 2 ] Gellf, G. and J.Henry (1976). Experimental and theoretical study of djffusion, convection and reaction phenomena for immobilized enzyme systems. *In* "Analysis and Control of Immobilized Enzyme System," ed. by D.Thomas and J.-P. Kernevez, pp.253-274, North-Holland, Amsterdam.

GENERAL

1. Crank. J. (1956). "The Mathematics of Diffusion," Clarendon Press, Oxford.
2. Ames, E.F. (1965). "Nonlinear Partial Differential Equations in Engineering," Academic Press, New York.
3. Jones, D.S. and B.D. Sleeman(1983). "Differential Equations and Mathematical Biology," George Allen & Unwin, London.

# CHAPTER 2

# Approximation Methods for Analysis of Rate Equations

In enzyme kinetics the Michaelis constant and maximum velocity are regarded as the most fundamental quantities to represent the kinetic property of an enzyme. A great number of biochemical experiments have been performed accumulating exhaustive knowledge on such kinetic parameters of enzymes. Kinetic characterization of an enzyme commonly begins with an attempt of their determination once it has been isolated. Time courses of enzymatic reactions obtained experimentally under various conditions are treated with the well-known Michaelis equation to estimate their values.

On the other hand, the dynamic behavior of enzymatic reaction can be represented by the rate equation as derived in Chapter 1 based on the deterministic treatment of the reaction process. It is naturally desirable that the time course from an experiment be directly and readily compared with that from the rate equation to estimate the reaction scheme and kinetic parameters. This quantitative (or analytical) treatment of experimental data, however, has not been possible because the rate equation for enzymatic reaction is inherently nonlinear, so that the relation of kinetic parameters with the feature of a time course cannot be expressed analytically. Some analytical relationships have thus been developed by appropriate approximations of the rate equation and has been widely used to qualitatively characterize the kinetic behavior.

In this chapter we discuss several approximation methods which can be applied to the rate equation to yield the analytical relationships commonly employed for treatment of experimental data. The quasi-steady-state approximation is first described for the rate equation in the form of ordinary differential equation. The rapid-equilibrium and linear approximations

---

This chapter was written by Yoichi Aso and Katsuya Hayashi.

are also explained. These approximation methods will be discussed in more detail in Chapter 4 in conjunction with the probable conditions for their application and the comparison with results from the computer simulation of some basic enzymatic reactions. For the rate equation in the form of partial differential equation, the linear and steady-state approximations are introduced in preparation for the analysis described in Chapter 7.

## 2.1 Quasi-Steady-State Approximation

### 1. Derivation of Steady-State Expression

Let us consider a reaction of the simplest Michaelis-Menten type as in Scheme 2.1. E, S, ES and P represent the enzyme, substrate, enzyme-

$$E + S \underset{k_2}{\overset{k_1}{\rightleftharpoons}} ES \overset{k_3}{\longrightarrow} E + P$$

Scheme 2.1. Michaelis-Menten-type reaction

substrate complex and product, respectively. As described in Chapter 1, we can easily derive the rate equation for the scheme:

$$\left. \begin{aligned}
\frac{dE}{dt} &= -k_1 SE + (k_2 + k_3) C \\
\frac{dS}{dt} &= -k_1 SE + k_2 C \\
\frac{dC}{dt} &= k_1 SE - (k_2 + k_3) C \\
\frac{dP}{dt} &= k_3 C ,
\end{aligned} \right\} \tag{1}$$

where $C$ is the concentration of enzyme-substrate complex ES. The conservation law is expressed as

$$\left. \begin{aligned}
E + C &= E_0 + C_0 \\
S + C + P &= S_0 + C_0 + P_0 ,
\end{aligned} \right\} \tag{2}$$

where the subscript 0 indicates an initial concentration of the corresponding species. Throughout this section, we suppose that a reaction starts from an initial condition of $C_0 = P_0 = 0$. Hence, equation (2) is written as

$$\left. \begin{aligned}
E + C &= E_0 \\
S + C + P &= S_0 .
\end{aligned} \right\} \tag{3}$$

Considering that none of $E$, $S$ and $C$ are dependent upon $P$ and that $dE/dt = -dC/dt$ and $E = E_0 - C$ from equation (3), we can reduce equation (1) to

$$\left.\begin{aligned}
\frac{dC}{dt} &= k_1 S(E_0 - C) - (k_2 + k_3) C \\
\frac{dS}{dt} &= -k_1 S(E_0 - C) + k_2 C.
\end{aligned}\right\} \tag{4}$$

The quasi-steady-state approximation implies that measurement of concentration change should be done during the time range in which the changing rates of enzyme species, E and ES in this case, are almost zero. Namely, the approximation for the scheme is expressed mathematically as $dE/dt = 0$ and $dC/dt = 0$. Then, from equation (4) we have

$$\left.\begin{aligned}
\bar{C} &= \frac{k_1 E_0 S}{k_1 S + k_2 + k_3} \\
v_s &\doteq \frac{dS}{dt} = \frac{-k_1 k_3 E_0 S}{k_1 S + k_2 + k_3},
\end{aligned}\right\} \tag{5}$$

where $\bar{C}$ and $v_s$ represent $C$ and decreasing rate of $S$ in the quasi-steady state, respectively. If it may be assumed that $S$ is not far from $S_0$ (but $dS/dt$ is not zero), we can set $S \simeq S_0$ mathematically to obtain

$$\left.\begin{aligned}
\bar{E} &= \frac{(k_2 + k_3) E_0}{D} \\
\bar{C} &= \frac{k_1 S_0 E_0}{D} \\
v_s &= \frac{-k_1 k_3 S_0 E_0}{D} \\
v_p &= \frac{k_1 k_3 S_0 E_0}{D} \\
P(t) &= v_p t,
\end{aligned}\right\} \tag{6}$$

where $D = k_1 S_0 + k_2 + k_3$, and $v_p$ is a generating rate of $P$ in the quasi-steady state. Hereafter, $v_p$ is abbreviated as $v$, because we frequently deal with $v$ in this chapter. If $S_0$ is taken as infinity, limits of the relationship (6) become

$$\left.\begin{aligned}
&\lim_{S_0 \to \infty} \bar{E} = 0, \quad \lim_{S_0 \to \infty} \bar{C} = E_0, \\
&\lim_{S_0 \to \infty} v_s = -k_3 E_0, \quad V_{max} = \lim_{S_0 \to \infty} v = k_3 E_0,
\end{aligned}\right\} \tag{7}$$

where $V_{max}$ denotes a maximum velocity. By using $V_{max}$ and Michaelis constant $K_m$, which is defined as

$$K_m = \frac{k_2 + k_3}{k_1} , \qquad (8)$$

$v$ is expressed in the well-known steady-state expression,

$$v = \frac{V_{max}}{1 + \frac{K_m}{S_0}} . \qquad (9)$$

The values of $K_m$ and $V_{max}$ can be experimentally evaluated by determining $v$ for various $S_0$. It should be noted that we cannot get the time courses of reactions with the technique of quasi-steady-state approximation.

When we only assume $S \simeq S_0$, and discard the quasi-steady-state assumption of $dC/dt = 0$, the differential equation for $C$ in equation (4) becomes linear ;

$$\frac{dC}{dt} = -k_a C + k_b , \qquad (10)$$

where $k_a = k_1 S_0 + k_2 + k_3$ and $k_b = k_1 E_0 S_0$. Solving equation (10) with the formula for linear differential equation, we obtain

$$C(t) = \frac{k_b}{k_a} (1 - e^{-k_a t}) . \qquad (11)$$

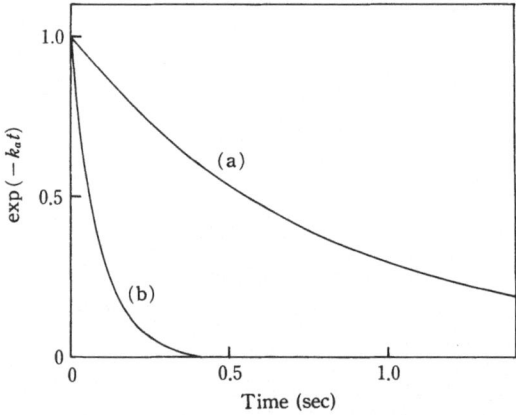

**Fig. 2. 1**  Transition to steady state. $k_1 = 10^5 \, \text{M}^{-1} \text{sec}^{-1}$, $k_2 = 1 \, \text{sec}^{-1}$, $k_3 = 0.1$ $\text{sec}^{-1}$. (a) : $S_0 = 1.0 \mu \text{M}$, (b) : $S_0 = 100 \mu \text{M}$.

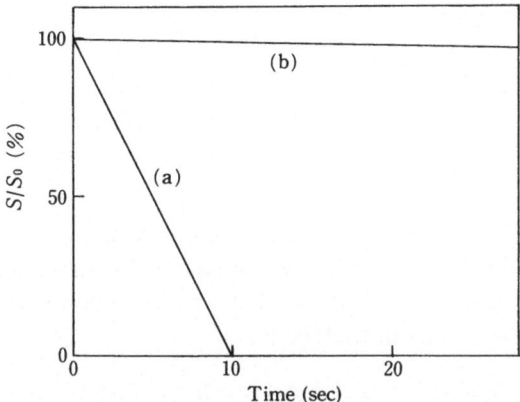

**Fig. 2. 2** Consumption of substrate at maximum velocity. The values of kinetic parameters and notation of (a) and (b) are the same as in Fig. 2. 1.

The exponential term on the right-hand side quickly converges to zero if the value of $k_a$ is relatively large. At that time, $C$ becomes $k_b/k_a$, which is equal to $\bar{C}$ in equation (6). Figure 2.1 reveals how fast $\exp(-k_a t)$ converges to zero with higher $S_0$. Assuming that the substrate is consumed at a maximum velocity of $k_3 E_0$ so that $S(t) = S_0 - k_3 E_0 t$, we have the time courses of $S$ in Fig. 2.2. It is shown that $S \simeq S_0$ may be a reasonable assumption when $S_0 \gg k_3 E_0$. Hence, we can say that the quasi-steady-state aproximation is valid for the simplest Michaelis-Menten mechanism.

## 2. Steady-State Expressions for Relatively Complex Systems

Let us consider the enzymatic reaction with two enzyme-substrate complexes of Scheme 2.2. The rate equation for the scheme is written as

$$\frac{dE}{dt} = -k_1 SE + k_2 C_1 + k_5 C_2$$

$$\frac{dS}{dt} = -k_1 SE + k_2 C_1$$

$$\mathrm{E + S} \underset{k_2}{\overset{k_1}{\rightleftarrows}} \mathrm{ES_1} \underset{k_4}{\overset{k_3}{\rightleftarrows}} \mathrm{ES_2} \overset{k_5}{\longrightarrow} \mathrm{E + P}$$

Scheme 2. 2.  Reaction with two enzyme-substrate complexes

$$\frac{dC_1}{dt} = -(k_2+k_3)C_1+k_1SE+k_4C_2$$

$$\frac{dC_2}{dt} = -(k_4+k_5)C_2+k_3C_1 \qquad (12)$$

$$\frac{dP}{dt} = k_5C_2,$$

where $C_1$ and $C_2$ represent the concentrations of $ES_1$ and $ES_{2'}$, respectively. If the quasi-steady state (*i.e.*, $dE/dt=0$, $dC_1/dt=0$ and $dC_2/dt=0$) and $S \simeq S_0$ are again assumed, equation (12) can be reduced to simultaneous linear algebraic equations in matrix form,

$$\begin{bmatrix} -k_1S_0 & k_2 & k_5 \\ k_1S_0 & -(k_2+k_3) & k_4 \\ 1 & 1 & 1 \end{bmatrix} \begin{bmatrix} \bar{E} \\ \bar{C}_1 \\ \bar{C}_2 \end{bmatrix} = \begin{bmatrix} 0 \\ 0 \\ E_0 \end{bmatrix}, \qquad (13)$$

and expressions for the reaction rates,

$$v_s = \frac{dS}{dt} = -k_1S_0\bar{E}+k_2\bar{C}_1$$

$$v = \frac{dP}{dt} = k_5\bar{C}_2, \qquad (14)$$

where $E_0$ denotes the total concentration of enzyme species. We can solve equation (13) by one of several procedures, for example, Cramer's formula to obtain

$$\bar{E} = \frac{(k_2k_4+k_2k_5+k_3k_5)E_0}{D}$$

$$\bar{C}_1 = \frac{k_1(k_4+k_5)S_0E_0}{D} \qquad (15)$$

$$\bar{C}_2 = \frac{k_1k_3S_0E_0}{D},$$

where $D = k_2k_4+k_2k_5+k_3k_5+k_1(k_3+k_4+k_5)S_0$. Substitution of equation (15) into equation (14) yields

$$v = -v_s = \frac{k_1k_3k_5S_0E_0}{D}. \qquad (16)$$

Limits of equations (15) and (16) with $S_0$ taken as infinity become

$$\lim_{S_0 \to \infty} \bar{E} = 0$$

$$\left.\begin{aligned}
\lim_{S_0 \to \infty} \bar{C}_1 &= \frac{(k_4 + k_5) E_0}{Z} \\
\lim_{S_0 \to \infty} \bar{C}_2 &= \frac{k_3 E_0}{Z} \\
V_{max} &= \lim_{S_0 \to \infty} v = \frac{k_3 k_5 E_0}{Z},
\end{aligned}\right\}
\tag{17}$$

where $Z = k_3 + k_4 + k_5$. If the Michaelis constant $K_m$ is defined as

$$K_m = \frac{k_2 k_3 + k_2 k_5 + k_3 k_5}{k_1 Z}, \tag{18}$$

equation (16) leads to the steady-state expression,

$$v = \frac{V_{max}}{1 + \dfrac{K_m}{S_0}}, \tag{19}$$

which is identical to equation (9).

As discussed above, for the quasi-steady-state approximation, we need to solve linear algebraic equations with respect to concentrations of enzyme species instead of nonlinear differential equations. The solution is usually most time-consuming in derivation of steady-state expressions, especially when the reaction scheme becomes more complex. To avoid such laborious and error-prone work, several methods have been developed for graphically solving algebraic equations by hand or computer [1]. The King-Altman method [2] is now representative of these, due to its easy and straight-forward use in application. Procedure of this method is as follows:

1) A reaction scheme may be drawn as a polygon with vertices which correspond to enzyme species. The kinetic parameter for each reaction path is assigned to the side between the relevant vertices. If we take Scheme 2.2 as an example, we obtain a triangle as in Fig. 2.3.

2) The steady-state concentrations of E, $ES_1$ and $ES_2$ are expressed in the form of their ratios to $E_0$ (total concentration of enzyme species).

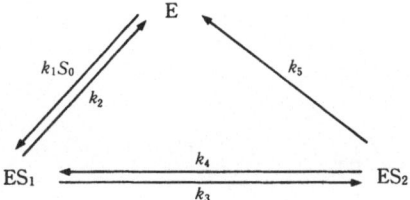

**Fig. 2.3** Graphical representation of reaction scheme (Scheme 2.2).

When we denote the numerators of ratios by $e$, $c_1$ and $c_2$ for $\bar{E}$, $\bar{C}_1$, and $\bar{C}_2$, respectively, and the denominator by $D$, we can write them as

$$\left. \begin{aligned} \bar{E} &= \frac{eE_0}{D} \\ \bar{C}_1 &= \frac{c_1E_0}{D} \\ \bar{C}_2 &= \frac{c_2E_0}{D}. \end{aligned} \right\} \tag{20}$$

Since $E_0 = \bar{E} + \bar{C}_1 + \bar{C}_2$, we have the relationship,

$$D = e + c_1 + c_2, \tag{21}$$

between the numerators and denominator. Determination of $e$, $c_1$ and $c_2$ by the procedure below leads to the expression of steady-state concentrations of the species.

3) In order to get numerators, we select a set of paths for each enzyme species. Each path reaches the corresponding vertex by going through the appropriate sides just once. All these paths are included in the set except for closed-loop paths. In this example, we obtain three sets of paths as in Fig. 2.4.

4) Making a summation of the products of kinetic parameters included in each path of a set, we can determine every numerator. For this example, we have from Fig. 2.4.

(a)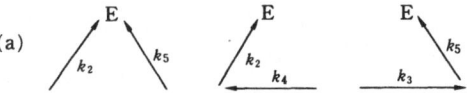

(b)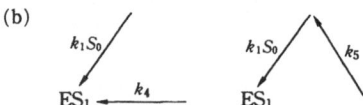

(c)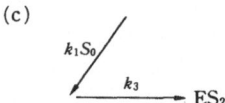

**Fig. 2.4** Selection of paths to derive numerators. (a) : for E ; (b) : for $\mathrm{ES}_1$ ; (c) : for $\mathrm{ES}_2$.

$$\left.\begin{array}{l} e = k_2 k_4 + k_3 k_5 + k_2 k_5 \\ c_1 = k_1 k_4 S_0 + k_1 k_5 S_0 \\ c_2 = k_1 k_3 S_0. \end{array}\right\} \tag{22}$$

5) We thus obtain the steady-state expression (15) from equations (20), (21) and (22). Expression for $v$ is easily derived by substituting $\bar{C}_2$ into $v = k_5 \bar{C}_2$.

In this procedure we need not be concerned with how to solve algebraic equations, but only with the construction of several graphs according to the rules. For a more complex scheme, however, it becomes rather difficult to correctly find all the paths. The computer procedure is devised to perform the King-Altman method [1]. Algebraic equations themselves can also be solved by computer to derive the steady-state expressions.

## 2.2 Rapid-Equilibrium and Linear Approximations

### 1. Rapid-Equilibrium Approximation

For the rapid-equilibrium approximation, we assume the existence of a rapid-equilibration step whereby the formation process of enzyme-substrate complex instantly reaches an equilibrium state. In the case of Scheme 2.2, for example, equilibrium constants $K_1 = k_2/k_1$ and $K_2 = k_4/k_3$ are defined for the processes of rapid equilibration. If $S \simeq S_0$, we obtain, instead of equation (13),

$$\begin{bmatrix} S_0 & -K_1 & 0 \\ 0 & 1 & -K_2 \\ 1 & 1 & 1 \end{bmatrix} \begin{bmatrix} \tilde{E} \\ \tilde{C}_1 \\ \tilde{C}_2 \end{bmatrix} = \begin{bmatrix} 0 \\ 0 \\ E_0 \end{bmatrix}, \tag{23}$$

where $\tilde{E}$, $\tilde{C}_1$ and $\tilde{C}_2$ are the concentrations at equilibrium. The solution of equation (23) is given by

$$\left.\begin{array}{l} \tilde{E} = \dfrac{K_1 K_2 E_0}{A} \\[2mm] \tilde{C}_1 = \dfrac{K_2 S_0 E_0}{A} \\[2mm] \tilde{C}_2 = \dfrac{S_0 E_0}{A}, \end{array}\right\} \tag{24}$$

where $A = K_1 K_2 + (1 + K_2) S_0$. Since $v = k_5 \tilde{C}_2$, we have

$$v = \frac{k_5 E_0}{1 + K_2 + K_1 K_2 / S_0} \tag{25}$$

and then

$$V_{\max} = \lim_{S_0 \to \infty} v = \frac{k_5 E_0}{1 + K_2}. \tag{26}$$

From the definition of $V_{\max}$ and $K = K_1 K_2 / (1 + K_2)$, we obtain

$$v = \frac{V_{\max}}{1 + \dfrac{K}{S_0}}, \tag{27}$$

which is an identical expression to equation (19). In a similar manner, application of the rapid-equilibrium approximation to Scheme 2.1 can lead easily to the relationship,

$$v = \frac{V_{\max}}{1 + \dfrac{K}{S_0}}, \tag{28}$$

where $K = k_1 / k_2$.

Next, as a typical example of rapid-equilibrium approximation, we consider a reaction mechanism in which an enzyme has several active sites. Let us start with the simple mechanism of Scheme 2.3, where enzyme R is

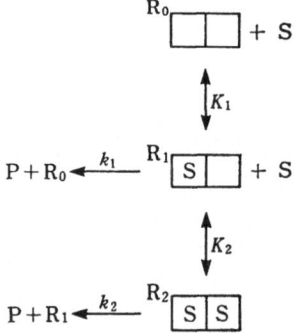

Scheme 2.3. Enzyme with two active sites

a dimer with two active sites, one for each subunit. As the enzyme species are to be at equilibrium, their concentrations are determined from a linear algebraic equation,

$$\begin{bmatrix} 2S_0 & -K_1 & 0 \\ 0 & S_0 & -2K_2 \\ 1 & 1 & 1 \end{bmatrix} \begin{bmatrix} \tilde{R}_0 \\ \tilde{R}_1 \\ \tilde{R}_2 \end{bmatrix} = \begin{bmatrix} 0 \\ 0 \\ E_0 \end{bmatrix}. \tag{29}$$

This equation is solved to yield

$$\left. \begin{aligned} \tilde{R}_0 &= \frac{E_0}{A} \\[2mm] \tilde{R}_1 &= \frac{2\alpha_1 E_0}{A} \\[2mm] \tilde{R}_2 &= \frac{\alpha_1\alpha_2 E_0}{A}, \end{aligned} \right\} \tag{30}$$

where $A = 1 + 2\alpha_1 + \alpha_1\alpha_2$, and $\alpha_1 = S_0/K_1$ and $\alpha_2 = S_0/K_2$. The velocities of $v$ and $V_{max}$ and their ratio $r$ ($= v/V_{max}$) are expressed as

$$\left. \begin{aligned} v &= k_1\tilde{R}_1 + 2k_2\tilde{R}_2 \\[2mm] &= \frac{2\alpha_1(k_1 + k_2\alpha_2)E_0}{A} \\[2mm] V_{max} &= \lim_{S_0\to\infty} v = \lim_{\alpha_1,\alpha_2\to\infty} v = 2k_2 E_0 \\[2mm] r &= \frac{\alpha_1\left(\dfrac{k_1}{k_2} + \alpha_2\right)}{A}. \end{aligned} \right\} \tag{31}$$

If $K_r = K_1 = K_2$, that is, $\alpha = \alpha_1 = \alpha_2$, and $k = k_1 = k_2$ are assumed, the quantities in equation (31) are reduced to

$$\left. \begin{aligned} v &= \frac{2k\alpha E_0}{1+\alpha} \\[2mm] V_{max} &= 2kE_0 \\[2mm] r &= \frac{\alpha}{1+\alpha}. \end{aligned} \right\} \tag{32}$$

Thus, we have

$$v = \frac{V_{max}}{1 + \dfrac{K_r}{S_0}}, \tag{33}$$

which is identical to equation (28). It follows that we cannot distinguish Scheme 2.1 from Scheme 2.3 when the rapid-equilibrium approximation is applied.

Scheme 2.3 may be extended to Scheme 2.4 including an $n$-meric enzyme with $n$ active sites (one for each subunit). Solving the linear algebraic equation derived from the rapid-equilibrium approximation in Scheme 2.4,

$$R_0 + S$$
$$\uparrow K_1$$
$$P + R_0 \xleftarrow{k_1} R_1 + S$$
$$\uparrow K_2$$
$$P + R_1 \xleftarrow{k_2} R_2 + S$$
$$\vdots$$
$$P + R_{n-2} \xleftarrow{k_{n-1}} R_{n-1} + S$$
$$\uparrow K_n$$
$$P + R_{n-1} \xleftarrow{k_n} R_n$$

Scheme 2.4. Enzyme with $n$ active sites

$$\begin{bmatrix} -n\alpha_1 & 1 & 0 & \cdots\cdots\cdots\cdots\cdots\cdots\cdots\cdots\cdots & 0 \\ \cdots\cdots\cdots\cdots\cdots\cdots\cdots\cdots\cdots\cdots\cdots\cdots \\ \cdots\cdots\cdots\cdots\cdots & 0 & -(n-i+1)\alpha_i/i & 1\cdots 0 \\ \cdots\cdots\cdots\cdots\cdots\cdots\cdots\cdots\cdots\cdots\cdots\cdots \\ 1 & 1 & 1 \cdots 1 & 1 & 1\cdots 1 \end{bmatrix} \begin{bmatrix} \tilde{R}_0 \\ \cdots \\ \tilde{R}_{i-1} \\ \cdots \\ \tilde{R}_n \end{bmatrix} = \begin{bmatrix} 0 \\ \cdots \\ 0 \\ \cdots \\ E_0 \end{bmatrix}, \quad (34)$$

where $\alpha_i = S_0/K_i$ $(i = 1,2,\cdots,n)$, we can obtain the equilibrium concentrations of enzyme species in a general form:

$$\tilde{R}_q = \frac{x_q E_0}{A}, \quad q = 0,1,2,\cdots,n, \tag{35}$$

where $A = 1 + \sum\limits_{m=1}^{n} x_m, x_i = {}_nC_i \prod\limits_{j=1}^{i} \alpha_j$, $x_0 = 1$, and ${}_nC_i = n!/[i!(n-i)!]$. The velocities of $v$ and $V_{\max}$ and their ratio $r$ are given as

$$\left. \begin{array}{l} v = \sum\limits_{i=1}^{n} i k_i \tilde{R}_i \\ V_{\max} = n k_n E_0 \\ r = \dfrac{\sum\limits_{i=1}^{n} i k_i x_i}{n k_n A}. \end{array} \right\} \tag{36}$$

If $\alpha = \alpha_1 = \alpha_2 = \cdots = \alpha_n$ and $k = k_1 = k_2 = \cdots = k_n$ may be assumed again, equation (36) becomes concise expressions,

$$\left. \begin{array}{l} v = kE_0 \sum\limits_{i=1}^{n} i \, {}_nC_i \alpha^i/(1+\alpha)^n \\ V_{\max} = nkE_0 \\ r = \sum\limits_{i=1}^{n} i \, {}_nC_i \alpha^i/\{n(1+\alpha)^n\}. \end{array} \right\} \tag{37}$$

Since the sum of combinations in equation (37) is calculated to be

$$\sum_{i=1}^{n} i_n C_i \alpha^i = \frac{n\alpha \sum_{i=1}^{n} (n-1)!\, \alpha^{i-1}}{(i-1)!\,(n-1)!}$$

$$= n\alpha\left(1+\sum_{i=1}^{n-1} {}_{n-1}C_i \alpha^i\right)$$

$$= n\alpha\,(1+\alpha)^{n-1},$$

$$(38)$$

equation (37) is rewritten as

$$\left.\begin{aligned} v &= \frac{\alpha V_{\max}}{1+\alpha} \\ V_{\max} &= nkE_0 \\ r &= \frac{\alpha}{1+\alpha}, \end{aligned}\right\}$$

$$(39)$$

which have the same forms as in equation (32).

Introducing another state for enzyme species, we modify Scheme 2.3 to Scheme 2.5, which is a well-known model of allosteric enzyme. In this

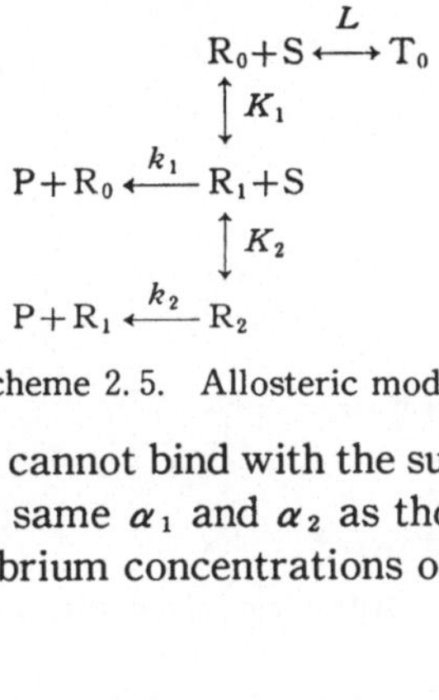

$$R_0 + S \xleftrightarrow{\;L\;} T_0$$
$$\uparrow K_1$$
$$P + R_0 \xleftarrow{\;k_1\;} R_1 + S$$
$$\uparrow K_2$$
$$P + R_1 \xleftarrow{\;k_2\;} R_2$$

Scheme 2.5. Allosteric model

model we assume that $T_0$ cannot bind with the substrate and that $T_0$ and $R_0$ are at equilibrium. If the same $\alpha_1$ and $\alpha_2$ as those defined for Scheme 2.3 are used again, the equilibrium concentrations of enzyme species are given by

$$\left.\begin{aligned} \tilde{T}_0 &= \frac{LE_0}{A} \\ \tilde{R}_0 &= \frac{E_0}{A} \\ \tilde{R}_1 &= \frac{2\alpha_1 E_0}{A} \\ \tilde{R}_2 &= \frac{\alpha_1 \alpha_2 E_0}{A}, \end{aligned}\right\}$$

$$(40)$$

where $A=L+1+2\alpha_1+\alpha_1\alpha_2$ and $L=\tilde{T}_0/\tilde{R}_0$. $v$ and $r$ are represented by

$$
\left.\begin{aligned}
v &= \frac{2\alpha_1(k_1+k_2\alpha_2)E_0}{A} \\[2mm]
r &= \frac{\alpha_1\left(\dfrac{k_1}{k_2}+\alpha_2\right)}{A},
\end{aligned}\right\}
\tag{41}
$$

which, with $\alpha=\alpha_1=\alpha_2$ and $k=k_1=k_2$, become

$$
\left.\begin{aligned}
v &= \frac{2k\alpha(1+\alpha)E_0}{D} \\[2mm]
r &= \frac{\alpha(1+\alpha)}{D},
\end{aligned}\right\}
\tag{42}
$$

where $D=L+(1+\alpha)^2$.

The treatment can be readily extended to an allosteric enzyme having $n$ protomers. Assuming that $\alpha=\alpha_1=\alpha_2=\cdots=\alpha_n$, that is, $K_R=K_1=K_2=\cdots=K_n$, and $k=k_1=k_2=\cdots=k_n$, we have a general allosteric model of

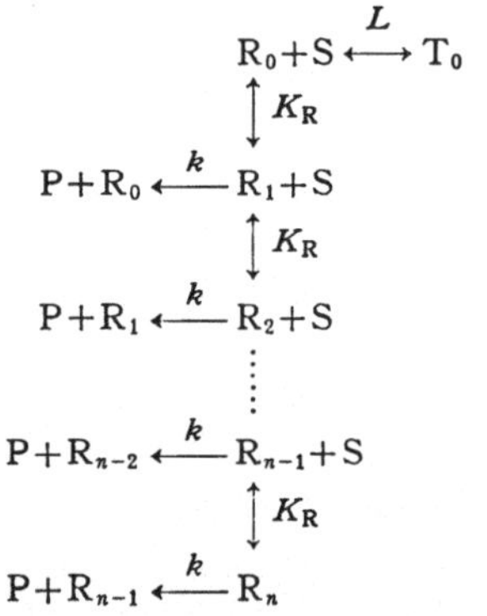

$$
\begin{array}{c}
\qquad\qquad\qquad\quad L \\
R_0+S \xleftrightarrow{\quad} T_0 \\
\uparrow K_R \\
P+R_0 \xleftarrow{\;k\;} R_1+S \\
\uparrow K_R \\
P+R_1 \xleftarrow{\;k\;} R_2+S \\
\vdots \\
P+R_{n-2} \xleftarrow{\;k\;} R_{n-1}+S \\
\uparrow K_R \\
P+R_{n-1} \xleftarrow{\;k\;} R_n
\end{array}
$$

Scheme 2.6.   Allosteric model with $n$ protomers

Scheme 2.6. In the similar manner, we can obtain

$$
\left.\begin{aligned}
v &= \frac{nk\alpha(1+\alpha)^{n-1}E_0}{D} \\[2mm]
r &= \frac{\alpha(1+\alpha)^{n-1}}{D},
\end{aligned}\right\}
\tag{43}
$$

where $D=L+(1+\alpha)^n$. If the $r$'s in equations (39) and (43) are denoted by $r_M$ and $r_A$, respectively, we have

$$
\lim_{L\to 0} r_A = r_M.
\tag{44}
$$

Since $\alpha > 0$, we can derive the relationships,

$$\left.\begin{aligned}
\frac{\partial r_M}{\partial \alpha} &= \frac{1}{(1+\alpha)^2} > 0 \\[2mm]
\lim_{\alpha \to \infty} \frac{\partial r_M}{\partial \alpha} &= 0 \\[2mm]
\frac{\partial^2 r_M}{\partial \alpha^2} &= \frac{-2}{(1+\alpha)^3} < 0,
\end{aligned}\right\} \tag{45}$$

for $r_M$ and

$$\left.\begin{aligned}
\frac{\partial r_A}{\partial \alpha} &= \frac{a^{n-2}(L+nL\alpha+a^n)}{b^2} > 0 \\[2mm]
\lim_{\alpha \to \infty} \frac{\partial r_A}{\partial \alpha} &= 0 \\[2mm]
\frac{\partial^2 r_A}{\partial \alpha^2} &= \frac{a^{n-3}}{b^3}\{na^n(aL-2L-2nL\alpha-a^n) \\
&\quad + anL(L+a^{n-1}) \\
&\quad + (n-2)(L+a^n)(L+nL\alpha+a^n)\},
\end{aligned}\right\} \tag{46}$$

for $r_A$, where $a=1+\alpha$ and $b=L+(1+\alpha)^n$. Equations (45) and (46) reveal the characteristics of relationship between $\alpha$ and $r$, that is, the so-called $S$-$v$ relationship. It follows from equations (44) and (45) that the $S$-$v$ relationship for a scheme without a transition path between $T_0$ and $R_0$ becomes a simple saturation curve. On the other hand, equation (46) indicates that the $S$-$v$ relationship for a scheme with T–R transition can have an inflection point for appropriate values of $\alpha$ and $L$, since partial derivative of the second order, $\partial^2 r_A/\partial \alpha^2$, can be zero at a certain value of $\alpha$ ($\geqq 0$). For example, with $n=2$, the condition of $\partial^2 r_A/\partial \alpha^2 = 0$ is written as

$$\alpha^3 + 3(1+L)\alpha^2 + 3(1+L)\alpha + 1 - L^2 = 0 \tag{47}$$

to give a relationship between $\alpha_f$, the value of $\alpha$ at the inflection point, and the allosteric constant $L$, equilibrium constant for T–R transition. Figure 2.5 illustrates the relationship from equation (47) in a plot of $\log_{10} L$ against $\alpha_f$. Figure 2.6 shows the $\alpha$-$r$ curve for $L=1024$. It should be noted from the analysis that allostericity is due to the transition between $T_0$ and $R_0$ in this model. We also realize that the rapid-equilibrium approximation results in the problem of how to solve simultaneous linear algebraic equations.

## 2. Linear Approximation

In general, the rate equation for an enzymatic reaction is expressed by a nonlinear ordinary differential equation. As stated above, it is impossible to

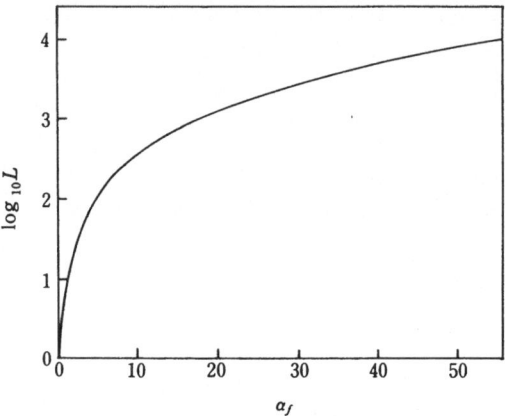

**Fig. 2. 5** Relationship between position of inflection and allosteric constant.

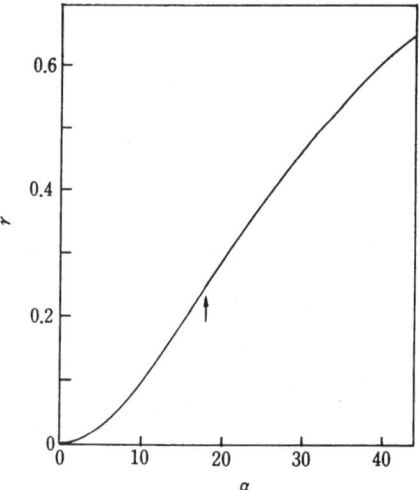

**Fig. 2. 6** $\alpha$-$r$ profile at $L = 1024$. Arrow indicates the position of inflection.

get an analytic solution (*i.e.,* a solution expressed by analytic functions) of a rate equation except for special cases. Instead of an analytic solution, it has recently become relatively easy to obtain the solution of a differential equation as a sequence of numerical values (called a numerical solution), by highly developed methods of numerical integration with a computer as described in Chapter 3.

Analysis of the rate equation by approximation methods, however, is still attractive from some aspects, especially because an analytic solution can provide us with global properties of the solution, that is, characteristics of the reaction. The method of linear approximation is often employed for this purpose. If, under appropriate assumptions, we can reduce a nonlinear term to a linear one in the neighborhood of a specified state, a nonlinear differential equation is approximately represented by a linear differential equation which provides the desired analytic solution. This process is called linearization. The procedure of approximation by linearization is referred to as the method of linear approximation. In fact, linearization under an assumption of $S \simeq S_0$ is used for the analysis in the above discussions, and is one of the techniques of linear approximation. In the following, we discuss the chemical relaxation method [3] as another example of linear approximation.

When a reaction system in an equilibrium state is perturbed, for example, by an abrupt change in temperature, the system undergoes an equilibration process to reach a new equilibrium state. Let us consider a simple reaction

$$E + M \underset{k_2}{\overset{k_1}{\rightleftharpoons}} EM$$

Scheme 2.7. Equilibration process

of Scheme 2.7. The rate equation for this scheme is given by

$$\left.\begin{aligned}
\frac{dE}{dt} &= -k_1 ME + k_2 C \\
\frac{dM}{dt} &= -k_1 ME + k_2 C \\
\frac{dC}{dt} &= k_1 ME - k_2 C,
\end{aligned}\right\} \tag{48}$$

where $C$ is the concentration of the complex EM. $E$, $M$ and $C$ during the equilibration process are expressed by

$$\left.\begin{aligned}
E &= \tilde{E} + \delta E \\
M &= \tilde{M} + \delta M \\
C &= \tilde{C} + \delta C,
\end{aligned}\right\} \tag{49}$$

where $\tilde{E}$, $\tilde{M}$ and $\tilde{C}$ are the respective concentrations in the new equilibrium state, and symbol $\delta$ indicates the deviation from the equilibrium concentration. Signs of $\delta E$, $\delta M$ and $\delta C$ depend on a direction in which the equilibrium is perturbed. According to the conservation law and the definition of equilibrium, we have

$$\delta E = \delta M = -\delta C \tag{50}$$

and

$$k_1 \tilde{M} \tilde{E} = k_2 \tilde{C}. \tag{51}$$

Since $\tilde{E}$, $\tilde{M}$ and $\tilde{C}$ are constants, substitution of equation (49) into equation (48) and use of equations (50) and (51) lead to

$$\left. \begin{array}{l} \dfrac{d(\delta E)}{dt} = -k\delta E - k_1(\delta E)^2 \\[2mm] \dfrac{d(\delta M)}{dt} = -k\delta M - k_1(\delta M)^2 \\[2mm] \dfrac{d(\delta C)}{dt} = -k\delta C + k_1(\delta C)^2, \end{array} \right\} \tag{52}$$

where $k = k_1(\tilde{E} + \tilde{M}) + k_2$. These are nonlinear differential equations. However, if we can suppose that $\delta E$, $\delta M$ and $\delta C$ have small enough values, the nonlinear terms, $(\delta E)^2$, $(\delta M)^2$ and $(\delta C)^2$, must be negligible in comparison with other terms. By such linearization, we can reduce equation (52) to a linear differential equation,

$$\frac{d}{dt} \begin{bmatrix} \delta E \\ \delta M \\ \delta C \end{bmatrix} = \begin{bmatrix} -k & 0 & 0 \\ 0 & -k & 0 \\ 0 & 0 & -k \end{bmatrix} \begin{bmatrix} \delta E \\ \delta M \\ \delta C \end{bmatrix}, \tag{53}$$

which is a set of independent differential equations. The solution is obtained easily in a general form as

$$\delta X = \delta X_0 \exp(-kt), \tag{54}$$

where $\delta X_0$ expresses the overall difference between the values of $X$ in initial and final equilibrium states. Equation (54) implies that, when a reaction at equilibrium is disturbed by slight perturbation, the deviation from the new equilibrium concentration decreases exponentially to finally be zero. This procedure is called chemical relaxation [3]. The rate of relaxation depends on coefficient $k$, that is, rate constants $k_1$ and $k_2$ and equilibrium concentrations. Hence, we can determine rate constants $k_1$ and $k_2$ by the chemical relaxation method, in which we measure the equilibrium concentrations and velocity of relaxation (or relaxation time $\tau [= 1/k]$) for various perturbations.

The exertion speed of perturbation is correlated to the equilibration rate (relaxation rate). If we were to perturb the system at a rate slower than that of equilibration, we could never observe the relaxation process. In such case, the equilibration process just follows the perturbation. The relaxation process is now made observable by the development of the fast-reaction technique (*i.e.*, abrupt perturbation and quick measurement). Since the chemical relaxation method provides us with detailed information on every step of the reaction mechanism, it might be much more effective in enzyme kinetics than the quasi-steady-state approximation.

Relevant to Chapter 1, we here formulate the linear-approximation method in a general form. The rate equation containing $n$ chemical species can be represented by

$$\frac{dX_i}{dt} = f_i(X_1, X_2, \cdots, X_n, t), \quad i = 1, 2, \cdots, n. \tag{55}$$

If $X_i$ can be expressed in the form of $X_i = C_i + \delta X_i$ with a constant $C_i$ and a deviation $\delta X_i$, equation (55) is written as

$$\frac{d(C_i + \delta X_i)}{dt} = f_i(C_1 + \delta X_1, C_2 + \delta X_2, \cdots, C_n + \delta X_n, t), \tag{56}$$

$$i = 1, 2, \cdots, n.$$

The left-hand side of equation (56) becomes $d\delta X_i / dt$ and the right-hand side can be expanded in a Taylor series around $C_i$:

$$f_i(C, t) + \sum_{j=1}^{n} a_{ij}\delta X_i + O\{(\delta X_i)^2\}, \tag{57}$$

where $a_{ij}$'s are the components of the Jacobian matrix defined below. If we can neglect the higher-order term $O\{(\delta X_i)^2\}$, and choose $C_i$'s to satisfy $f_i(C, t) = 0$, which is possible in a case where the reaction is in an equilibrium state with $C_i$'s as equilibrium concentrations, we obtain from equation (55) a linear differential equation in matrix form as

$$\frac{d}{dt}\begin{bmatrix} \delta X_1 \\ \delta X_2 \\ \vdots \\ \delta X_n \end{bmatrix} = J \begin{bmatrix} \delta X_1 \\ \delta X_2 \\ \vdots \\ \delta X_n \end{bmatrix}. \tag{58}$$

$J$ is called the Jacobian matrix, and has the components,

$$a_{ij} = \frac{\partial f_i}{\partial X_j}\bigg|_{X=C}, \tag{59}$$

where $X = C$ means that $[X_1, X_2, \cdots, X_n] = [C_1, C_2, \cdots, C_n]$. For equation (48), the Jacobian matrix is given as

$$J = \begin{bmatrix} -k_1\tilde{M} & -k_1\tilde{E} & k_2 \\ -k_1\tilde{M} & -k_1\tilde{E} & k_2 \\ k_1\tilde{M} & k_1\tilde{E} & -k_2 \end{bmatrix}. \tag{60}$$

Any time the approximation methods described in this chapter are used, careful analysis must be made so that the objective reaction system actually satisfies the conditions to substantiate the assumption on which the approximation method is based. In fact, validity of the approximation, which can be assured by simple reactions in closed system, sometimes becomes uncertain when applied to complex and/or open systems. It should be remembered that the dynamic characteristics of enzymatic reaction are rarely consistent with one's intuition.

## 2.3 Approximate Solution of Partial Differential Equation

### 1. Analytic Solution for First-Order Reaction
A partial differential equation for the diffusion process is represented by

$$\frac{\partial C(x,t)}{\partial t} = D \frac{\partial^2 C(x,t)}{\partial x^2}, \tag{61}$$

where $C(x,t)$ is the concentration of the diffusing substance and $D$ is the diffusion coefficient. The solution of equation (61) is given by

$$C(x,t) = \frac{A}{2(\pi Dt)^{1/2}} e^{-x^2/(4Dt)}, \tag{62}$$

where $A$ represents the total amount of diffusing substance [4].

We now consider a first-order reaction,

$$A \xrightarrow{k} B$$

of diffusing reactant A. The rate equation is written as

$$\frac{\partial C_1}{\partial t} = D \frac{\partial^2 C_1}{\partial x^2} - kC_1, \tag{63}$$

where $C_1(x,t)$ expresses the concentration of A. When it is assumed that A is kept at a constant concentration $C_0$ at boundary (0) and permeates through boundary (1) at a rate proportional to its concentration, the initial and boundary conditions are represented by

$$
\begin{aligned}
&C_1(x,0) = 0, \quad x > 0 \\
&C_1(0,t) = C_0 \\
&\frac{\partial C_1(1,t)}{\partial x} = hC_1(1,t),
\end{aligned}
\tag{64}
$$

where $h$ is a constant.

Let $C(x,t)$ in equation (62) express the solution of the diffusion process of A without the reaction. Then, the solution of $C_1(x,t)$ in equation (63) under the conditions of (64) is given [4] by

$$C_1(x,t) = k \int_0^t C(x,\tau) e^{-k\tau} d\tau + C(x,t) e^{-kt}. \tag{65}$$

## 2. Approximate Solution of Bimolecular Reaction [5]

We consider the reaction–diffusion system shown in Section 1.2.3. to examine the concentration change of reactants in the system. If the concentrations of A and B are represented by $C_1(x,t)$ and $C_2(x,t)$, respectively, equations (58), (59) and (60) in Section 1.2 can be rewritten as

$$\begin{aligned}
\frac{\partial C_1}{\partial t} &= D_A \frac{\partial^2 C_1}{\partial x^2} - kC_1 C_2 \\
\frac{\partial C_2}{\partial t} &= D_B \frac{\partial^2 C_2}{\partial x^2} - kC_1 C_2
\end{aligned} \tag{66}$$

$$\begin{aligned}
&C_1(x,0) = 0, \quad C_2(x,0) = C_{20} \\
&C_1(0,t) = C_1^*, \quad \frac{\partial C_2(0,t)}{\partial x} = 0 \\
&C_1(\infty,t) = 0, \quad C_2(\infty,t) = C_{20}.
\end{aligned} \tag{67}$$

Here, we introduce new dimensionless parameters:

$$\begin{aligned}
&\alpha = \frac{C_1}{C_1^*}, \quad \beta = \frac{C_2}{C_{20}}, \quad \theta = kC_1^* t, \\
&\xi = \left( \frac{kC_1^*}{D_A} \right)^{\frac{1}{2}} x, \quad \Delta = \frac{D_B}{D_A}, \quad \Gamma = \frac{C_{20}}{C_1^*}.
\end{aligned} \tag{68}$$

By using the new parameters, equations (66) and (67) can be transformed to

$$\begin{aligned}
\frac{\partial \alpha}{\partial \theta} &= \frac{\partial^2 \alpha}{\partial \xi^2} - \Gamma \alpha \beta \\
\frac{\partial \beta}{\partial \theta} &= \Delta \frac{\partial^2 \beta}{\partial \xi^2} - \alpha \beta
\end{aligned} \tag{69}$$

$$\begin{aligned}
\theta = 0 & \begin{cases} \alpha = 0, & \xi > 0 \\ \beta = 1, & \xi \geq 0 \end{cases} \\
\theta > 0 & \begin{cases} \alpha = 1, & \partial \beta / \partial \xi = 0, \quad \xi = 0 \\ \beta = 0, & \beta \to 1, \quad \xi \to \infty. \end{cases}
\end{aligned} \tag{70}$$

This nonlinear partial differential equation cannot be solved analytically. Therefore, we need to solve equation (69) approximately, assuming some special conditions. Some examples of approximate solution are shown in the following.

a) Let us assume that $\theta$ (reaction time) is very small and that the extent of reaction is negligible. Under these conditions, $\alpha$ and $\beta$ are regarded as $\alpha \to 0$ and $\beta = 1$, respectively, and then equation (69) is reduced approximately to a diffusion equation as

$$\frac{\partial \beta}{\partial \theta} = \Delta \frac{\partial^2 \beta}{\partial \xi^2},$$ (71)

which has the same solution as equation (62).

b) Let us assume that $D_B$ is extremely large and $\Delta \to \infty$. In this case, the consumed $\beta$ owing to the reaction is rapidly restored at a high diffusion rate and the consumption of $\beta$ may be negligible. This also corresponds to the extremely large value of $\beta$ in comparison with $\alpha$, allowing the bimolecular reaction to be regarded as pseudo first-order reaction. Thus, equation (69) may be transformed to

$$\frac{\partial \alpha}{\partial \tau} = \frac{\partial^2 \alpha}{\partial \eta^2} - \alpha,$$ (72)

where $\tau = \Gamma \theta$ and $\eta = \Gamma^{1/2} \xi$.

Equation (72) is the rate equation for first-order reaction and has the same solution as equation (65) with $k=1$.

## 3. Steady-State Approximation

We consider here the enzyme system represented by equations $(61) \sim (66)$ in Section 1.2 and assume that the system is in a stationary (steady) state. Since in the stationary state

$$\frac{\partial S}{\partial t} = \frac{\partial P_i}{\partial t} = \frac{\partial E_i}{\partial t} = \frac{\partial C_i}{\partial t} = 0, \quad i=1,2,$$

at any position in the region, the following equations are obtained:

$$\left. \begin{array}{l} D_S \dfrac{d^2 S}{dx^2} = k_{11} S E_1 - k_{-11} C_1 \\[2mm] D_{P_1} \dfrac{d^2 P_1}{dx^2} = -k_{12} C_1 + k_{21} P_1 E_2 - k_{-21} C_2 \\[2mm] D_{P_2} \dfrac{d^2 P_2}{dx^2} = -k_{22} C_{22}, \end{array} \right\}$$ (73)

$$\left. \begin{array}{l} -k_{11} S E_1 + k_{-11} C_1 + k_{12} C_1 = 0 \\[1mm] -k_{21} P_1 E_2 + k_{-21} C_2 + k_{22} C_2 = 0. \end{array} \right\}$$ (74)

With simple manipulation we have

$$\left. \begin{array}{l} \dfrac{d^2 S}{dx^2} = \dfrac{1}{D_S} \dfrac{k_{12} S E_{10}}{K_{m1} + S} \\[2mm] \dfrac{d^2 P_1}{dx} = -\dfrac{1}{D_{P_1}} \left( \dfrac{k_{12} S E_0}{K_{m1} + S} - \dfrac{k_{22} P_1 E_{20}}{K_{m2} + P_1} \right) \\[2mm] \dfrac{d^2 P_2}{dx^2} = -\dfrac{1}{D_{P_2}} \dfrac{k_{22} P_1 E_{20}}{K_{m2} + P_1} \end{array} \right\}$$ (75)

$$E_1 = \frac{E_{10}}{1 + \frac{S}{K_{m1}}}, \quad E_2 = \frac{E_{20}}{1 + \frac{P_1}{K_{m2}}}$$

$$K_{m1} = \frac{k_{-11} + k_{12}}{k_{11}}, \quad K_{m2} = \frac{k_{-21} + k_{22}}{k_{21}}.$$

$$\tag{76}$$

The initial and boundary conditions are the same as for equations (65) and (66) in Section 1.2. Equation (75) is a system of nonlinear ordinary differential equations. The numerical solution of such a system will be described in Section 3.4.

## References

[1] Lam, C.F. and N. Schatz (1978). Automatic generation of steady state enzyme rate equation and the associated constraint equations. *J. Biochem.*, **84**, 585-595.

[2] King, E.L. and C. Altman (1956). A schematic method of deriving the rate laws for enzyme-catalyzed reactions. *J. Phys. Chem.*, **60**, 1375-1378.

[3] Hammes, G.G. and P.R. Schimmel (1970). Rapid reactions and transient states. *In* "The Enzymes," 3rd ed., ed. by P.D. Boyer, Vol.2, pp.67-114, Academic Press, New York.

[4] Crank, J. (1956). "The Mathematics of Diffusion," Clarendon Press, Oxford.

[5] Ames, E.F. (1965). "Nonlinear Partial Differential Equations in Engineering," Academic Press, New York.

## CHAPTER 3

# Numerical Methods for Solution of Rate Equations

The temporal evolution of enzymatic reactions, which we commonly encounter in *in vivo* or *in vitro* processes, can be expressed by rate equations in the form of ordinary or partial differential equations, as derived in Chapter 1. The kinetic properties of enzymes can be obtained from treatment of the experimental data and rate equations with the approximation methods described in Chapter 2. On the other hand, for the elucidation of more detailed mechanisms in enzymatic reaction, the analysis of dynamic behavior of enzymatic reactions naturally requires the procedure for solution of the rate equations, that is, solution of either initial value problems of nonlinear ordinary differential equations for homogeneous system or boundary value problems of partial differential equations for inhomogeneous system. An initial value problem of the rate equation in homogeneous and closed system corresponds to an experimental process of enzymatic reaction starting with mixing the substrate and enzyme. In inhomogeneous system with spatial distribution of enzymes and metabolites, the rate equation contains both reaction and diffusion terms to express a reaction-diffusion system with a certain boundary condition.

We can analytically obtain solutions of linear ordinary differential equations in a rather straightforward manner using the established procedures. Nonlinear equations involved in the rate equations for enzymatic reactions, however, cannot generally offer the analytic solutions because of the essential nonlinear-term representing the process of formation of the enzyme-substrate complex. We inevitably employ the computational procedure of numerical integration for the solution. In this chapter the numerical methods for solution of rate equations are described and applied to

---

This chapter was written by Naoto Sakamoto and Katsuya Hayashi.

some enzymatic reactions in homogeneous and inhomogeneous systems.

Most program libraries provide the Runge–Kutta–Gill method for numerical solution of ordinary differential equations. Unfortunately, application of this method to the rate equations for enzymatic reactions often gives rise to various difficulties. In some cases consumption of too much computer-time (CPU time) makes the solution virtually impossible. In other cases the correct solution cannot be obtained at the first step of numerical integration, leading to negative values for the concentrations of chemical species. Sometimes the variation in rate constants and initial conditions yields no solutions for equations which have numerical solutions under a certain condition.

This chapter is concerned with the cause of these difficulties and the general numerical method which overcomes them to suit the dynamic analysis of enzymatic reactions. In Section 3.1 the principle of numerical integration of ordinary differential equations is described. In Section 3.2 a numerical method is introduced which is suitable for solution of the rate equations. The most difficult problem in application of numerical methods to rate equations had been the "stiffness," but research on the numerical solution of stiff differential equations has advanced rapidly since the end of the 1960s. Several effective methods have been proposed [1]. We here consider mostly the Gear method, which is accepted as the most general and efficient procedure [1]. In Section 3.3 the method is applied to reactions of the Michaelis–Menten-type and allosteric (MWC dimeric model) enzymes. The chapter closes with the introduction of numerical methods for partial differential equations and their application to some simple enzyme systems.

## 3.1  Principle of Numerical Solution of Ordinary Differential Equations

In enzymatic reactions, for which the effects of spatial distribution and molecular diffusion in enzymes and metabolites are not taken into consideration, the system is assumed to be homogeneous and its temporal evolution is represented by a system of ordinary differential equations, as described in the preceding chapters. In the case of the simple Michaelis–Menten-type reaction in Scheme 3.1, the differential equations given in Chapter 1,

$$E + S \underset{k_{-1}}{\overset{k_1}{\rightleftarrows}} ES \xrightarrow{k_2} E + P$$

$$k_1 = 10^6 \, M^{-1} sec^{-1}, \quad k_{-1} = 100 \, sec^{-1}, \quad k_2 = 10 \, sec^{-1}$$

Scheme 3.1.  Michaelis–Menten-type reaction

$$
\left.
\begin{aligned}
\frac{dS(t)}{dt} &= -k_1 SE + k_{-1} C \\
\frac{dE(t)}{dt} &= -k_1 SE + k_{-1} C + k_2 C \\
\frac{dC(t)}{dt} &= k_1 SE - k_{-1} C - k_2 C \\
\frac{dP(t)}{dt} &= k_2 C,
\end{aligned}
\right\}
\tag{1}
$$

are regarded as the rate equation in homogeneous and closed system and are often solved under an initial condition of $S(t_0) = S_0$, $E(t_0) = E_0$ and $C(t_0) = P(t_0) = 0$, in which $C$ represents the concentration of the enzyme-substrate complex ES and $t_0$ denotes the initial time.

With the generalization of such a system, we now consider the numerical solution of an initial value problem,

$$
\left.
\begin{aligned}
\frac{dy}{dt} &= f(y, t) \\
y(t_0) &= y_0,
\end{aligned}
\right\}
\tag{2}
$$

where $y$ and $f(y,t)$ express the concentration and reaction term of a chemical species, respectively. The initial time and concentration are respectively denoted by $t_0$ and $y_0$.

The numerical methods are classified into two general types. One type is devised to approximate the correct solution $y(t)$ of equation (2) by a linear combination of orthogonal functions. The second type is called the difference method to evaluate approximate solutions $\{y_i\}$ to $\{y(t_i)\}$ at discrete points $\{t_i\}$. The former type has long been used to solve linear ordinary differential equations. The computational load, however, is too heavy to be applied to nonlinear or multi-variable problems. The remarkable development of digital computers allows the latter type to process any problem with a reasonable expenditure of CPU time. Consequently, we consider here the difference method suitable for solution of general nonlinear problems by computer.

The difference method yields an approximation $y_{n+1}$ to the exact solution $y(t_{n+1})$ in terms of some of the previously evaluated values $\{y_k$ at $t_k\}$ ($k = 0, 1, \cdots, n$). In general, the step size of time $h_{n+1} (= t_{n+1} - t_n)$ varies at each step, but we assume here a constant step-size $h$ for the sake of common and simple discussion. In addition, the initial time is set to $t_0 = 0$, and then we have $t_{n+1} = (n+1)h$.

The approximation of $y_{n+1}$ by the difference method introduces two kinds of errors. One kind is a truncation error, which arises from the variation in value of $f$ on the right-hand side of equation (2) during an interval $h$, for $f$ is assumed to have a fixed value $f(y_n, t_n)$ in the difference method. This error is dependent on the method employed, such as the Euler method or the Runge-Kutta method. Most program libraries usually provide the Runge-Kutta-Gill method of the fourth order, which means that the truncation error is of the order of $h^5$, that is, $O(h^5)$ for the method. The second kind of error is a round-off error, which is due to the restriction in a computer whereby a number is expressed by limited digits. Computers commonly used for scientific computation have at most eight significant digits for a decimal number. Hence, the difference method always contains both truncation and round-off errors. The truncation error can be decreased by reducing the step size $h$. This procedure, however, increases the computation time and the round-off error. These situations can be explained by the basic difference method of Euler, which is described in the following.

## 1.  Euler Method [2]

The Euler method, which is the simplest of the difference methods, is important in the sense that it is the basis for many other methods and is suitable for understanding the principle of numerical integration of ordinary differential equations. With an initial value $y(0) = y_0$ given in equation (2), the derivative $y'$ at $t=0$ is calculated from equation (2), that is, $y'(0) = f(y_0, 0)$. Using these values, we obtain $y(t_1) = y(h)$ from the expansion of $y(h)$ in a Taylor series around $t=0$ :

$$y(h) = y(0) + y'(0) h + \frac{1}{2} y''(0) h^2 + \cdots\cdots + \frac{1}{n!} y^{(n)}(0) h^n + R_n$$
$$\equiv y(0) + y'(0) h + T_e ,$$

(3)

where

$$R_n = \frac{y^{(n+1)}(\theta h)}{(n+1)!} h^{n+1}, \quad 0 < \theta < 1$$
$$T_e = \frac{1}{2} y''(0) h^2 + \cdots\cdots + \frac{1}{n!} y^{(n)}(0) h^n + R_n.$$

The Euler method uses the first two terms on the right-hand side of equation (3) to evaluate the approximation $y_1$ to $y(h)$ ;

$$y(h) \simeq y_1 = y_0 + hy'(0) = y_0 + hf(y_0, 0) .$$

(4)

The third term $T_e$ on the right-hand side of equation (3) is the truncation error for the Euler method. In a similar manner, the approximation $y_{n+1}$ at the time $t_{n+1}$ is obtained from the previously calculated value $y_n$ using the formula,

$$y_{n+1} = y_n + hf(y_n, t_n). \tag{5}$$

For the purpose of explanation, we now apply the Euler method of equation (5) to one of the simplest examples:

$$\frac{dy}{dt} = \lambda y, \quad \lambda < 0, \quad y_0 = 1, \tag{6}$$

which represents a first-order reaction. Equations (5) and (6) yield

$$y_{n+1} = y_n + h\lambda y_n = (1 + h\lambda) y_n. \tag{7}$$

The step size is always the first consideration in obtaining the numerical solution $\{y_k\}$ $(k = 1, 2, \cdots, n)$ from equation (7), because the truncation and round-off errors affect the evaluation at subsequent steps. The numerical solution $y_n$ at the $n$th step consists of the analytic (or exact) solution $y(t_n)$ and error term $\varepsilon_n$:

$$y_n = y(t_n) + \varepsilon_n. \tag{8}$$

The numerical solution $y_{n+1}$ in the following step is obtained by using $y_n$, producing the error term of $(1 + h\lambda)\varepsilon_n$, which is called a propagation error:

$$\begin{aligned}
y_{n+1} &= (1 + h\lambda)\{y(t_n) + \varepsilon_n\} \\
&= (1 + h\lambda) y(t_n) + (1 + h\lambda)\varepsilon_n.
\end{aligned} \tag{9}$$

Selection of step size $h$ such as $|(1 + h\lambda)\varepsilon_n| > |\varepsilon_n|$ has the error term increase at each step. The stable calculation thus requires the following condition of the step size:

$$|1 + h\lambda| \leq 1. \tag{10}$$

The concept of absolute stability of the difference formula is important in relation to the selection of step size $h$. We define "absolute stability" as follows: a difference formula is absolutely stable for a given step-size $h$ and a given differential equation if the propagation error $\varepsilon_m$ in all subsequent values $y_m (m > n)$ is smaller than the error $\varepsilon_n$, which happens to be contained in the numerical solution $y_n$. This definition depends on differential equations. It is common that the absolutely stable region for $h$ is defined for the "test equation" of equation (6); in general, $\lambda$ in equation (6) can be a complex number, and the absolutely stable region is defined in the complex $h\lambda$-plane as

$$|\varepsilon_m| \leq |\varepsilon_n| \quad \text{if} \quad m > n.$$

From inequality (10) the region of absolute stability for the Euler method is inside a unit circle centered at $(-1, 0)$ in the complex $h\lambda$-plane, as shown in Fig. 3.1. Figure 3.2 illustrates an example of application of the Euler

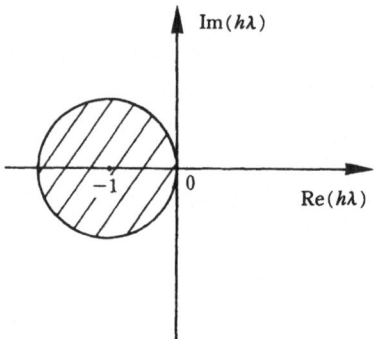

**Fig. 3. 1**  Region of absolute stability in the Euler method.

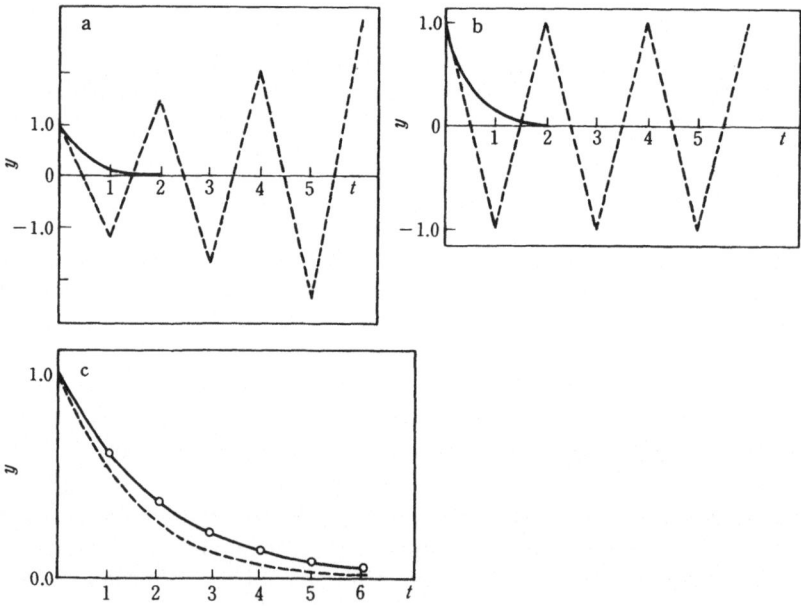

**Fig. 3. 2**  Numerical solution by the Euler method. Differential equation :
$dy/dt = -1000\,y$, initial condition : $y_0 = 1$. Solid line : analytic solution $y = \exp(-1000\,t)$ ; broken line : numerical solution. The unit for
the abscissa $(h)$ : a : $h = 2.2 \times 10^{-3}$ ; b : $h = 2.0 \times 10^{-3}$ ; c : $h = 0.5 \times 10^{-3}$.

method to equation (6) with $\lambda = -1000$ and various step-size $h$.  In (a) of
Fig. 3.2 the solution diverges rapidly because $|1 + h\lambda| = 1.2(>1)$.  In (b),
where $|1 + h\lambda| = 1$, the solution leads to a sustained oscillation.  For these

two cases the method yields a numerical solution quite different from the exact solution. On the other hand, in (c), where $|1+h\lambda|=0.5(<1)$, the method provides a good approximation to the exact solution. As in (b), the condition for absolute stability, if satisfied, cannot always assure the good performance of the method. Thus, the stable solution to equation (6) for negative $\lambda$ requires the bound for $h$:

$$h\leq|2/\lambda|. \tag{11}$$

In contrast to the (forward) Euler method of equation (5), the backward Euler method is absolutely stable for arbitrary $\lambda$. The formula is given by

$$y_{n+1}=y_n+hf(y_{n+1}, t_{n+1}), \tag{12}$$

which is derived by replacing $f(y_n, t_n)$ on the right-hand side of equation (5) with $f(y_{n+1}, t_{n+1})$. Equation (12) indicates that $y_{n+1}$ cannot be expressed explicitly if $f$ is nonlinear. This kind of formula is called the implicit method, while the forward Euler method is one of the explicit methods. Application of the backward Euler method to the test equation leads to the formula for solution:

$$y_{n+1}=y_n+h\lambda y_{n+1},$$

or

$$y_{n+1}=(1-h\lambda)^{-1}y_n. \tag{13}$$

Hence, the backward Euler method is absolutely stable for any negative $\lambda$ because $|(1-h\lambda)^{-1}|<1$. In fact, the region of absolute stability is the area outside of a unit circle as shown in Fig. 3.3. The results of the solution to equation (6) with $\lambda=-1000$ by the backward Euler method are demon-

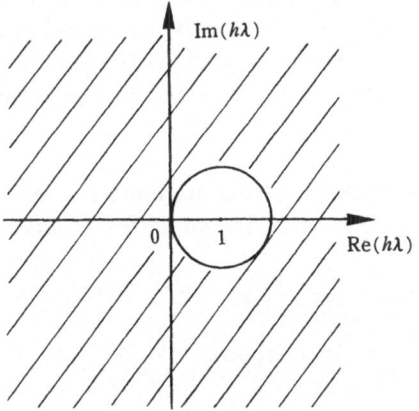

**Fig. 3. 3** Region of absolute stability in the backward Euler method.

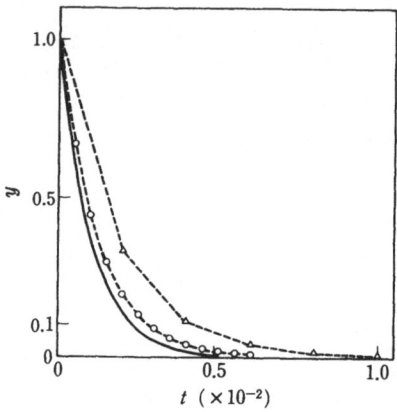

**Fig. 3. 4** Numerical solution by the backward Euler method. The differential equation, initial condition and connotation of the solid line are the same as in Fig. 3. 2. Broken line: numerical solution with $h=5.0\times 10^{-4}(\bigcirc)$; $h=2.0\times10^{-3}(\triangle)$.

strated in Fig. 3.4. Although the accuracy of the solution becomes worse as the step size $h$ increases, the solution is always a reasonable approximation to the analytic solution, $y = \exp(-1000t)$. Thus, the backward Euler method extends the region of absolute stability, providing the basis for the numerical methods suitable for rate equations of enzymatic reactions.

## 2. One-Step Methods [2]

The numerical method for ordinary differential equations wherein the approximation $y_{n+1}$ at the subsequent step can be evaluated only with $y_n$ is called the one-step method. The Euler method is the simplest of the one-step methods. The one-step method is expressed in a general form,

$$y_{n+1} = y_n + h\phi(y_n, t_n, h),\tag{14}$$

where $\phi(y_n, t_n, h)$ is the increment function of Henrici. To obtain $\phi$ in an explicit form, we expand $y(t_n+h)$ in a Taylor series around $t_n$:

$$y(t_n+h) = y(t_n) + hy'(t_n) + \frac{h^2}{2}y''(t_n)$$
$$+\cdots\cdots+\frac{h^n}{n!}y^{(n)}(t_n)+O(h^{n+1}).\tag{15}$$

Using equation (2), we have

$$y(t_n+h) = y(t_n) + hf(y_n, t_n) + \frac{h^2}{2} f'(y_n, t_n)$$

$$+ \cdots\cdots + \frac{h^n}{n!} f^{(n-1)}(y_n, t_n) + O(h^{n+1}) \tag{16}$$

$$\equiv y(t_n) + h\Delta(y_n, t_n).$$

It is naturally seen that the increment function $\phi$ should coincide with $\Delta$ as accurately as possible. In alternation, it apparently seems best to perform the numerical calculation by direct use of equation (16). However, that is not practical for two reasons. First, inclusion of total differentials $f^{(n-1)} = d^{n-1}f/dt^{n-1}$ as well as function $f$ requires a larger memory-area for programming. Secondly, the calculation of $f^{(n-1)}$ is rather difficult. In fact, the derivative of first order is given by

$$\frac{df(y(t), t)}{dt} = \frac{\partial f}{\partial y} \frac{dy}{dt} + \frac{\partial f}{\partial t} = f\frac{\partial f}{\partial y} + \frac{\partial f}{\partial t}, \tag{17}$$

and the derivatives of higher orders become too complex to compute.

Some methods are devised for approximation of $\Delta$ with $\phi$. The simplified Runge-Kutta method is derived by approximating the first-order derivative of $f$ in $\Delta$ with the sum of the values of $f$ at two different pairs of arguments $(y, t)$ and $(y', t')$. In this method the increment function $\phi$ is given by

$$\phi = a_1 k_1 + a_2 k_2, \tag{18}$$

where

$$k_1 = f(y, t), \quad k_2 = f(y + phf(y, t), t + qh).$$

The coefficients $a_1$, $a_2$, $p$ and $q$ are determined so that $\phi$ coincides with $\Delta$ up to the order $O(h^2)$. That is, the first-order derivative of $f$ is approximated by

$$\frac{h}{2} f'(y, t) = (a_1 - 1) f(y, t) + a_2 f(y + phf(y, t), t + qh) + O(h^2). \tag{19}$$

Expanding the right-hand side of equation (19) in a Taylor series and matching the terms of an equal order of $h$ with use of equation (17) yield the coefficients in parametric form:

$$a_1 = 1 - \alpha, \quad a_2 = \alpha, \quad p = q = \frac{1}{2\alpha}. \tag{20}$$

The parameter $\alpha$ is chosen to suit a given problem. The Heun method is for $\alpha = 1/2$ and the modified Euler method is for $\alpha = 1$. These methods are also called the Runge-Kutta method of the second order since the increment function $\phi$ is the approximation to $\Delta$ within an error of $O(h^2)$.

The Runge-Kutta method of the fourth order employs the evaluation of

$\phi$ with $k_1$, $k_2$, $k_3$ and $k_4$, *i.e.*, the values of $f$ at four points. The eight coefficients used in the method are determined similarly to the derivation of equation (20). The various methods in the family of the Runge-Kutta method of the fourth order have been devised to suit given problems. The classical Runge-Kutta method developed earliest is given as follows:

$$\phi = \frac{1}{6}(k_1 + 2k_2 + 2k_3 + k_4),\qquad(21)$$

where

$$k_1 = f(y, t)$$
$$k_2 = f\left(y + \frac{1}{2}hk, t + \frac{1}{2}h\right)$$
$$k_3 = f\left(y + \frac{1}{2}hk_2, t + \frac{1}{2}h\right)$$
$$k_4 = f(y + hk_3, t + h).$$

The Runge-Kutta method, which has a truncation error of order $O(h^5)$, is of very high accuracy. Its region of absolute ability, however, is as limited as that of the Euler method indicated in Fig. 3.1. The method thus has no liberty to choose a large step-size at will; this will be described in detail in Section 3.3. The Runge-Kutta-Gill method commonly offered in most program libraries is one of the Runge-Kutta methods which was developed for the British computer EDSAC in 1951.

## 3. Linear Multi-Step Methods [2]

The one-step methods discussed in the preceding sections require only a knowledge of the solution at one mesh-point $(y_{n-1})$ and the differential equation to compute the value of $y_n$ at the subsequent $n$th step. When we have performed the numerical integration up to the $(n-1)$th step, the values of $(y_0, y_1, \cdots, y_{n-1})$ and $(hy_0', hy_1', \cdots, hy_{n-1}')$, where $hy_i' = hf(y_i, t_i)$, are available for computation at the subsequent steps. Natural extension of the one-step methods thus leads to the methods called linear multi-step methods which evaluate the value at a point using information from a number of previous points. The $k$-step methods can obtain $y_n$ from a knowledge of the differential equation and the class of values $\hat{y}_{n-1}$ such as

$$\hat{y}_{n-1} = [y_{n-1}, y_{n-2}, \cdots\cdots, y_{n-k}, hy_{n-1}', \cdots\cdots, hy_{n-k}'].\qquad(22)$$

The linear multi-step methods are most suitable for problems to be solved with high accuracy and speed. One drawback of these methods is the requirement of the starting values of $\hat{y}_k = [y_{k-1}, y_{k-2}, \cdots, y_0, hy_{k-1}', hy_{k-2}', \cdots, hy_0']$, which have to be calculated by a certain one-step method like one of the Runge-Kutta methods.

The general expression of the $k$-step method is given as follows:

$$y_n = \sum_{i=1}^{k} \alpha_i y_{n-i} + h \sum_{i=0}^{k} \beta_i y'_{n-i}. \tag{23}$$

The forward and backward Euler methods correspond to the specific cases in equation (23). If $\beta_0 \neq 0$, equation (23) is called an implicit formula, the right-hand side of which contains a term $y'_n = f(y_n, t_n)$ to make the direct computation of $y_n$ impossible. In the case of $\beta_0 = 0$, equation (23) is an explicit formula to yield $y_n$ directly.

A notable example of explicit methods is the Adams-Bashforth method, and its $k$-step formula is given by

$$y_n = y_{n-1} + h \sum_{i=1}^{k} \beta_{ki} y'_{n-i} . \tag{24}$$

The coefficients $\beta_{ki}$ are determined from an expansion of the right-hand side of equation (24) in a Taylor series around $t_n$ to equate the terms of equal orders up to $h^k$ on both sides. The $k$-step Adams-Bashforth method is a formula of $k$th order with truncation error of the order $O(h^{k+1})$. The method, which requires only one evaluation of $f$ for each step, is more efficient for problems involving complex calculation of $f$ than the high-order one-step methods like the Runge-Kutta method of the fourth order, which needs four evaluations of $f$ for each step. It should be mentioned that the Adams-Bashforth method is rarely used as an independent formula but is rather employed as predictor in the predictor-corrector methods introduced below, because the region of absolute stability, which is an important factor in the numerical methods for the rate equations of enzymatic reactions, quickly becomes smaller as the order $k$ increases.

A well-known example of implicit methods is the Adams-Moulton method, which is represented by

$$\begin{aligned} y_n &= y_{n-1} + h \sum_{i=0}^{k-1} \beta_i y'_{n-i} \\ &= y_{n-1} + h \sum_{i=1}^{k-1} \beta_i y'_{n-i} + \beta_0 h f(y_n, t_n). \end{aligned} \tag{25}$$

The $(k-1)$-step formula has a truncation error of the order $O(h^{k+1})$, thus belonging to the methods of order $k$. In general, equation (25) is a nonlinear equation of $y_n$, from which $y_n$ cannot be solved analytically. Iteration of the predictor-corrector process is a common procedure for solving it. The predictor process is to evaluate the starting value $y_{n,(0)}$ for the formula (25). An explicit method like the Adams-Bashforth method is often employed for this process. The corrector process is expressed by

$$y_{n,(m)} = y_{n-1} + h \sum_{i=1}^{k-1} \beta_i y'_{n-i} + \beta_0 h f(y_{n,(m-1)}, t_n), \tag{26}$$

which is derived from equation (25). By starting from $y_{n,(0)}$ evaluated as the predictor, the corrector process is repeated to get convergence of $y_{n,(m)}$. In practical computation, the corrector formula of $k$th order corresponds to the predictor formula of $k$th or $(k-1)$th order, which evaluates the starting value close to the convergent solution, and hence two or three iterations suffice for the corrector process.

The Adams-Moulton method of first order is the backward Euler method, whose region of absolute stability is shown in Fig. 3.3. The region for the second-order method occupies the half plane of negative $\mathrm{Re}(h\lambda)$ in the complex $h\lambda$-plane. If the order $k$ becomes higher than three, however, the region gets limited, even though it is still about ten times larger than in the Adams-Bashforth method. It is therefore concluded that the Adams-Moulton method cannot be the general method for dynamic analysis of enzymatic reactions.

## 3.2 Numerical Method for Rate Equations

The numerical methods for ordinary differential equations described in the preceding sections pose three difficult problems in their application to the rate equations for enzymatic reactions. First, the rate equations constitute systems of differential equations, as seen in equation (1) for a simple Michaelis-Menten-type reaction. Secondly, they are nonlinear in most cases. The third and most serious problem is concerned with the stiffness of the rate equations. In fact, stiff differential equations commonly result from kinetic representation of chemical reaction systems, giving rise to great difficulty in the numerical analysis. In this section, we consider these problems in order to develop a numerical procedure suitable for the rate equations of enzymatic reactions.

### 1. Treatment of System of Nonlinear Differential Equations

A system of ordinary differential equations is written as

$$\frac{dy(t)}{dt} = f(y, t), \tag{27}$$

where $y$ and $f$ are vectors of $[y^1, y^2, \cdots, y^n]$ and $[f^1, f^2, \cdots, f^n]$, respectively. Application of the backward Euler method of equation (12) to the system yields a nonlinear vector equation:

$$y_{n+1} = y_n + hf(y_{n+1}, t_{n+1}). \tag{28}$$

This can be solved by iterative substitution similar to equation (26) in the Adams-Moulton method. As seen from an expansion of equation (28) as

$$y_{n+1} = y_n + hf(y_n, t_n)\left[1 + h\frac{\partial f}{\partial y} + h^2\left(\frac{\partial f}{\partial y}\right)^2 + \cdots\cdots\right],$$

the solution converges rapidly if the step size $h$ is small and the norm $\|h\partial f/\partial y\| \ll 1$. This method, however, is not appropriate for the rate equations of enzymatic reactions because the condition for convergence is satisfied only with extremely small $h$.

Instead, Gear recommends employment of the Newton method [2]. Denoting an approximation to $y_{n+1}$ by $y_{(0)}$ and a correction to $y_{(0)}$ by $\eta_{(0)}$, respectively, we have

$$y_{n+1} = y_{(0)} + \eta_{(0)},\tag{29}$$

which is substituted into equation (28) to yield the following relationship using an expansion of $f$ in a Taylor series:

$$y_{(0)} + \eta_{(0)} = y_n + hf(y_{(0)} + \eta_{(0)}, t_{n+1})$$
$$= y_n + hf(y_{(0)}, t_{n+1}) + hJ\eta_{(0)},$$

where $\eta_{(0)}$ is assumed to be small enough to ignore the terms higher than second order. $J$ is the Jacobian matrix, $(i, j)$ component of which is $\partial f^i/\partial y^j$. The relationship is rearranged as follows:

$$\eta_{(0)} = (I - hJ(y_{(0)}, t_{n+1}))^{-1}\{y_n + hf(y_{(0)}, t_{n+1}) - y_{(0)}\},\tag{30}$$

where $I$ is a unit matrix. Thus, a new approximation $y_{(1)}(= y_{(0)} + \eta_{(0)})$ to $y_{n+1}$ is determined. The similar process of equations (29) and (30) is repeated with $y_{n+1} = y_{(1)} + \eta_{(1)}$ until the convergence criterion is fulfilled. This procedure is called the Newton method. The method becomes inefficient if the calculation of Jacobian matrix $J$ consumes much CPU time. Fortunately, this is not the case for enzymatic reactions, in which $J$ generally varies very gradually as demonstrated in Section 3.3. Therefore, $J$ need not be reevaluated at each step of the numerical computation and this assures high efficiency with the Newton method.

## 2. Stiff Stability

The difficulty due to stiffness of differential equations is related to the absolute stability of the numerical method. In order to explain the situation, let us consider the behavior around the time $t_n$ in the exact solution $y(t)$ of equation (27). When an approximate solution $y_n$ is obtained, the right-hand side of equation (27) is expanded in a Taylor series with respect to $(y(t) - y_n)$ to retain the first-order term as follows:

$$\frac{dy(t)}{dt} = f(y_n, t_n) + J(y_n, t_n)(y(t) - y_n),\tag{31}$$

where $J$ is the Jacobian matrix same as in equation (30). Equation (31) is an extended form of vector expression of equation (6), $\lambda$ of which corresponds to the eigenvalues $\lambda_i$ of $J(y_n, t_n)$. Therefore, the general form of solution $y(t)$ is given by

$$y(t) - y_n = b + \sum_i c_i \omega_i e^{\lambda_i(t-t_n)}, \tag{32}$$

where $\omega_i$ is the eigenvector of $J$ corresponding to eigenvalue $\lambda_i$. The coefficients $b$ and $c_i$ are determined by

$$\left. \begin{array}{l} Jb + f(y_n, t_n) = 0 \\ b + \sum_i c_i \omega_i = 0. \end{array} \right\} \tag{33}$$

The numerical solution of $y(t)$ from equation (32) requires the calculation of every term of $\exp(\lambda_i(t-t_n))$. For almost any enzymatic reaction $\lambda_i$'s are negative real numbers, and this is assumed in the following discussion. With explicit methods such as the Euler, Runge-Kutta and Adams-Bashforth methods, the absolute stability requires the condition,

$$h \leq |\lambda_i|^{-1}, \tag{34}$$

to be valid. The stable computation of equation (32), which is a linear combination of $\exp(\lambda_i(t-t_n))$, depends on $\lambda_M$, maximum value of $|\lambda_i|$, in the sense that the step size $h$ is limited by

$$h \leq \lambda_M^{-1}. \tag{35}$$

On the other hand, if the relationship between $\lambda_M$ and $\lambda_m$, minimum value of $|\lambda_i|$, is such that

$$\lambda_M / \lambda_m \gg 1, \tag{36}$$

$\exp(-\lambda_m(t-t_n))$ is virtually constant during the time period $h$ under the condition (35), for we have

$$e^{-\lambda_m(t_{n+1}-t_n)} - e^{-\lambda_m(t_n-t_n)} = e^{-\lambda_m h} - 1 \simeq e^{-\lambda_m/\lambda_M} - 1 \simeq 0. \tag{37}$$

The differential equations for which the relationship (36) is valid are called stiff. The numerical solution of stiff differential equations has to be performed with very small step-size. Consequently, the computation of the time course of the chemical species involving the term $\exp(-\lambda_m t)$ requires considerable CPU time. It also happens in many enzymatic reactions that computers cannot appropriately handle the numbers of virtual null in the change of $\exp(-\lambda_m t)$, resulting in inaccurate and meaningless values of negative concentration and overflows.

The numerical methods for stiff differential equations were first considered by Curtis and others in the early 1950s [1]. The exponentially fitted Runge-Kutta method was proposed by Fowler and Warten [3] in 1967 as a nonlinear multi-step formula. Subsequently, in 1968 Gear developed the stiffly stable method of linear multi-step implicit formula, generating the remarkable advancement in this field [1].

The stiff stability of a method is defined as follows [2] : In the complex $h\lambda$-plane, if the region $R_1$ [$\text{Re}(h\lambda) < D$] for a real number $D$ is the region of absolute stability, and accuracy is established in the region $R_2$ [$D < \text{Re}(h\lambda) < \alpha, |\text{Im}(h\lambda)| < \theta$] for real numbers $\alpha$ and $\theta$, a method is stiffly stable.

The definition implies that a component including an eigenvalue $\lambda$ varies for one step of the numerical calculation by a quantity of $\exp(\text{Re}(h\lambda))$, which decays rapidly for $\text{Re}(h\lambda) < D(<0)$, and that the computation process can neglect the component without sacrificing the accuracy. The absolute stability has only to be required to prevent the enlargement of errors. On the other hand, around the origin in the complex $h\lambda$-plane the method must be stable and accurate so that the step size $h$ locates in the regions $R_1$ or $R_2$. The advantage of the stiff stability will again be explained in Section 3.3 with examples of numerical computation for enzymatic reactions.

## 3. Gear Method

The stiffly stable method proposed by Gear [4] is a predictor-corrector method as follows: for the formula of order $q$,

$$\text{Predictor} : y_{n,(0)} = \alpha_1 y_{n-1} + \cdots\cdots + \alpha_q y_{n-q} + \eta_1 h y'_{n-1}$$
$$\text{Corrector} : y_{n,(m+1)} = \alpha_1^* y_{n-1} + \cdots\cdots + \alpha_q^* y_{n-q} + \eta_0^* h f(y_{n,(m)}, t_n). \tag{38}$$

The truncation error at convergence of the corrector is given by

$$h^{q+1} y^{(q+1)}/(q+1) + O(h^{q+2}),$$

where $y^{(q+1)}$ denotes the $(q+1)$th derivative of $y$. The coefficients $\alpha_i$, $\alpha_1^*$, $\eta_1$ and $\eta_0^*$ are evaluated in the text [2]. The formula for $q=1$ corresponds to the backward Euler method.

The method makes use of the already computed values of $[y_{n-1}, h y'_{n-1}, y_{n-2}, \cdots, y_{n-q}]$ to determine the predictor $y_{n,(0)}$ for $y_n$ with the $q$th-degree polynomial approximation given in equation (38). As the initial condition specifies only $y_0$ and $h y'_0$, the computation starts with the first-order formula. The order can increase subsequently, but the stiff stability is assured only for the formulas with $q < 6$.

The convergent solution of the corrector equation is obtained using the Newton method by the iteration from $y_{n,(0)}$ evaluated in the predictor. That is, as in equation (30), we have

$$y_{n,(m+1)} = y_{n,(m)} + \{I - h\eta_0^* J(y_{n,(m)}, t_n)\}^{-1}$$
$$\times \left\{ \sum_{i=1}^{q} \alpha_i^* y_{n-i} + h\eta_0^* f(y_{n,(m)}, t_n) - y_{n,(m)} \right\}. \tag{39}$$

The Gear method proceeds with the iteration of equation (39) without evaluation at each step of the Jacobian $J$, which is known to generally vary very slowly. If the convergence of equation (39) should not be attained after three iterations, $J$ would then be computed at the step (time), and the iteration would resume.

The method thus provides the stable numerical solution of a stiff differential equation for relatively large step-size $h$. In practice, equation (38) is not employed in the form as given because programming becomes difficult for a change in step size $h$ during computation. Instead, the formulation is done in the Nordsieck form. Introduction of a matrix $Z_n = [y_n, hy_n', \cdots, h^q y_n^{(q)}/q!]^t$ makes it possible to transform the predictor-corrector formula (38) to the following form:

$$Z_{n, (0)} = QBQ^{-1}Z_{n-1} \tag{40}$$

$$Z_{n, (m+1)} = Z_{n, (m)} + lF(Q^{-1}Z_{n, (m)})$$
$$= Z_{n, (m)} + lF(Z_{n, (m)}), \tag{41}$$

where

$$B = \begin{pmatrix} \alpha_1 & \eta_1 & \alpha_2 & \cdots\cdots & \alpha_{q-1} & \alpha_q \\ \gamma_1 & \delta_1 & \gamma_2 & \cdots\cdots & \gamma_{q-1} & \gamma_q \\ 1 & 0 & 0 & \cdots\cdots & 0 & 0 \\ 0 & 0 & 1 & 0\cdots & 0 & 0 \\ \vdots & \vdots & 0 & \ddots & \vdots & \vdots \\ 0 & 0 & 0 & \cdots 1 & 0 \end{pmatrix}$$

$$\gamma_i = (\alpha_i - \alpha_i^*)/\eta_0^*, \quad \delta_1 = \eta_1/\eta_0^*$$
$$l = Q[\eta_0^* \ 1 \ 0 \cdots\cdots 0]^t$$
$$F(Z_{n(m)}) = hf(y_{n(m)}, t_n) - hy_{n(m)}'.$$

The matrix $Q$ is determined from the relationship,

$$Z_n = Q[y_n, hy_n', y_{n-1}, \cdots\cdots, y_{n-q+1}] \tag{42}$$

Then, equation (39) is written as

$$Z_{n, (m+1)} = Z_{n, (m)} + lWF(Z_{n, (m)}), \tag{43}$$

where

$$W = (l_1 - hl_0 \partial f/\partial y)^{-1}.$$

An algorithm for the Gear method is thus derived which retains $Z_n$ for each step of the computation to process equation (40) as the predictor and equation (43) as the corrector, respectively. As mentioned above, the computation starts with $Z_0 = [y_0, hy_0']^t$.

The general-purpose program for the Gear method was coded by Gear

[2]. Its characteristics are summarized as follows.

1) The computation starts using the backward Euler method, which corresponds to the formula for $q = 1$ in the Gear method, thus automatically providing the starting values for the linear multi-step procedure.

2) A large step-size $h$ can be taken since the stability is assured up to the fifth-order formula.

3) The results accumulated so far have proved its applicability to various problems.

Subsequent research on numerical methods for stiff differential equations has been performed in the sense of extending the Gear method. At present, this method is still regarded as the best due to its general applicability, ease of use and economy.

## 3.3 Application to Enzymatic Reactions — Examples and Remarks

In this section, the numerical methods described in the preceding sections are applied to the rate equations for fundamental and important reactions of Michaelis-Menten-type and allosteric enzymes. Particular remarks on the application are pointed out as a comparison of the Runge-Kutta-Gill method of the fourth order and the Gear method. As mentioned above, the former method is commonly employed in many cases of numerical integration of ordinary differential equation, while the latter is known to suit the stiff equation.

We first consider the rate equation for a Michaelis-Menten-type reaction given in equation (1). The numerical integration performed with the rate constants of

$$k_1 = 10^6 \, M^{-1} \, sec^{-1}, \quad k_{-1} = 100 \, sec^{-1}, \quad k_2 = 10 \, sec^{-1} \tag{44}$$

and the initial and total concentrations of $S(t_0) = 10.0 \, mM$ and $E_0 = 10 \, \mu M$ yields the time course of the progress in reaction as illustrated in Fig. 3.5. The reaction is in a transient phase from the start to $10^{-4}$ sec, during which the enzyme-substrate complex is formed. Then, a steady state during which the enzyme species do not change their concentrations is maintained for some period. The substrate is consumed steadily until its complete consumption ceases the reaction.

The time course in Fig. 3.5 is obtained by application of the Gear method with an expense of about 3 sec of CPU time in a medium-sized computer. In contrast, the Runge-Kutta-Gill method of the fourth-order spends more than 3 min of CPU time in integration of the same rate equations for the first 4.5 sec in the progress of reaction. An extremely long CPU time would

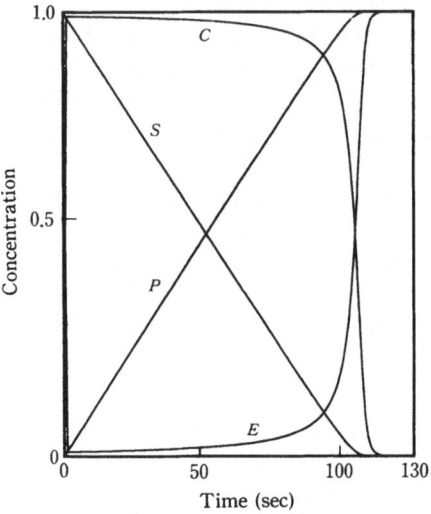

**Fig. 3. 5** Time course of the Michaelis-Menten-type reaction represented by equation (1). Computation by the Gear method. Initial condition: $S = 10$mM, $E = 10\mu$M, $C = P = 0$. Full scale (1.0) on the ordinate: $E = C = 10\mu$M, other chemical species $= 10$mM.

be required to continue the computation until the end of reaction. In order to explain the reason for this difference, we examine the eigenvalues of the Jacobian matrix $J$. The matrix for equation (1) is written as

$$J = \begin{bmatrix} -k_1 E(t) & -k_1 S(t) & k_{-1} & 0 \\ -k_1 E(t) & -k_1 S(t) & k_{-1}+k_2 & 0 \\ k_1 E(t) & k_1 S(t) & -k_{-1}-k_2 & 0 \\ 0 & 0 & k_2 & 0 \end{bmatrix}, \tag{45}$$

and its characteristic equation is given by

$$|\lambda I - J| = \lambda^2 \{\lambda^2 + (k_1 S(t) + k_1 E(t) + k_{-1} + k_2)\lambda + k_1 k_2 E(t)\} = 0, \tag{46}$$

where $\lambda$ represents the eigenvalue. Equation (46) has a null double root, for equation (1) implies the conservation equations,

$$\left. \begin{array}{l} S(t) + P(t) + C(t) = S(t_0) \\ E(t) + C(t) = E_0 . \end{array} \right\} \tag{47}$$

Around the time $t$, the concentration in each chemical species varies according to

$$C_{0t} + C_{Mt}e^{-\lambda_M t} + C_{mt}e^{-\lambda_m t} , \tag{48}$$

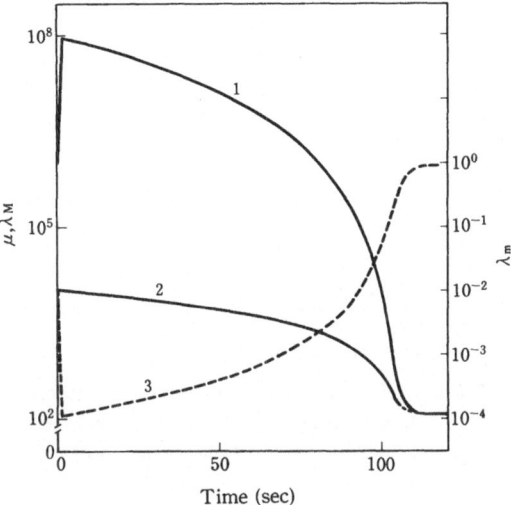

**Fig. 3.6** Time course of the eigenvalues of equation (1) (Michaelis-Menten-type reaction). 1: $\mu = \lambda_M/\lambda_m$; 2: $\lambda_M$; 3: $\lambda_m$.

where $-\lambda_M$ and $-\lambda_m$ ($\lambda_M, \lambda_m > 0$) are two other eigenvalues of $J$, and $C_{0i}$, $C_{Mi}$ and $C_{mi}$ are appropriate constants. It is seen from equation (46) that $\lambda_M$ and $\lambda_m$ are not constants but certain functions of time. Now that the temporal change in concentration of the chemical species in equation (46) is known, the temporal variation in $\lambda_M$ and $\lambda_m$ can be calculated by sub-stitution. The result is demonstrated in Fig. 3.6, indicating that at the start of reaction $\lambda_M = 10^4$ and $\lambda_m = 10^{-2}$, and then $\lambda_m$ decreases rapidly to $10^{-4}$. The degree of stiffness of the rate equation is represented by a value of $\mu = \lambda_M/\lambda_m$, which reaches a value of $10^8$ just after the start of reaction. When the reaction is in the steady state, neither $\lambda_M$ nor $\lambda_m$ vary noticeably.

The numerical integration of this type of equations by the Euler method and the Runge-Kutta method of the fourth-order is imposed by the limi-tation of very small step-size as

$$h \simeq \lambda_M^{-1} \simeq 10^{-4}. \tag{49}$$

In fact, the computation proceeds in the Runge-Kutta-Gill and Gear methods, respectively, as shown in Fig. 3.7. The iteration by 60 times in the Runge-Kutta-Gill method computes only 0.013 sec in the progress of reaction, while an equal number of iterations in the Gear method corre-sponds to 1.43 sec in the progress. The Gear method is less efficient than the Runge-Kutta-Gill method until 55 iterations, but after 60 iterations it

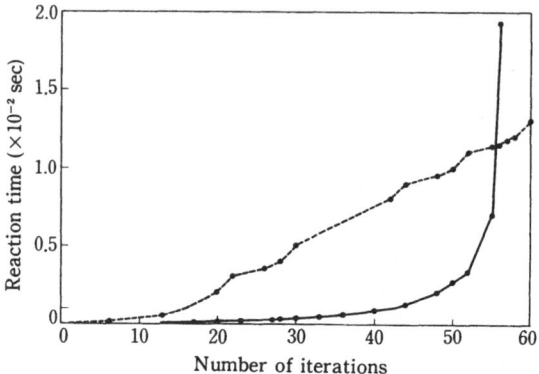

**Fig. 3. 7** Comparison of progress in numerical integration. Initial condition: $S = 10$ mM, $E = 10 \mu$M, $C = P = 0$. Solid line: Gear method; broken line: Runge-Kutta-Gill method of the fourth order.

processes the computation with a step size larger than 1 sec. This difference results in $10^6$ and only $10^2$ iterations in the Runge-Kutta-Gill and Gear methods, respectively, for the progress of about 100 sec in the reaction displayed in Fig. 3.5.

As seen in equation (49), the step-size for the Runge-Kutta-Gill method is limited to a very small value throughout the reaction, because $\lambda_M$ is very large in many enzymatic reactions. The term $\exp(-\lambda_M t)$ in equation (48), however, contributes to the temporal change in concentration of chemical species at the early stage of reaction, actually only in the transient phase, becoming almost negligible as the reaction proceeds. This is clearly indicated in Table 3.1 for the evaluation of the value of $\exp(-\lambda_M t)$. This situation is similar to the case of the backward Euler method as shown in

**Table 3. 1**  Values of $\exp(-\lambda_M t)$ of the Michaelis-Menten-type reaction

| Reaction time (sec) | $\lambda_M$ | $\lambda_M t$ | $e^{-\lambda_M t}$ |
|---|---|---|---|
| 0 | $1.012 \times 10^4$ | 0 | 1 |
| $3.54 \times 10^{-5}$ | $1.012 \times 10^4$ | 0.3582 | 0.6990 |
| $5.41 \times 10^{-4}$ | $1.010 \times 10^4$ | 5.467 | $4.224 \times 10^{-3}$ |
| $1.11 \times 10^{-3}$ | $1.010 \times 10^4$ | 10.16 | $3.869 \times 10^{-5}$ |
| 1.434 | $1.006 \times 10^4$ | $1.443 \times 10^4$ | $10^{-6267}$ |
| 5.434 | $9.563 \times 10^3$ | $5.197 \times 10^4$ | $10^{-22570}$ |
| 50.43 | $5.133 \times 10^3$ | $2.589 \times 10^6$ | $10^{-1124300}$ |
| 100.4 | $4.386 \times 10^2$ | $4.404 \times 10^4$ | $10^{-19126}$ |
| 115.4 | $1.192 \times 10^2$ | $1.376 \times 10^4$ | $10^{-5796}$ |

Fig. 3.4. The absolute stability is the only concern in having the round-off and truncation errors propagate without divergence. On the other hand, the accuracy for the term $\exp(-\lambda_m t)$ is treated rigorously.

The next example is application to a more complex reaction of an allosteric enzyme. Scheme 3.2 presents an MWC dimeric model, in which substrate S interacts only with the enzyme species in R-state to be con-

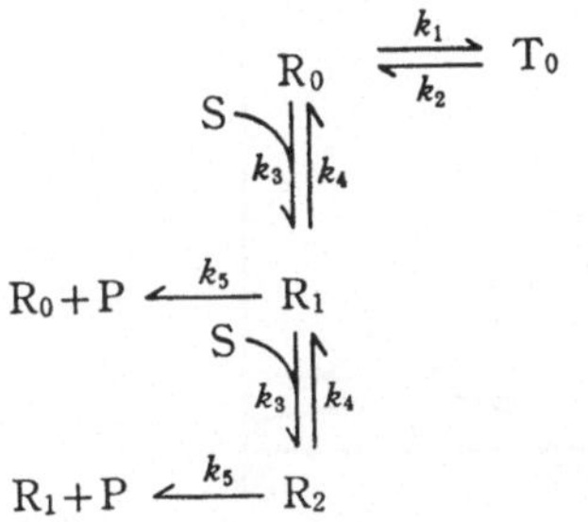

$$k_1 = 500 \, \text{sec}^{-1}$$
$$k_2 = 0.5 \, \text{sec}^{-1}$$
$$k_3 = 10^7 \, \text{M}^{-1} \text{sec}^{-1}$$
$$k_4 = 500 \, \text{sec}^{-1}$$
$$k_5 = 1 \, \text{sec}^{-1}$$

Scheme 3.2.  MWC dimeric model of allosteric reaction

verted to product P from $R_1$ and $R_2$ species. The rate equation for the reaction is written as follows:

$$\frac{dS}{dt} = -2k_3 R_0 S + k_4 R_1 - k_3 R_1 S + 2k_4 R_2$$

$$\frac{dP}{dt} = k_5 R_1 + 2k_5 R_2$$

$$\frac{dR_0}{dt} = -2k_3 R_0 S + k_4 R_1 + k_5 R_1 - k_1 R_0 + k_2 T_0$$

$$\frac{dR_1}{dt} = 2k_3 R_0 S - k_4 R_1 - k_5 R_1 - k_3 R_1 S + 2k_4 R_2 + 2k_5 R_2$$

$$\frac{dR_2}{dt} = k_3 R_1 S - 2k_4 R_2 - 2k_5 R_2$$

$$\frac{dT_0}{dt} = k_1 R_0 - k_2 T_0 .$$

(50)

The time course of the reaction with an initial condition of $S(t_0) = 10.0 \, \text{mM}$, $R_0(t_0) = 10 \, \text{nM}$, $T_0(t_0) = 9.99 \, \mu\text{M}$ and null for other chemical species is shown in Fig. 3.8. The reaction is in the transient phase for about 20 sec after the start, and then stays nearly in the steady state for about 300 sec to reach the end.

The eigenvalues of Jacobian matrix $J$ for equation (50) are evaluated to obtain

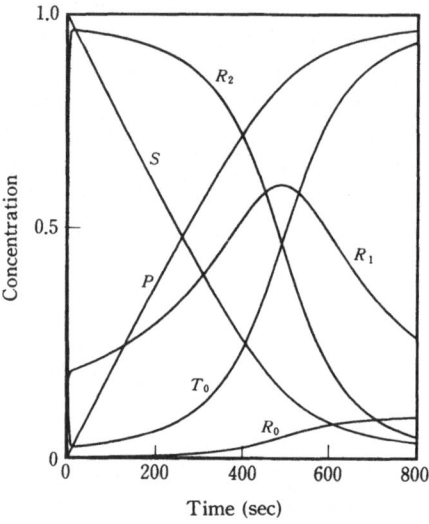

**Fig. 3. 8**  Time course of the allosteric reaction of Scheme 3. 2. Initial condition: $S=10$mM, $R_0=10$nM, $T_0=9.99\mu$M; other chemical species$=$ 0. Full scale (1.0) on the ordinate: $S=P=10$mM, $R_0=0.1\mu$M, $R_1=0.5\mu$M, $R_2=T_0=10\mu$M.

$$\left.\begin{array}{l} \lambda_M=\max_i|\mathrm{Re}(\lambda_i)| \\ \lambda_m=\min_i|\mathrm{Re}(\lambda_i)|\,. \end{array}\right\} \tag{51}$$

As demonstrated in Fig. 3.9, $\lambda_M$ decreases monotonously from $2\times10^5$ to $10^4$, implying that the Runge-Kutta-Gill method would require $10^7\sim10^8$ iterations if the computation could be carried out until the end of reaction. In actual computation, however, the solution of equation (50) leads to an overflow in the computer as well as unrealistic negative concentrations in the chemical species. This is because the solution is strongly affected by the high degree of stiffness of the equation which is represented by the very large value of $\mu\,(=\lambda_M/\lambda_m)$, as seen in Fig. 3.9.

On the other hand, the Gear method needs only several hundred iterations to perform the computation for the time course in Fig. 3.8. This is due to the evaluation of $\exp(-\lambda_M t)$ by retaining only the absolute stability and ignoring the accuracy, as mentioned above. This never means, however, that the computation of temporal change in concentration of chemical species is run without accuracy. In fact, good accuracy for the time course in Fig. 3.8 is maintained throughout the progress of reaction, since the conservation of concentrations is exactly valid, for example, at 704 sec so that

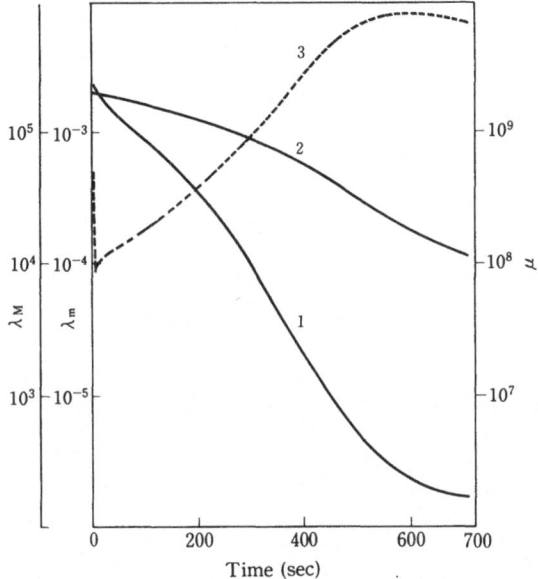

**Fig. 3.9**  Time course of the eigenvalues of allosteric reaction. 1: $\mu = \lambda_M/\lambda_m$; 2: $\lambda_M$; 3: $\lambda_m$.

$$S + R_1 + 2R_2 + P = 10.0000 \text{ mM}$$
$$R_0 + R_1 + R_2 + T_0 = 10.0000 \text{ } \mu\text{M}.$$

(52)

## 3.4  Numerical Method for Partial Differential Equations [5]

### 1.  Difference Expression of Derivative

Let the region in $x$ be divided into a number of equal intervals $h$ and the time into equal intervals $k$ as shown in Fig. 3.10, and the variable $u(x, t)$ be represented as $u(i,j)$ or $u_{i,j}$ using the lattice-points. Taylor-series expansions of $u(i,j)$ with respect to $x$ and $t$ are written as

$$u_{i+1,j} = u_{i,j} + hu'_{i,j} + \frac{1}{2} h^2 u''_{i,j} + \cdots\cdots$$

(53)

$$u_{i,j+1} = u_{i,j} + k\dot{u}_{i,j} + \frac{1}{2} k^2 \ddot{u}_{i,j} + \cdots\cdots,$$

(54)

where $u'$ and $\dot{u}$ indicate $\partial u/\partial x$, and $\partial u/\partial t$, respectively. From equations (53) and (54) we obtain a difference expression of derivative as

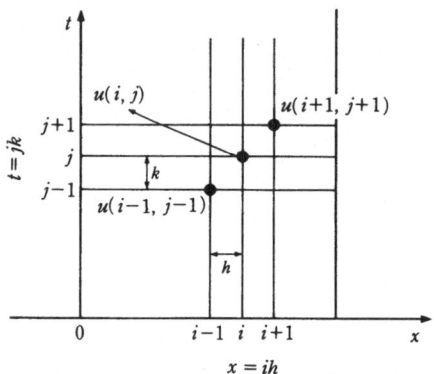

**Fig. 3. 10**  Lattice for numerical solution.

$$\frac{\partial u}{\partial x}\bigg|_{i,j} = \frac{1}{h}(u_{i+1,j} - u_{i,j}) + O(h) \tag{55}$$

$$\frac{\partial u}{\partial t}\bigg|_{i,j} = \frac{1}{k}(u_{i,j+1} - u_{i,j}) + O(k). \tag{56}$$

In general, there are several types of difference expression of the derivative. Equation (55) is called forward difference. Similarly, we can define the backward and central differences as

$$\frac{\partial u}{\partial x}\bigg|_{i,j} = \frac{1}{h}(u_{i,j} - u_{i-1,j}) + O(h) \tag{57}$$

and

$$\frac{\partial u}{\partial x}\bigg|_{i,j} = \frac{1}{2h}(u_{i+1,j} - u_{i-1,j}) + O(h), \tag{58}$$

respectively. From a summation of the Taylor-series expansions of $u_{i+1,j}$ and $u_{i-1,j}$, that is,

$$u_{i+1,j} = u_{i,j} + hu'_{i,j} + \frac{1}{2}h^2 u''_{i,j} + \cdots\cdots$$

and

$$u_{i-1,j} = u_{i,j} - hu'_{i,j} + \frac{1}{2}h^2 u''_{i,j} - \cdots\cdots,$$

we obtain the difference expression of second-order derivative as

$$\frac{\partial^2 u}{\partial x^2}\bigg|_{i,j} = \frac{1}{h^2}(u_{i+1,j} - 2u_{i,j} + u_{i-1,j}) + O(h^2). \tag{59}$$

## 2. Explicit Method

Dimensionless rate equation for a reaction-diffusion system may be generally represented by

$$\frac{\partial u}{\partial t} = \frac{\partial^2 u}{\partial x^2} + \lambda \phi (u, x, t)$$

$$u(0, t) = f(t), \quad \frac{\partial u(1, t)}{\partial x} = 0 \tag{60}$$

$$u(x, 0) = g(x), \quad 0 \leq x \leq 1,$$

where $\phi(u, x, t)$ is the reaction term, and $\lambda$ is a constant. Using equations (56) and (59) and ignoring $O(k)$ and $O(h^2)$, we can transform equation (60) into the difference equation:

$$u_{i,j+1} = r(u_{i+1,j} - u_{i-1,j}) + (1 - 2r) u_{i,j} + \lambda r h^2 \phi (ih, jk, u_{i,j})$$

$$u_{i,0} = g(ih), \quad u_{0,j} = f(jk), \tag{61}$$

where $r = k/h^2$, $t = jk$ and $x = ih$ $(i = 1, 2, \cdots, N-1; \ N = 1/h)$, and $i = 0, N$ for the boundary. From the boundary conditions we have at $i = N$ or $x = 1$

$$\left. \frac{\partial u}{\partial x} \right|_{i,j} = \frac{1}{2h} (u_{N+1,j} - u_{N-1,j}) = 0. \tag{62}$$

Therefore, it is convenient to introduce an auxiliary point of $i = N+1$ to put

$$u_{N+1,j} = u_{N-1,j}.$$

Thus, the value of $u$ at $x = ih$ and $t = (j+1)k$ is calculated from the values $u_{i+1,j}$, $u_{i,j}$ and $u_{i-1,j}$ at the previous time $t = jk$.

## 3. Implicit Method

We consider a reaction-diffusion system represented by the dimensionless rate equation,

$$\frac{\partial u}{\partial t} = \frac{\partial^2 u}{\partial x^2} + f(u, x, t)$$

$$u(x, 0) = 1, \quad 0 \leq x \leq 1 \tag{63}$$

$$\frac{\partial u(1, t)}{\partial x} = 0, \quad u(0, t) = 1.$$

If the values of $u(x, t)$ at $t = (j+1)k$ and $t = jk$ are used for the difference expression of $\partial^2 u / \partial x^2$ and the weighted average is taken by introducing the relaxation parameter $\omega$, the expression,

$$\left. \frac{\partial^2 u}{\partial x^2} \right|_{i,j} = \frac{\omega}{h} (u_{i+1,j+1} - 2u_{i,j+1} + u_{i-1,j+1})$$

$$+ \frac{(1+\omega)}{h^2} (u_{i+1,j} - 2u_{i,j} + u_{i-1,j}), \tag{64}$$

is obtained. Use of equations (64) and (56) yields transformation of equation (63) into

$$r\omega u_{i+1,j+1} + [1-2r(1-\omega)] u_{i,j+1} - r\omega u_{i-1,j+1}$$
$$= r(1-\omega) u_{i+1,j} + [1-2r(1-\omega)] u_{i,j} + r(1-\omega) u_{i-1,j} \quad (65)$$
$$+ kf(u_{i,j}, ih, jk), \quad r = k/h^2.$$

Thus, the linear combination of three unknown values at $t = (j+1)k$ can be calculated by four known values at the previous time $t = jk$.

For $\omega = 0$, equation (65) becomes the explicit expression. For $\omega = 1/2$, it is called the Cranck-Nicolson formula and for $\omega = 1$, the O'Brien-Hyman-Kaplan formula. The calculation of equation (65) is stable for all values of $u$ as long as $\omega \geq 1/2$.

## 4. Gauss Elimination Method

Equation (65) can be expressed in the form of

$$a_i u_{i-1,j+1} + b_i u_{i,j+1} + c_i u_{i+1,j+1} = d_i. \quad (66)$$

The coefficients $a_i$, $b_i$ and $c_i$ are known, and all other known coefficients and $u$ are included in $d_i$. Denoting $u_{i,j+1}$ by $u_i$, we write equation (66) as

$$b_1 u_1 + c_1 u_2 = d_1$$
$$a_2 u_1 + b_2 u_2 + c_2 u_3 = d_2$$
$$\vdots \quad (67)$$
$$a_{n-1} u_{n-2} + b_{n-1} u_{n-1} + c_{n-1} u_n = d_{n-1}$$
$$a_n u_{n-1} + b_n u_n = d_n$$

or

$$\begin{bmatrix} b_1 & c_1 & 0 & 0 & \cdots & 0 & 0 \\ a_2 & b_2 & c_2 & 0 & \cdots & 0 & 0 \\ \cdots & & & & & & \\ 0 & 0 & 0 & 0 & \cdots & a_n & b_n \end{bmatrix} \begin{bmatrix} u_1 \\ u_2 \\ \cdots \\ u_n \end{bmatrix} = \begin{bmatrix} d_1 \\ d_2 \\ \cdots \\ d_n \end{bmatrix}. \quad (67')$$

A triple-diagonal matrix equation can be solved by the Gauss elimination method. Each equation multiplied by a constant is successively subtracted from the next equation. This manipulation gives upper double-diagonal matrix. The new coefficients $a_i'$, $b_i'$, $c_i'$ and $d_i'$ are given by

$$a_i' = 0, \quad i = 2, 3, \cdots, n$$
$$b_i' = 0, \quad i = 2, 3, \cdots, n \quad (68)$$
$$c_1' = c_1/b_1, \quad d_1' = d_1/b_1$$

$$c_{i+1}' = \frac{c_{i+1}}{b_{i+1} + a_{i+1} c_i'}, \quad d_{i+1}' = \frac{d_{i+1} - a_{i+1} d_i'}{b_{i+1} - a_{i+1} c_i'}, \quad i = 1, 2, \cdots, n-1. \quad (69)$$

In the new equations, the last one becomes

$$u_n = d_n' \quad (70)$$

and the $(n-1)$th and $i$th equations are

$$u_{n-1} = d_n' - c_{n-1}' u_n \tag{71}$$

and

$$u_i = d_{i+1}' - c_i' u_{i+1} , \quad i = n-1, n-2, \cdots, 1, \tag{72}$$

respectively. By using equation (72), the solution $u_i$ can be obtained starting from $u_0$ and $u_{n+1}$ at the boundaries.

## 5. Over-Relaxation Method

We consider the system represented by equation (63). With equations (56) and (64) applied, the Cranck-Nicolson formula is obtained as

$$\frac{1}{k}(u_{i,j+1} - u_{i,j}) = \frac{1}{2} h^{-2} (u_{i+1,j+1} - 2u_{i,j+1} + u_{i-1,j+1})$$
$$+ \frac{1}{2} h^{-2} (u_{i+1,j} - 2u_{i,j} + u_{i-1,j}) + kf(u_{i,j}, ih, jk) . \tag{73}$$

The solution $u_{i,j+1}$ of equation (73) is given by

$$u_{i,j+1} = u_{i,j} + \frac{r}{2}(u_{i+1,j+1} - 2u_{i,j+1} + u_{i-1,j+1}$$
$$+ u_{i+1,j} - 2u_{i,j} + u_{i-1,j}) + f(u_{i,j}, ih, jk) . \tag{74}$$

Changing the notation as $u_{i+1,j+1} \to u_{i+1}, u_{i,j+1} \to u_i$ and $u_{i-1,j+1} \to u_{i-1}$, we reduce equation (73) to

$$u_i = \frac{r}{2}(u_{i-1} - 2u_i + u_{i+1}) + a_i$$
$$a_i = u_{i,j} + \frac{r}{2}(u_{i+1,j} - 2u_{i,j} + u_{i-1,j}) + kf(u_{i,j}, ih, jk) , \tag{75}$$

where $a_i$ is a known term.

Let $u_i^{(1)}$ and $u_i^{(n)}$ be the first and $n$th approximations of $u_i$, respectively, and we assume that all approximations are calculated starting from $u_i^{(1)}$ to $u_i^{(n)}$. Then, from equation (75) the iterative formula can be derived as

$$u_i^{(n+1)} = \frac{r}{2}(u_{i-1}^{(n)} - 2u_i^{(n)} + u_{i+1}^{(n)}) + a_i . \tag{76}$$

The iteration should be continued until the convergence of $u_i^{(n)}$ to $u_i$.

Because the convergence of equation (76) is usually unsatisfactory, a little modification is necessary for the procedure. As $u_{i-1}^{(n+1)}$ is already known in the calculation of equation (76), an implicit iterative formula,

$$u_i^{(n+1)} = \frac{r}{2}(u_{i-1}^{(n+1)} - 2u_i^{(n+1)} + u_{i+1}^{(n)}) + a_i , \tag{77}$$

can be used, instead of equation (76). Equation (77) is transformed to

$$u_i^{(n+1)} = \frac{r}{2(1+r)}(u_{i-1}^{(n+1)} + u_{i+1}^{(n)}) + \frac{a_i}{1+r}$$

$$= u_i^{(n)} + \left\{\frac{r}{2(1+r)}(u_{i-1}^{(n+1)} + u_{i+1}^{(n)}) + \frac{a_i}{1+r} - u_i^{(n)}\right\}, \tag{78}$$

which can be rewritten with introduction of a parameter $\lambda$ as

$$u_i^{(n+1)} = u_i^{(n)} + \lambda \left\{\frac{r}{2(1+r)}(u_{i-1}^{(n+1)} + u_{i+1}^{(n)}) + \frac{a_i}{1+r} - u^{(n)}\right\}$$

$$= \lambda \left\{\frac{r}{2(1+r)}(u_{i-1}^{(n+1)} + u_{i+1}^{(n)}) + \frac{a_i}{1+r}\right\} - (\lambda - 1) u_i^{(n)}. \tag{79}$$

$\lambda$, which is called an over-relaxation parameter, is found to have the optimal value,

$$\lambda = \frac{2}{1+(1-\mu^2)^{1/2}}, \quad \mu = \frac{r}{1+r} \cos \frac{\pi}{n}. \tag{80}$$

The over-relaxation method employing equation (79) yields a satisfactory result.

## 6. Boundary Conditions of Gradient Form

We assume the following type of boundary conditions:

$$\frac{\partial C(x,t)}{\partial x}\bigg|_{x=0} = a, \quad \frac{\partial C(x,t)}{\partial x}\bigg|_{x=0} = bC(0,t)$$

$$\frac{\partial C(x,t)}{\partial x}\bigg|_{x=1} = d, \quad \frac{\partial C(x,t)}{\partial x}\bigg|_{x=1} = eC(1,t),$$

where $a$, $b$, $d$ and $e$ are constants and $C$ represents the concentration of a reactant with $x \in [0,1]$. The gradient of $C$ at $x=1$ and $t=jk$ may be expressed as

$$\frac{\partial C(x,t)}{\partial x}\bigg|_{x=1} = \frac{1}{h}\left(\nabla_1 + \frac{1}{2}\nabla_2 + \frac{1}{3}\nabla_3 + \frac{1}{4}\nabla_4\right)C(N+1), \tag{81}$$

where $\nabla_i$ is the backward difference operator, and $C(N,j)$ is abbreviated by $C(N)$. $N$ corresponds to the point at $x=1$, and $N+1$ is an auxiliary point on the $x$-axis. The $\nabla_i C(N+1)$ are given by

$$\begin{aligned}
\nabla_1 C(N+1) &= C(N+1) - C(N) \\
\nabla_2 C(N+1) &= C(N+1) - 2 C(N) + C(N-1) \\
\nabla_3 C(N+1) &= C(N+1) - 3 C(N) + 3 C(N-1) - C(N-1) \\
\nabla_4 C(N+1) &= C(N+1) - 4 C(N) + 6 C(N-1) - 4 C(N-2) \\
&\quad + C(N-3).
\end{aligned} \tag{82}$$

Substitution of equation (82) into equation (81) leads to

$$\frac{\partial C\,(x,t)}{\partial x}\bigg|_{x=1} = \frac{1}{h}\left\{\frac{15}{12}C\,(N+1)-4\,C\,(N)+3\,C\,(N-1)\right.$$
$$\left.-\frac{4}{3}C\,(N-2)+\frac{1}{4}C\,(N-3)\right\}. \tag{83}$$

The difference equation of a rate equation such as equation (79) can be formulated up to $i=N+1$ to obtain $C\,(N+1,j)$. This value can be substituted into $C\,(N+1)$ $[=C\,(N+1,j)]$ in equation (83) for numerical calculation.

## 7. Steady-State Approximation of Reaction-Diffusion System

As described in Section 2.3, the quasi-steady-state approximation to the rate equation for the enzyme system (Scheme 1.19) can yield the following nonlinear ordinary differential equation:

$$\frac{d^2S}{dx^2} = \frac{1}{D_S}\frac{k_{12}SE_{10}}{K_{m1}+S}$$
$$\frac{d^2P_1}{dx^2} = -\frac{1}{D_{P_1}}\left(\frac{k_{12}SE_{10}}{K_{m1}+S}-\frac{k_{22}P_1E_{20}}{K_{m2}+P_1}\right) \tag{84}$$
$$\frac{d^2P_2}{dx^2} = -\frac{1}{D_{P_2}}\frac{k_{22}P_1E_{20}}{K_{m2}+P_1}.$$

The boundary conditions are given by

$$S(x,0)=0,\quad x>0,\quad P_1(x,0)=P_2(x,0)=0,$$
$$S(0,t)=S_0,\quad \frac{\partial P_1(0,t)}{\partial x}=aP_1(0,t),$$
$$\frac{\partial P_2(0,t)}{\partial x}=aP_2(0,t),\quad \frac{\partial S(1,t)}{\partial x}=bS(1,t), \tag{85}$$
$$\frac{\partial P_1(1,t)}{\partial x}=dP_1(1,t),\quad \frac{\partial P_2(1,t)}{\partial x}=dP_2(1,t).$$

Equation (84) can be solved numerically. It should be noted that the terms on the right-hand side of equation (84) contain $E_{10}$ and/or $E_{20}$ vanishing at all points on the $x$-axis except for $L_1$ and $L_2$ at which $E_1$ and $E_2$ are fixed respectively. Let $P_1$ and $P_2$ be denoted by $P$ and $Q$, respectively, to simplify the notation, and the region in $x$ be divided into $N$ equal intervals with $\Delta x = h$. Thus, by using equation (59), the difference equations for equation (84) may be written as

$$S_{i-2}-2S_{i-1}+S_i=0,\quad i=3,4,\cdots,m$$
$$S_{m-1}-2S_m+S_{m+1}=\frac{h^2}{D_S}\frac{k_{12}E_{10}S_m}{K_{m1}+S_m},\quad m=L_1/h \tag{86}$$
$$S_{i-1}-2S_i+S_{i+1}=0,\quad i=m+1,m+2,\cdots,N$$

$$P_{i-2} - 2P_{i-1} + P_i = 0, \quad i = 3, 4, \cdots, m$$

$$P_{m-1} - 2P_m + P_{m+1} = \frac{h^2}{D_P} \frac{k_{12} E_{10} S_m}{K_{m1} + S_m}, \quad m = L_1/h$$

$$P_{i-1} - 2P_i + P_{i+1} = 0, \quad i = m+1, m+2, \cdots, n \tag{87}$$

$$P_{n-1} - 2P_n + P_{n+1} = \frac{h^2}{D_P} \frac{k_{22} E_{20} P_n}{K_{m2} + P_n}, \quad n = L_2/h$$

$$P_{i-1} - 2P_i + P_{i+1} = 0, \quad i = n+1, n+2, \cdots, N$$

$$Q_{i-2} - 2Q_{i+1} + Q_i = 0, \quad i = 3, 4, \cdots, n$$

$$Q_{n-1} - 2Q_n + Q_{n+1} = \frac{h^2}{D_Q} \frac{k_{22} E_{20} P_n}{K_{m2} + P_n}, \quad n = L_2/h \tag{88}$$

$$Q_{i-1} - 2Q_i + Q_{i+1} = 0, \quad i = n+1, n+2, \cdots, N.$$

Equations (87) and (88) are solved iteratively from the boundary conditions. The difference equations for the boundary conditions are the same as those given by equation (83).

## References

[1] Garfinkel, D., C.B. Marbach and N.Z. Shapiro (1977). Stiff differential equations. *Annu. Rev. Biophys. Bioeng.*, **6**, 525-542.
[2] Gear, C.W. (1971). "Numerical Initial Value Problems in Ordinary Differential Equations," Prentice-Hall, Englewood Cliffs.
[3] Fowler, M.E. and R.M. Warten (1967). A numerical integration technique for ordinary differential equations. *IBM J. Res. Develop.*, **11**, 537-543.
[4] Gear, C.W. (1971). The automatic integration of ordinary differential equations. *Commun. A.C.M.*, **14**, 176-179.
[5] Smith, G.D. (1965). "Numerical Solution of Partial Differential Equations," Oxford University Press, London.

# CHAPTER 4

# Analysis of Enzymatic Reactions in Closed System

Individual enzymes are certainly the most fundamental and important constituents of an enzyme system. The knowledge of dynamic behavior of individual enzymatic reactions hence provides the basis for the dynamic analysis of enzyme systems. In enzyme kinetics the Michaelis-Menten-type reactions have been studied in detail, for instance, with respect to the relationship between reaction rate and substrate concentration and the effects of inhibitors and activators. The characteristics of allosteric enzymes have been explained by various models with regard to the cooperative mechanism. The kinetic properties of enzymes thus obtained, however, are generally based on the treatment of experimental data with quasi-steady-state and rapid-equilibrium approximations.

In fact, the behavior of enzymatic reactions in nonsteady states is little known, and the validity of the approximations has not been examined from the aspect of dynamic analysis. In this chapter, therefore, we perform the dynamic analysis of reactions of both Michaelis-Menten-type and allosteric enzymes in closed and homogeneous system. Although the procedure for dynamic analysis described in the preceding chapters is applicable to enzyme systems in both open and closed systems, we first treat the enzymatic reactions taking place in closed system, which are dealt with commonly in *in vitro* or experimental systems.

In Section 4.1 the rate equations for Michaelis-Menten-type reactions with or without effectors are numerically solved to reveal the dynamic behavior of reactions. The validity of application of the quasi-steady-state approximation is examined by a comparison of the results from computer simulation and approximation. The effects of rate constants and effectors

---

This chapter was written by Naoto Sakamoto, Masahiro Okamoto and Yukihiro Eguchi.

on the validity are analyzed. The behavior of lysozyme is also studied as an example of a reaction of enzyme with subsites. In Section 4.2 the time courses in reactions of the most basic MWC and KNF models for dimeric allosteric enzymes are obtained by computer simulation and analyzed to derive the characteristics of dynamic behavior of allosteric enzymes in closed system. Simulation is also performed for reactions of MWC tetrameric model and monomeric enzyme with cooperativity.

## 4.1 Reactions of Michaelis-Menten-Type Enzymes

In the practical kinetics for enzymatic reactions, suitable approximation methods such as the quasi-steady-state and rapid-equilibrium approximations have been generally employed to obtain the kinetic parameters like Michaelis constant $K_m$ and maximum velocity $V_{max}$. The successful application of the approximation method depends strongly on the experimental conditions under which the time course of enzymatic reaction is measured to estimate the reaction rate. The kinetic parameters such as $K_m$ and $V_{max}$ determined by the approximation method may provide meaningless values and erroneous characterization unless the experimental conditions are properly arranged.

This section is concerned with the experimental conditions which allow us to apply the quasi-steady-state approximation to Michaelis-Menten-type reactions. Computer simulation is performed with variation in conditions for the standard reaction, reactions with inhibitor and activator, and reaction of enzyme with subsites. These results lead to the characterization of the several factors affecting the validity of application of the quasi-steady-state approximation.

### 1. Standard Reaction

In recent years it has been realized that enzymatic reactions generally involve a sequential formation of the enzyme-substrate complexes as shown in Scheme 4.1. We consider here the typical three-step enzymatic reaction

$$E+S \rightleftharpoons ES_1 \rightleftharpoons ES_2 \rightleftharpoons \cdots\cdots \rightleftharpoons ES_n \longrightarrow E+P$$

Scheme 4.1. Sequential formation of enzyme-substrate complexes in enzymatic reaction

$$E+S \underset{k_{-1}}{\overset{k_{+1}}{\rightleftharpoons}} ES_1 \underset{k_{-2}}{\overset{k_{+2}}{\rightleftharpoons}} ES_2 \overset{k_{+3}}{\longrightarrow} E+P$$

Scheme 4.2. Standard reaction of Michaelis-Menten-type enzyme of Scheme 4.2 as the standard reaction of Michaelis-Menten-type enzyme, in which $k_{+1}$ and $k_{-1}$ are the rate constants for formation of the complex $ES_1$, $k_{+2}$ and $k_{-2}$ for isomerization between the complexes $ES_1$ and $ES_2$, and $k_{+3}$ for formation of the product. $E(t)$, $S(t)$, $C_1(t)$, $C_2(t)$ and $P(t)$ represent the concentrations of enzyme E, substrate S, $ES_1$, $ES_2$ and product P, respectively, as a function of time. In ordinary cases, we can obtain experimentally the time course of the change in substrate or product concentration. It is very difficult, however, to determine from those measurements (*i. e.*, $S(t)$ or $P(t)$) the rate constants ($k_{+1} \sim k_{+3}$) by suitable techniques (see Section 8.2). Incidentally, when the time courses of $C_1(t)$ and $C_2(t)$ are obtained by fast-reaction technique, it is possible to determine the rate constants precisely with the chemical relaxation method.

The rate equation for Scheme 4.2 in the form of simultaneous differential equations may be written as

$$\left.\begin{aligned}
\frac{dE(t)}{dt} &= -k_{+1}E(t)S(t) + k_{-1}C_1(t) + k_{+3}C_2(t) \\
\frac{dS(t)}{dt} &= -k_{+1}E(t)S(t) + k_{-1}C_1(t) \\
\frac{dC_1(t)}{dt} &= k_{+1}E(t)S(t) - (k_{-1}+k_{+2})C_1(t) + k_{-2}C_2(t) \\
\frac{dC_2(t)}{dt} &= k_{+2}C_1(t) - (k_{-2}+k_{+3})C_2(t) \\
\frac{dP(t)}{dt} &= k_{+3}C_2(t).
\end{aligned}\right\} \qquad (1)$$

For the simulation the values of rate constants are chosen as

$$\begin{aligned}
&k_{+1} = 5 \times 10^5 \text{ M}^{-1}\text{ sec}^{-1}, \quad k_{-1} = 5 \text{ sec}^{-1}, \\
&k_{+2} = 10^3 \text{ sec}^{-1}, \quad k_{-2} = 10^2 \text{ sec}^{-1}, \quad k_{+3} = 0.12 \text{ sec}^{-1}
\end{aligned} \qquad (2)$$

with reference to the experimentally obtained values in enzymatic reactions of lysozyme and chymotrypsin. The rate equation can be solved numerically by the modified Runge-Kutta-Gill method. The computation speed (not CPU time but number of computational steps) is greatly shortened within a reasonable error range by the modified Fowler-Warten method [1] or by the Gear method [2]. These methods provide the integration formulas suitable for stiff systems, which arise from the values of rate constants or concentrations of chemical species greatly different from one another as seen in equation (2).

In Scheme 4.2 the conditions of $dC_1(t)/dt = 0$ and $dC_2(t)/dt = 0$ are essential for assupmtion of the quasi-steady state in the reaction. In the actual reaction, the quasi-steady state as mathematically defined would not be attained, even if the reaction should take place under the most perfect conditions. In the present study, however, we assume for the sake of convenience that a quasi-steady state is established when the trajectory of

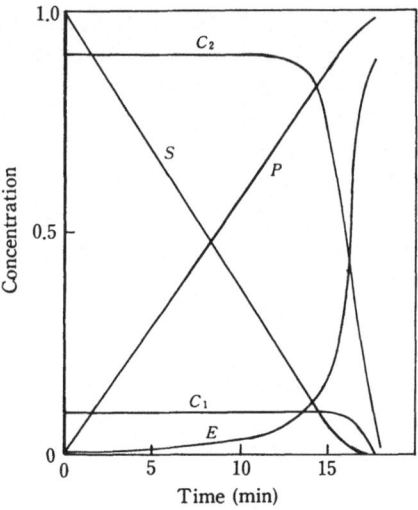

**Fig. 4. 1**  Time course of the standard reaction. (I) : $a=100$. Full scale (1.0) on the ordinate : $S=P=100\mu$M, $E=C_1=C_2=1.0\mu$M.

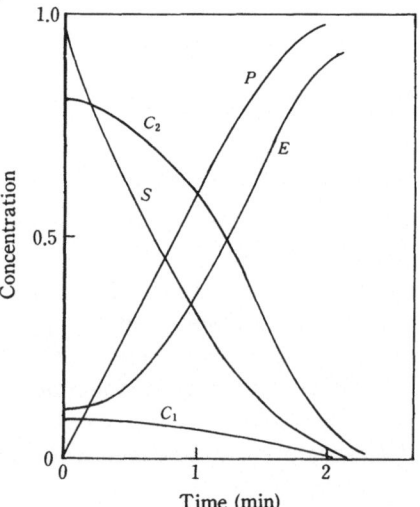

**Fig. 4. 2**  Time course of the standard reaction. (II) : $a=10$. Full scale (1.0) on the ordinate : $S=P=10\mu$M, $E=C_1=C_2=1.0\mu$M.

$C_2(t)$ is apparently parallel to the horizontal axis (time axis). Because the calculated value of $C_1(t)$ contains a computational error due to its very small value, it is not used for the judgment of quasi-steady state. Furthermore, the period in which the quasi-steady-state approximation can be applied is defined as that in which the absolute value of deviation of $C_2(t)$ is within 5% of the steady-state value. This period is called steady-state time, and the ratio $R_T$ is defined as percentage of the steady-state time in the whole reaction time.

Typical time courses of Scheme 4.2 from simulation are shown in Figs. 4.1 and 4.2. The initial condition is given by $E_0 = 1.0\,\mu$M, $S_0 = aE_0$ and $C_{10} = C_{20} = P_0 = 0$, where $a$ is a constant specifying a ratio of $S_0/E_0$. In Fig. 4.1, $C_2(t)$ has a nearly constant value during the period from 0.1 sec to 8 min. Therefore, $R_T$ is roughly estimated to be 72.5%. In contrast, in Fig. 4.2, $R_T$ becomes almost zero because the quasi-steady state is not attained under the specified condition.

### Effect of $k_{+3}$ on the Steady-State Time

As revealed in Table 4.1, the change in $k_{+3}$ alone with the other rate constants fixed does not vary the value of $R_T$. On the other hand, it causes a change in the whole reaction time, as shown in Fig. 4.3. The value of $k_{+3}$ affects the overall rate of the enzymatic reaction.

### Effect of Substrate-Enzyme Ratio on the Steady-State Time

It is obvious from the time courses of $C_2(t)$ shown in Figs. 4.1 and 4.2 that the substrate-enzyme ratio $a$ $(= S_0/E_0)$ plays the primary role in the appearance of quasi-steady state. The time courses of $C_2(t)$ obtained with variation in $a$ are shown in Fig. 4.4. The rate constants used for the simulation are as follows:

$$k_{+1} = 5 \times 10^5\,\text{M}^{-1}\text{sec}^{-1}, \quad k_{-1} = 5\,\text{sec}^{-1},$$
$$k_{+2} = 10^3\,\text{sec}^{-1}, \quad k_{-2} = 10^2\,\text{sec}^{-1}, \quad \Bigg\} \qquad (3)$$
$$k_{+3} = 0.16\,\text{sec}^{-1}.$$

The values of $R_T$ estimated from the curves in Fig. 4.4 are listed in Table 4.2. It follows that $a$ larger than 50 is necessary for application of the quasi-steady-state approximation. However, for a reaction with $a$ of 30, the first one-third of the whole reaction time may allow application of

**Table 4.1** Effect of $k_{+3}$ on $R_T$

| $k_{+3}(\text{sec}^{-1})$ | $R_T (\%)$ |
|---|---|
| 0.12 | 72.5 |
| 0.16 | 73.0 |
| 0.20 | 75.0 |

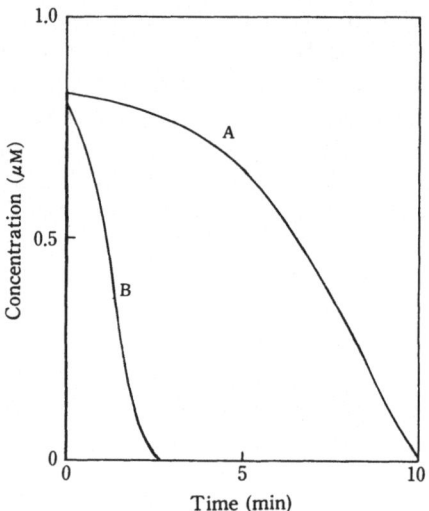

**Fig. 4. 3**  Effect of $k_{+3}$ on the time course of $C_2(t)$. The value of $k_{+3}$ is:
A: $0.03\,\mathrm{sec^{-1}}$; B: $0.16\,\mathrm{sec^{-1}}$; $a=10$.

the quasi-steady-state approximation, if a rather large error in computation can be accepted.

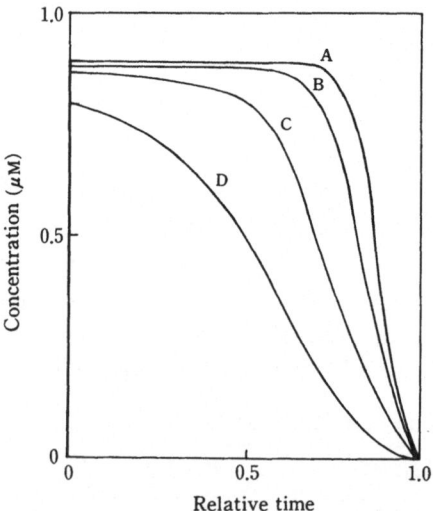

**Fig. 4. 4**  Time courses of $C_2(t)$ with variation in $a$. The substrate-enzyme ratio $a$ is: A: 100; B: 50; C: 30; D: 10. The abscissa (time) is normalized with respect to the whole reaction time.

**Table 4.2**  Effect of $a$ on $R_T$

| $a$ | $R_T$ (%) |
|------|-----------|
| 100  | 72.5      |
| 50   | 57.0      |
| 30   | 0         |
| 10   | 0         |

### Effect of $k_{+1}$ and $k_{-1}$ on the Steady-State Time

The effect of variation in $k_{+1}$ on the appearance of quasi-steady state is demonstrated in Fig. 4.5 and Table 4.3. For the simulation with various $k_{+1}$, the other kinetic parameters are fixed as $k_{-1}=5$, $k_{+2}=10^3$, $k_{-2}=10^2$, $k_{+3}=0.16$ (in $sec^{-1}$) and $a=50$. Thus, it is clearly seen that $k_{+1}$ strongly governs the steady state. Moreover, the extent of the change in steady-state time with variation in $k_{+1}$ depends on the starting value of $k_{+1}$. As shown in Fig. 4.5, the changes in $k_{+1}$ from $5\times10^6$ to $5\times10^5$ and from $5\times10^5$ to $5\times10^4$ (in $M^{-1} sec^{-1}$) give rise to quite different time courses of $C_2(t)$. This is due to the nonlinearity in the reaction. In relation to this aspect, the combined contribution of $k_{+1}$ and $k_{-1}$ to the quasi-steady state is examined under the fixed value of $10^5\ M^{-1}$ for equilibrium constant $K_1$. Table 4.4

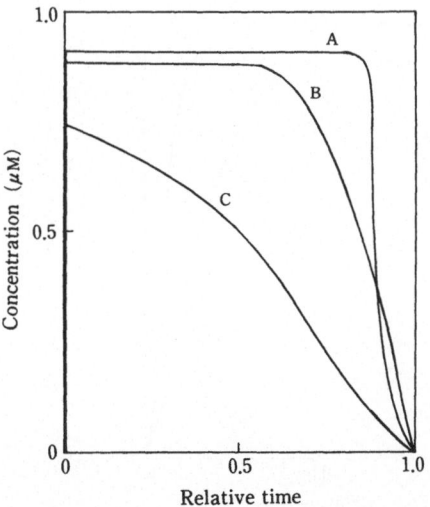

**Fig. 4.5**  Time courses of $C_2(t)$ with variation in $k_{+1}$. The value of $k_{+1}$ is: A: $5\times10^6 M^{-1} sec^{-1}$; B: $5\times10^5 M^{-1} sec^{-1}$; C: $5\times10^4 M^{-1} sec^{-1}$. The abscissa (time) is normalized with respect to the whole reaction time.

**Table 4. 3**  Effect of $k_{+1}$ on $R_T$

| $k_{+1}(M^{-1}sec^{-1})$ | $R_T$ (%) |
|---|---|
| $5\times10^6$ | 85 |
| $5\times10^5$ | 57 |
| $5\times10^4$ | 0 |

**Table 4. 4**  Effect of equilibrium constant $K_1(k_{+1}/k_{-1})$ on $R_T$

| $k_{+1}(M^{-1}\,sec^{-1})$ | $k_{-1}(sec^{-1})$ | $K_1\,(M^{-1})$ | $R_T$ (%) |
|---|---|---|---|
| $5\times10^5$ | 5 | $10^5$ | 57 |
| $5\times10^4$ | 0.5 | $10^5$ | 0 |

reveals that the combination of rate constants for forward and reverse paths in the equilibration process strongly affects the steady-state time.

**Effect of $k_{+2}$ and $k_{-2}$ on the Steady-State Time**

Since both forward and reverse reactions at the second step $(ES_1 \rightleftarrows ES_2)$ are of first order, $k_{+2}$, $k_{-2}$ and equilibrium constant $K_2$ seem to affect the steady-state time in a rather simpler manner than those at the first step. Therefore, the effect is examined only with the rate constant $k_{-2}$. The other

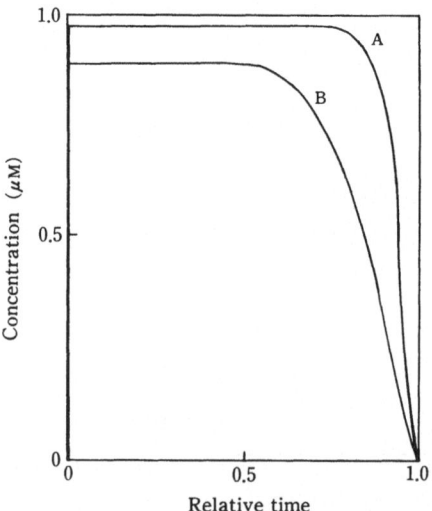

**Fig. 4. 6**  Time courses of $C_2(t)$ with variation in $k_{-2}$. The value of $k_{-2}$ is: A: $10\,sec^{-1}$; B: $10^2\,sec^{-1}$. The abscissa (time) is normalized with respect to the whole reaction time.

kinetic parameters are fixed as $k_{+1} = 5 \times 10^5 \text{ M}^{-1} \text{ sec}^{-1}$, $k_{-1} = 5 \text{ sec}^{-1}$, $k_{+2} = 10^3$ $\text{sec}^{-1}$, $k_{+3} = 0.16 \text{ sec}^{-1}$ and $a = 50$. As shown in Fig. 4.6, ten-fold increase in $k_{-2}$ causes a small decrease in $R_T$. It should be noted that a small value of $k_{-2}$ leads to elongation of the steady-state time, in contrast to the effects of other rate constants.

## Factors Affecting the Steady-State Time

The most influential factor is the ratio of initial concentration of substrate to total concentration of enzyme (*i. e.,* ratio $a$). An enzymatic reaction with $a$ less than 50 cannot establish the quasi-steady state during the whole reaction time, even if the rate constants are chosen in various combinations. The rate constants for the first step governing the formation and dissociation of the complex $ES_1$ are also relevant to the appearance of quasi-steady state. The rate constants for the second step affect only the elongation of steady-state time, provided that the values of $k_{+1}$ and $k_{-1}$ are appropriate for the occurrence of quasi-steady state. The value of $k_{+3}$ has an effect only on the whole reaction time. This infers that, when the original enzymatic reaction does not exhibit the quasi-steady state, the appearance of quasi-steady state may not be expected by the slowdown of the reaction with control of the milieu conditions such as the reaction temperature.

In this section, the values of rate constants are chosen with reference to the experimentally obtained values for hydrolytic enzymes. In general, however, the rate constants of enzymatic reactions have a wide range of values. Therefore, the results from the present study might not be applicable to specify the experimental conditions for a reaction having quite different rate constants from those used here. Nevertheless, the simulation for a specific reaction certainly enables us to examine the experimental conditions and time range in which the quasi-steady-state approximation is applicable to the reaction.

## 2. Reaction with Inhibitor

The reversible inhibition, which reduces the enzyme activity through reversible interaction of enzyme with a substance called inhibitor, generally operates with three major mechanisms: competitive, noncompetitive and uncompetitive inhibitions. The competitive inhibition has the effect of increasing the apparent value of $K_m$ for the substrate. The inhibitor is structurally similar to the substrate, competing for the active site. In the noncompetitive inhibition, the inhibitor, which is structurally different from the substrate, does not affect the binding of substrate to the active site (*i. e.,* no change in $K_m$) but causes a decrease in $V_{max}$. The uncompetitive inhibition affects both the apparent $K_m$ and $V_{max}$. The inhibitor can bind only to the enzyme-substrate complex and not to the free enzyme. In the following, we examine the effects of the three types of inhibition on the time course of reaction and the appearance of quasi-steady state [3].

**Competitive Inhibition**

In order to study the effect of competitive inhibition on the appearance of quasi-steady state, a three-step enzymatic reaction of Michaelis-Menten type in Scheme 4.2 is taken as the standard reaction. We consider the

$$
\begin{array}{ccc}
E + S & \underset{k_{-1}}{\overset{k_{+1}}{\rightleftharpoons}} ES_1 & \underset{k_{-2}}{\overset{k_{+2}}{\rightleftharpoons}} ES_2 \overset{k_{+3}}{\longrightarrow} E + P \\
+ & & \\
I & & \\
k_{-i} \updownarrow k_{+i} & & \\
EI & &
\end{array}
$$

Scheme 4.3.  Competitive inhibition

typical model for competitive inhibition in Scheme 4.3. The values of rate constants for interaction of enzyme with inhibitor are given as

$$
k_{+i}=5\times10^5\,\text{M}^{-1}\text{sec}^{-1}, \quad k_{-i}=5\,\text{sec}^{-1}, \quad K_I=k_{-i}/k_{+i}=10^{-5}\text{M}. \tag{4}
$$

The inhibition constant $K_I$ is defined as $k_{-i}/k_{+i}$. The other kinetic parameters are equal to those in equation (2). With an initial condition of $E_0=1.0\,\mu\text{M}$, $S_0=100\,\mu\text{M}$ and $I_0=1.0\,\text{mM}$, the simulation yields the time course of every chemical species in the system (Fig. 4.7). The whole

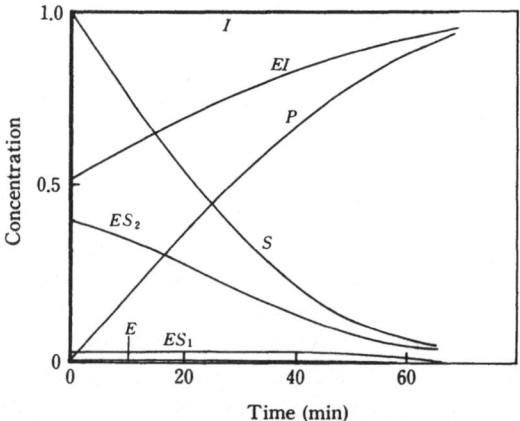

**Fig. 4.7** Time course of Scheme 4.3. Initial condition: $E_0=1.0\,\mu\text{M}$, $S_0=100\,\mu\text{M}$, $I_0=1.0\,\text{mM}$. Full scale (1.0) on the ordinate: $S=P=100\,\mu\text{M}$, $E=ES_1=ES_2=EI=1.0\,\mu\text{M}$, $I=1.0\,\text{mM}$.

**Table 4.5**  Effect of competitive inhibition on whole reaction time and steady-state time

| $S_0$ ($\mu$M) | $I_0$ ($\mu$M) | Whole reaction time (min) | Steady-state time(min) | $R_T$ (%) |
|---|---|---|---|---|
| 100 | 0 | 16.3 | 11.8 | 72.5 |
| " | 10 | 16.9 | 11.0 | 65.0 |
| " | 100 | 22.1 | 5.6 | 26.1 |
| " | 1000 | 65.6 | 3.0 | , 4.9 |
| 10 | 0 | 2.7 | 0.5 | 18.3 |
| " | 1 | 2.8 | 0.5 | 17.6 |
| " | 10 | 3.5 | 0.4 | 10.8 |
| " | 100 | 6.2 | 0.3 | 5.0 |

reaction time is 66 min, which is approximately four times as long as that in the standard reaction. On the other hand, the steady-state time decreases to a quarter of that in the standard reaction. These results indicate that the competitive inhibition may affect not only the whole reaction time but also the steady-state time.

We next examine the effects of inhibitor concentration ($I_0$) on the whole reaction time and steady-state time. The results are summarized in Table 4.5. Increase in $I_0$ leads to elongation of the whole reaction time, while the steady-state time becomes shorter, especially with lower $S_0$. Therefore, it is assumed that $I_0$ is a critical factor for application of the approximation to competitive inhibition. This can be verified by investigating how the inhibition constant $K_I$ obtained from the data of computed time courses by the quasi-steady-state approximation differs from the true value (10 $\mu$M) of $K_I$. We use the Woolf plot of $S/v$ vs. $S$ for the linear regression technique. The quasi-steady-state approximation to competitive inhibition (Scheme 4.3) yields the equation for the Woolf plot:

$$\frac{S}{v} = \frac{k_{+2}+k_{-2}+k_{+3}}{k_{+2}k_{+3}E_0} S + \frac{k_{-1}k_{-2}+k_{-1}k_{+3}+k_{+2}k_{+3}}{k_{+1}k_{+2}k_{+3}E_0}\left(1+\frac{I}{K_I}\right), \qquad (5)$$

where $S$ represents the substrate concentration, $I$ the concentration of inhibitor and $v$ the rate of product formation, respectively. The inhibition constant $K_I$ can be estimated from the intercept $B$ of the line on the ordinate; we get a relationship,

$$B = \frac{k_{-1}k_{-2}+k_{-1}k_{+3}+k_{+2}k_{+3}}{k_{+1}k_{+2}k_{+3}E_0}\left(1+\frac{I}{K_I}\right), \qquad (6)$$

from which we have

**Table 4.6** Inhibition constant ($K_I$) of competitive inhibition obtained by Woolf plot

| $I_0$ ($\mu$M) | $t$ (sec) | $K_I$ ($\mu$M) | $r$ |
|---|---|---|---|
| 100 | 30 | 8.20 | 0.999 |
| " | 60 | 6.95 | 0.999 |
| " | 120 | 4.88 | 0.997 |
| 1000 | 10 | 9.82 | 0.999 |
| " | 30 | 9.68 | 0.999 |
| " | 60 | 9.41 | 0.999 |
| 100000 | 10 | 9.96 | 0.978 |
| " | 30 | 9.99 | 0.990 |
| " | 60 | 9.93 | 0.296 |

True value of $K_I$ is $10\mu$M.

$$K_I = \frac{(k_{-1}k_{-2}+k_{-1}k_{+3}+k_{+2}k_{+3})I}{k_{+1}k_{+2}k_{+3}E_0B-(k_{-1}k_{-2}+k_{-1}k_{+3}+k_{+2}k_{+3})} . \tag{7}$$

The values of $K_I$ obtained from equations (6) and (7) for various values of $I_0$ are shown in Table 4.6, where $t$ indicates the time of measurement for the rate of enzymatic reaction and $r$ denotes the regression coefficient. It follows that the inhibition constant can be exactly obtained by the quasi-steady-state approximation at high $I_0$. On the other hand, in the presence of 0.1 M of inhibitor, the enzymatic reaction may not result in a significant value of $R_T$ owing to the slowdown effect of the inhibitor, that is, very long whole-reaction-time. It is obvious that, in order to obtain $K_I$ correctly, the measurements should be made in the early stage of reaction.

The accuracy in evaluation of the inhibition constant $K_I$ is checked by another procedure with the Dixon plot of $1/v$ vs. $I$. The relationship in the Dixon plot from the quasi-steady-state approximation for Scheme 4.3 is written as

$$\frac{1}{v} = \frac{k_{+2}+k_{-2}+k_{+3}}{k_{+2}k_{+3}E_0} + \frac{k_{-1}k_{-2}+k_{-1}k_{+3}+k_{+2}k_{+3}}{k_{+1}k_{+2}k_{+3}E_0}\left(1+\frac{I}{K_I}\right)\frac{1}{S} , \tag{8}$$

from which $K_I$ is obtained by the projection to the $I$-axis of the intersecting point of two lines drawn with two different values of $S$. Figure 4.8 shows the Dixon plot at $t=60$ sec. The values of $K_I$ thus evaluated at various measurement-times are listed in Table 4.7. Under these conditions the reaction is certainly in nonsteady state. Therefore, the values of $K_I$ in

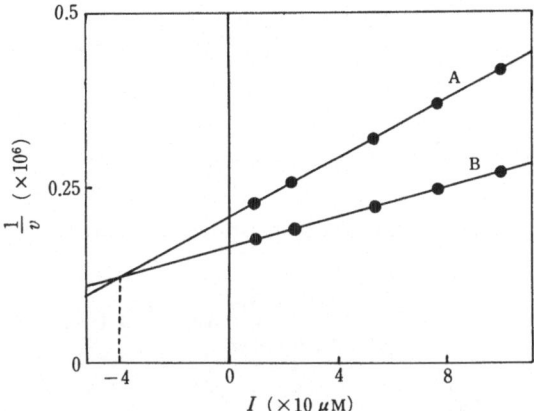

**Fig. 4.8** Dixon plot of Scheme 4.3 in a nonsteady state condition. $S_0$ is: A: 10$\mu$M, B: 20$\mu$M. The reaction rate is measured at $t = 60$ sec.

**Table 4.7** Inhibition constant ($K_1$) of competitive inhibition obtained by Dixon plot

| $t$ (sec) | $S_0$ ($\mu$M) | $r$ | $K_1$ ($\mu$M) |
|---|---|---|---|
| 10 | 10 | 0.999 | |
| $''$ | 20 | 0.999 | 15 |
| 30 | 10 | 0.999 | |
| $''$ | 20 | 0.999 | 21 |
| 60 | 10 | 0.998 | |
| $''$ | 20 | 0.999 | 40 |

Table 4.7 turn out to be several times as high as the true value (10 $\mu$M). Furthermore, there is a serious problem that the regression coefficient ($r$) is almost 1.0 in every case in Table 4.7. When a Michaelis–Menten-type reaction is exactly in the quasi-steady state, the data must be on a straight line in the Dixon plot. However, as revealed in Table 4.7 and Fig. 4.8, the inverse is not necessarily true. Hence, in the application of approximation methods to the enzymatic reactions with competitive inhibition, we should keep in mind that data on a straight line do not always lead to the correct evaluation of $K_1$.

**Noncompetitive Inhibition**

In noncompetitive inhibition, the inhibitor does not prevent the substrate

$$E + S \underset{k_{-1}}{\overset{k_{+1}}{\rightleftharpoons}} ES_1 \underset{k_{-2}}{\overset{k_{+2}}{\rightleftharpoons}} ES_2 \overset{k_{+3}}{\longrightarrow} P+E$$

$$+ \qquad\qquad\qquad +$$
$$I \qquad\qquad\qquad\qquad I$$

$$k_{-i} \left\|\, k_{+i} \right. \qquad\qquad\quad k_{-i} \left\|\, k_{+i} \right.$$

$$EI+S \underset{k_{-1}}{\overset{k_{+1}}{\rightleftharpoons}} EIS$$

Scheme 4. 4.   Noncompetitive inhibition (I)

$$E + S \underset{k_{-1}}{\overset{k_{+1}}{\rightleftharpoons}} ES_1 \underset{k_{-2}}{\overset{k_{+2}}{\rightleftharpoons}} ES_2 \overset{k_{+3}}{\longrightarrow} P+E$$

$$+ \qquad\qquad +$$
$$I \qquad\qquad\quad I$$

$$k_{-i} \left\|\, k_{+i} \right. \qquad\quad k_{-i} \left\|\, k_{+i} \right.$$

$$EI+S \underset{k_{-1}}{\overset{k_{+1}}{\rightleftharpoons}} EIS$$

Scheme 4. 5.   Noncompetitive inhibition (II)

from binding to an enzyme molecule, which interacts simultaneously with the inhibitor. We analyze the models for typical noncompetitive inhibition of Schemes 4.4 and 4.5. The rate constants are equal to those in equations (2) and (4). Figure 4.9 illustrates the time course of every chemical species in Scheme 4.4 from the simulation under an initial condition of $E_0 = 1.0\,\mu$M and $S_0 = I_0 = 100\,\mu$M. The time course with a similar profile is also obtained

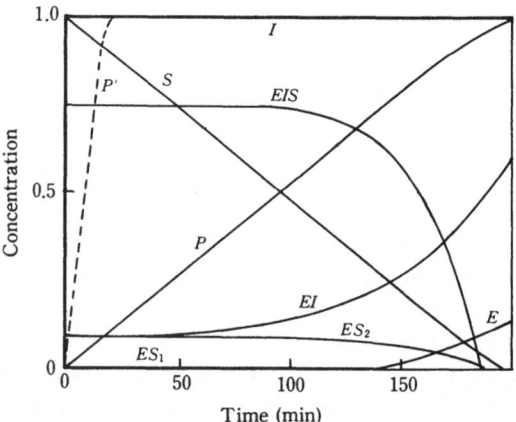

**Fig. 4. 9**   Time course of Scheme 4.4. Initial condition: $S_0 = I_0 = 100\mu$M, $E_0 = 1.0\mu$M.   Full scale (1.0) on the ordinate: $S = P = I = 100\mu$M, $E = ES_1 = ES_2 = EIS = 1.0\mu$M.

**Table 4.8** Effect of noncompetitive inhibition on whole reaction time and steady-state time

| $S_0$ ($\mu$M) | $I_0$ ($\mu$M) | Whole reaction time (min) | Steady-state time (min) | $R_T$ (%) |
|---|---|---|---|---|
| Scheme 4.4 | | | | |
| 100 | 0 | 16.3 | 11.8 | 72.5 |
| " | 10 | 32.5 | 18.2 | 55.9 |
| " | 100 | 185.4 | 39.4 | 21.2 |
| 10 | 0 | 2.7 | 0.5 | 18.3 |
| " | 1 | 3.2 | 0.4 | 12.9 |
| " | 10 | 8.7 | 0.4 | 4.7 |
| Scheme 4.5 | | | | |
| 100 | 0 | 16.3 | 11.8 | 72.5 |
| " | 10 | 18.2 | 12.4 | 68.2 |
| " | 100 | 37.1 | 16.4 | 44.1 |
| 10 | 0 | 2.7 | 0.5 | 18.3 |
| " | 1 | 2.9 | 0.5 | 16.4 |
| " | 10 | 3.1 | 0.4 | 13.2 |
| " | 100 | 13.0 | 0.4 | 1.4 |

for Scheme 4.5 under the same condition. In the figure, $P'$ (broken line) represents the time course of product concentration in the standard reaction (Scheme 4.2), demonstrating that the whole reaction time and steady-state time are elongated by about eleven times and thrice, respectively, longer than those of the standard reaction. The profile of the time course in Fig. 4.9 is very similar to that in Fig. 4.7. The effects of inhibitor concentration on the steady-state time for noncompetitive inhibition are summarized in Table 4.8. Increase in $I_0$ gives rise to elongation of the steady-state time at higher $S_0$, but not at lower $S_0$. This implies that, in the case of noncompetitive inhibition, kinetic parameters like $K_1$ can be determined exactly by means of the quasi-steady-state approximation, even if high $I_0$ is used. Furthermore, it is revealed that the inhibitor has a greater inhibitory effect in Scheme 4.4 than in Scheme 4.5.

## Uncompetitive Inhibition

In uncompetitive inhibition, the inhibitor has affinity only to the ES-complex to form an inactive EIS-complex. Typical models for uncompetitive inhibition are given in Schemes 4.6 and 4.7. The rate constants used are equal to those in equations (2) and (4). Figure 4.10 illustrates the

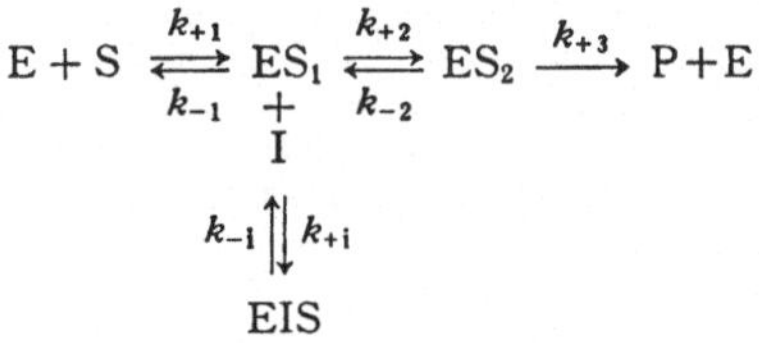

$$E + S \underset{k_{-1}}{\overset{k_{+1}}{\rightleftarrows}} ES_1 \underset{k_{-2}}{\overset{k_{+2}}{\rightleftarrows}} ES_2 \xrightarrow{k_{+3}} P + E$$

$$\begin{array}{c} + \\ I \end{array}$$

$$k_{-i} \updownarrow k_{+i}$$

$$EIS$$

Scheme 4.6.  Uncompetitive inhibition ( I )

$$E + S \underset{k_{-1}}{\overset{k_{+1}}{\rightleftarrows}} ES_1 \underset{k_{-2}}{\overset{k_{+2}}{\rightleftarrows}} ES_2 \xrightarrow{k_{+3}} P + E$$

$$\begin{array}{c} + \\ I \end{array}$$

$$k_{-i} \updownarrow k_{+i}$$

$$EIS$$

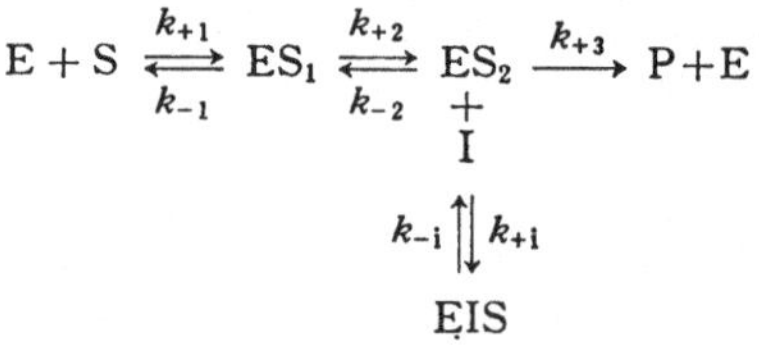

Scheme 4.7.  Uncompetitive inhibition (II)

time course of every chemical species in Scheme 4.7 from the simulation under an initial condition of $S_0/E_0 = 10$ and $I_0/S_0 = 10$. The broken line represents the time course of P in Scheme 4.2. It follows from the figure that the whole reaction time and steady-state time are both elongated by seven and 29 times, respectively, longer than those of Scheme 4.2. The effects of inhibitor concentration on the time course are summarized in Table 4.9. The whole reaction time and steady-state time are both

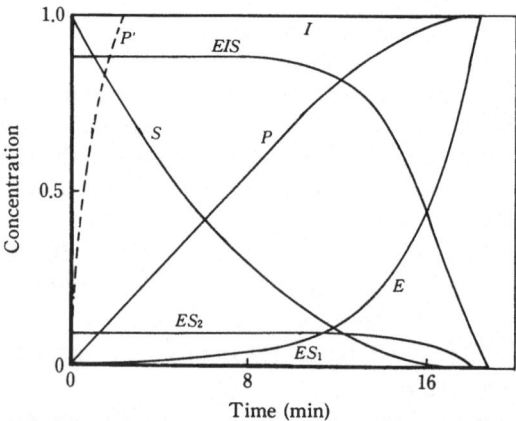

**Fig. 4.10**  Time course of Scheme 4.7. Initial condition : $S_0 = 10\mu$M, $E_0 = 1.0\mu$M, $I_0 = 100\mu$M. Full scale (1.0) on the ordinate : $S = P = 10\mu$M, $E = ES_1 = ES_2 = ESI = 1.0\mu$M, $I_0 = 100\mu$M.

**Table 4.9** Effect of uncompetitive inhibition on whole reaction time and steady-state time

| $S_0$ ($\mu$M) | $I_0$ ($\mu$M) | Whole reaction time (min) | Steady-state time(min) | $R_T$ (%) |
|---|---|---|---|---|
| Scheme 4.6 | | | | |
| 100 | 0 | 16.3 | 11.8 | 72.5 |
| " | 10 | 17.4 | 14.1 | 81.2 |
| " | 100 | 30.0 | 25.2 | 83.9 |
| 10 | 0 | 2.7 | 0.5 | 18.3 |
| " | 1 | 2.7 | 0.5 | 19.4 |
| " | 10 | 2.9 | 0.6 | 20.4 |
| " | 100 | 4.4 | 1.3 | 29.3 |
| Scheme 4.7 | | | | |
| 10 | 0 | 2.7 | 0.5 | 18.3 |
| " | 1 | 2.9 | 0.6 | 20.6 |
| " | 10 | 4.3 | 1.1 | 24.7 |
| " | 100 | 18.6 | 11.7 | 63.0 |

elongated with the increase in $I_0$, similarly to the case of noncompetitive inhibition. Therefore, it is concluded that the enzymatic reactions with uncompetitive inhibition can be analyzed in the usual manner with the quasi-steady-state approximation.

**Summary**

In the present study, a clear distinction is attempted between the three types of inhibition, *i. e.*, competitive, noncompetitive and uncompetitive inhibitions, in their dynamic behavior from the simulation rather than the approximation methods. Figures 4.7, 4.9 and 4.10 demonstrate that the temporal profiles of the three inhibitions are very similar. This finding indicates that the inhibition type cannot be identified simply with the temporal profile of the enzymatic reaction in the presence of inhibitor, and that the approximation methods widely used for the distinction of the inhibition type might become the most practical ones when the validity of approximation is assured by the simulation. With respect to the quasi-steady-state approximation, the effects of inhibitor concentration on steady-state time in the reactions with the three types of inhibition are summarized as follows; (1) In competitive inhibition, the increase in $I_0$ leads to reduction in steady-state time. (2) Even if the plot of $1/v$ *vs.* $1/S$ or $1/I$ yields a straight line, it cannot necessarily be inferred that the reaction is in quasi-steady state. (3) In noncompetitive and uncompetitive inhibitions, the steady-state time does not change with the increase in $I_0$. Therefore, the quasi-steady-state approximation can be applied without special precaution. Recently, Frere *et al.* [4] have discussed in detail the problems using linear-regression techniques for reactions in nonsteady state.

## 3. Reaction with Activator

The modes of enzyme activation may be put into two major classes: a) nonessential activation in which an enzyme exhibits some activity without activator, and b) essential activation in which the substrate-activator complex behaves as the true substrate. We here discuss only the nonessential activation [3]. The Michaelis-Menten-type reaction in Scheme 4.8 is postulated as the standard reaction, for which the rate constants used are given as

$$k_{+1} = 5 \times 10^5 \text{ M}^{-1} \text{ sec}^{-1}, \quad k_{-1} = 5 \text{ sec}^{-1}, \quad \text{and} \quad k_{+2} = 0.12 \text{ sec}^{-1}. \quad (9)$$

A general mechanism of nonessential activation is modeled in Schemes 4.9 and 4.10 with the rate constants of

$$\text{E} + \text{S} \underset{k_{-1}}{\overset{k_{+1}}{\rightleftharpoons}} \text{ES} \xrightarrow{k_{+2}} \text{P} + \text{E}$$

Scheme 4.8. Standard reaction for nonessential activation

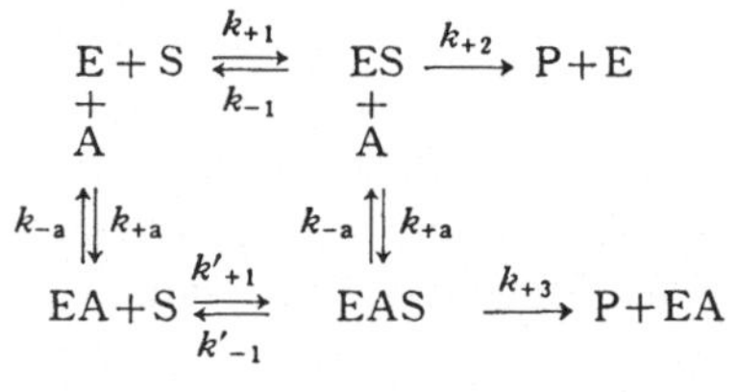

Scheme 4.9. Nonessential activation (I)

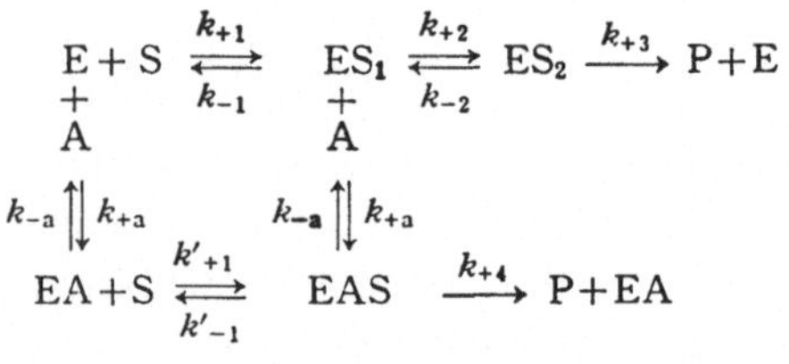

Scheme 4.10. Nonessential activation (II)

$$k'_{+1}=k_{+1}=5\times10^5\,\text{M}^{-1}\text{sec}^{-1}, \quad k'_{-1}=k_{-1}=5\,\text{sec}^{-1},$$
$$k_{+a}=5\times10^5\,\text{M}^{-1}\text{sec}^{-1}, \quad k_{-a}=5\,\text{sec}^{-1}, \tag{10}$$
$$k_{+3}=3k_{+2}=0.36\,\text{sec}^{-1}, \quad K_A=k_{-a}/k_{+a}=10^{-5}\text{M}.$$

The activation constant $K_A$ is defined by $k_{-a}/k_{+a}$. Figure 4.11 exhibits the

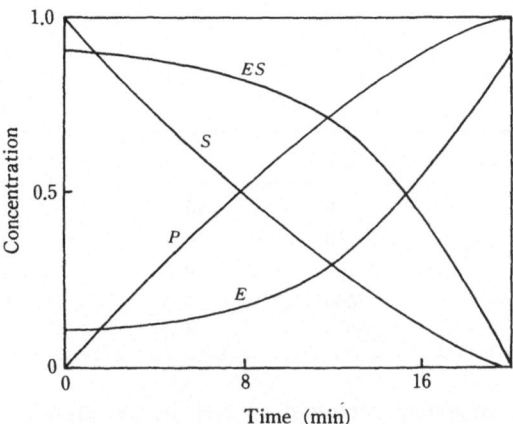

**Fig. 4. 11**  Time course of Scheme 4.8.  Initial condition: $S_0=100\mu\text{M}$, $E_0=1.0\mu\text{M}$.  Full scale (1.0) on the ordinate: $S=P=100\mu\text{M}$, $E=ES=1.0\mu\text{M}$.

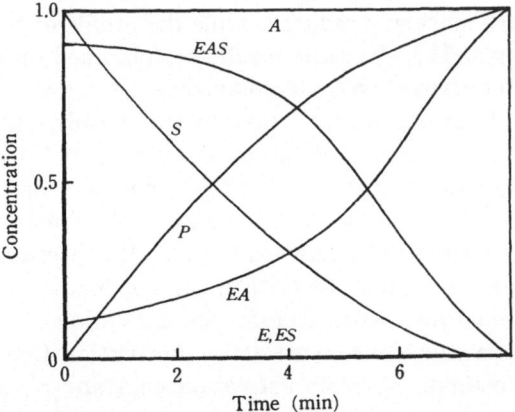

**Fig. 4. 12**  Time course of Scheme 4.9.  Initial condition: $S_0=100\mu\text{M}$, $E_0=1.0\mu\text{M}$, $A_0=100\,\text{mM}$.  Full scale (1.0) on the ordinate: $S=P=100\mu\text{M}$, $E=EA=ES=EAS=1.0\mu\text{M}$, $A=100\,\text{mM}$.

**Table 4. 10** Effect of activator concentration on whole reaction time and steady-state time

| $S_0$ ($\mu$M) | $A_0$ ($\mu$M) | Whole reaction time (min) | Steady-state time(min) | $R_T$ (%) |
|---|---|---|---|---|
| Scheme 4. 9 | | | | |
| 100 | 0 | 21. 0 | 5. 6 | 26. 7 |
| 100 | 10 | 10. 6 | 3. 0 | 27. 9 |
| 100 | 100 | 7. 5 | 2. 2 | 29. 0 |
| 100 | 1000 | 7. 1 | 1. 9 | 27. 1 |
| 100 | 100000 | 7. 0 | 1. 9 | 26. 7 |
| Scheme 4. 10 | | | | |
| 100 | 0 | 16. 3 | 11. 8 | 72. 5 |
| 100 | 10 | 14. 0 | 9. 0 | 64. 3 |
| 100 | 100 | 9. 2 | 3. 9 | 42. 1 |
| 100 | 1000 | 7. 3 | 2. 2 | 30. 7 |
| 100 | 100000 | 7. 0 | 1. 8 | 26. 0 |

time course of every chemical species in the standard reaction (Scheme 4.8) from the simulation under an initial condition of $E_0 = 1.0$ $\mu$M and $S_0 = 100$ $\mu$M. The whole reaction time and steady-state time are 21 min and 5.6 min, respectively. Figure 4.12 illustrates the time course of Scheme 4.9 from the simulation under a condition of $A_0/S_0 = 1000$. The other conditions are same as those in Fig. 4.11. The whole reaction time and steady-state time are markedly reduced, while the profile of time course is similar to that in Fig. 4.11. The same results are obtained for Scheme 4.10, which involves two enzyme-substrate complexes.

**Effect of Activator Concentration on the Steady-State Time**

The effects of initial concentration of activator ($A_0$) on the steady-state time are examined for reactions of Schemes 4.9 and 4.10. Results are summarized in Table 4.10. The increase in $A_0$ causes the reduction in the whole reaction time and steady-state time. It is presumed from this finding that the value of $A_0$ may be critical for application of the quasi-steady-state approximation to the nonessential activation.

**Effect of Activator Concentration on Activation Constant $K_A$**

We now investigate how the activation constant $K_A$ obtained by rapid-equilibrium approximation as in the case of inhibition differs from the true value of $K_A$ (10 $\mu$M). We use the Woolf plot of $S/v$ vs. $S$ for the linear regression technique. The rapid-equilibrium approximation leads to the equation of the Woolf plot for Scheme 4.9 [5]:

$$v = \frac{V_{max}\dfrac{S}{K_S} + \beta V_{max}\dfrac{AS}{\alpha K_A K_S}}{1 + \dfrac{S}{K_S} + \dfrac{A}{K_A} + \dfrac{AS}{\alpha K_A K_S}},$$

(11)

where $V_{max} = k_{+2}E_0$, $K_S = k_{-1}/k_{+1}$, $K_S' = k_{-1}'/k_{+1}'$, $\alpha = K_S'/K_S$ and $\beta = k_{+3}/k_{+2}$. In this analysis, we set $K_S' = 10\ \mu M$, $K_S = 10\ \mu M$, $\alpha = 1$ and $\beta = 3$, respectively. Equation (11) may be transformed to

$$\frac{1}{v} = \frac{1 + \dfrac{k_{+1}}{k_{-1}}S + \dfrac{A}{K_A} + \dfrac{k_{+1}}{k_{-1}}\dfrac{AS}{K_A}}{\dfrac{k_{+1}\cdot k_{+2}}{k_{-1}}E_0 S + \dfrac{k_{+1}\cdot k_{+3}}{k_{-1}}\dfrac{A}{K_A}E_0 S}.$$

(12)

By multiplying both sides by $S$, we have

$$\frac{S}{v} = \left(\frac{1 + \dfrac{A}{K_A}}{k_{+2} + \dfrac{k_{+3}A}{K_A}}\right)\frac{S}{E_0} + \frac{1 + \dfrac{A}{K_A}}{\dfrac{k_{+1}}{k_{-1}}\left(k_{+2} + \dfrac{k_{+3}A}{K_A}\right)E_0},$$

(13)

with which the activation constant $K_A$ can be obtained from the intercept of the line on the ordinate. That is, the intercept $B$ on the ordinate is given by

$$B = \frac{1 + \dfrac{A}{K_A}}{\dfrac{k_{+1}}{k_{-1}}\left(k_{+2} + \dfrac{k_{+3}A}{K_A}\right)E_0},$$

(14)

and hence, $K_A$ can be represented by

$$K_A = \frac{(k_{-1} - k_{+1}k_{+3}E_0 B)A}{k_{+1}k_{+2}E_0 B - k_{-1}}.$$

(15)

The relationship between $A_0$ and $K_A$ from equation (15) is shown in Table 4.11, where $t$ indicates the time of measurement for the reaction rate and $r$ denotes the coefficient of linear regression. Despite the fact that the reaction is in quasi-steady state at $t = 10$ sec or $t = 30$ sec, the obtained values of $K_A$ deviate extremely from the true value $(10\ \mu M)$, increasing sharply with increase in $A_0$. Furthermore, it should be noted that the regression coefficient of the Woolf plot is nearly unity in every case, implying that the data are exactly on a straight line.

It is thus concluded that $K_A$ in the nonessential activation cannot be determined by the rapid-equilibrium approximation unless a very low initial-concentration of activator, at least lower than the total concentration

**Table 4.11**  Activation constant $(K_A)$ obtained from Woolf plot

| $A_0$ ($\mu$M) | $t$ (sec) | $K_A$ ($\mu$M) | $r$ |
|:---:|:---:|:---:|:---:|
| 10 | 10 | 23.30 | 0.999 |
| " | 30 | 60.55 | 0.999 |
| 100 | 10 | 68.27 | 0.999 |
| " | 30 | 265.0 | 0.997 |
| 1000 | 10 | 565.5 | 0.999 |
| " | 30 | 2450 | 0.998 |

of the enzyme, is used. It should also be emphasized that linearity of the plot from the approximation does not guarantee an accurate evaluation of the kinetic parameters like $K_A$. As described above, the conditions allowing the accurate determination of kinetic parameters by the approximation methods are different and dependent on the reaction mechanisms. We should pay particular attention to this finding in performing the experimental determination.

## 4. Reaction of Enzyme with Subsites

The enzyme with several subsites in the substrate-binding site to catalyze the reaction of polymeric substrate generally yields a very complicated time course of reaction even in closed system. A typical example of such enzymes is lysozyme, which has six subsites and catalyzes the cleavage of glycosidic linkage in chitooligosaccharide, homooligomer of 2-acetamido-2-deoxy-D-glucopyranose (GlcNAc). In addition to hydrolysis of the substrate, the enzyme catalyzes the transglycosylation of glycosidic linkage[6]. Consequently, the time course of lysozyme catalysis with chitooligosaccharides exhibits a complicated pattern. Such a complicated enzymatic reaction does not allow the application of a simple kinetic analysis to the experimental data, but requires a much more mathematically sophisticated technique. We here analyze the behavior of lysozyme catalysis by computer simulation.

### Reaction Scheme of Lysozyme Catalysis

Figure 4.13 shows a symbolic representation of a typical reaction scheme for substrate chitopentamer (GlcNAc)$_5$ (1 : 1 complex). The troughs in the box represent the six subsites, A~F, each of which interacts with the unit residue GlcNAc of the substrate. M$_n$, E, C, B and A indicate the substrate, lysozyme, lysozyme-substrate complex, lysozyme-cleaved substrate complex and lysozyme-carbonium ion intermediate complex, respectively. The

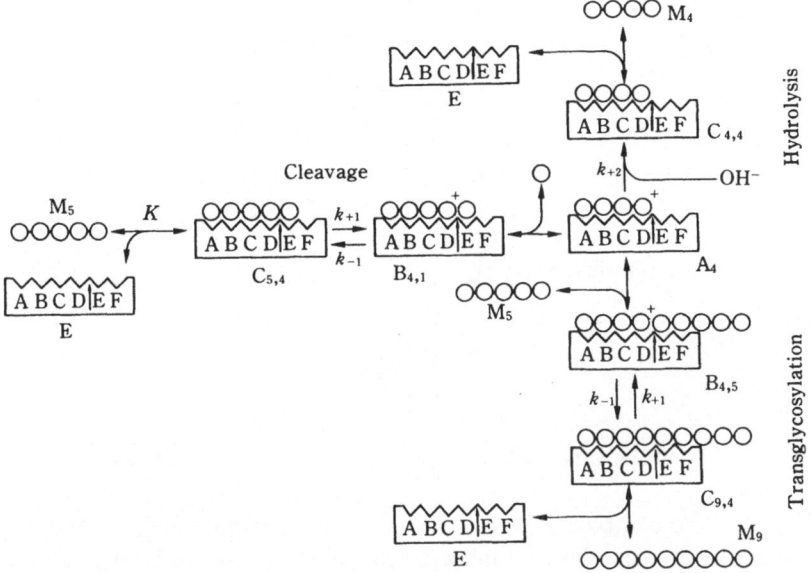

**Fig. 4.13** Schematic representation of lysozyme-catalyzed reaction with (GlcNAc)$_5$.

circle-filled and empty troughs denote the subsite occupied by GlcNAc residue and the free subsite, respectively. $K$ and $k$ are the equilibrium (binding) constant and the rate constant of the indicated step, respectively. The lysozyme-catalyzed reaction consists of three processes. We explain these processes using Fig. 4.13. In the first process, the cleavage of glycosidic linkage of the substrate occurs at the position indicated by the arrow, and C(1) carbonium-ion intermediate is formed. The reaction then branches in two directions, transglycosylation and hydrolysis processes. When lysozyme binds with the substrate (GlcNAc)$_5$ denoted by $M_5$, which occupies five subsites, A to E, the emzyme-substrate complex $C_{5,4}$ is formed. In the complex $C_{n,i}$, the first subscript $n$ expresses the size $n$ of the substrate (*i.e.*, $n$-mer of GlcNAc) and the second subscript $i$ indicates that the $i$th GlcNAc-residue numbered from the nonreducing terminal is located on subsite D (see illustration in Fig. 4.14). For example, $C_{4,3}$ means that in the complex (GlcNAc)$_4$ binds to lysozyme so that the third residue is located on subsite D.

Lysozyme cleaves the glycosidic linkage situated between subsites D and E. Hence, $C_{5,4}$ is cleaved into a carboniumion intermediate (GlcNAc)$_4$$^+$ and (GlcNAc)$_1$. (GlcNAc)$_4$$^+$ remains bound to subsites A, B, C and D, and (GlcNAc)$_1$ to subsite E. This complex is denoted by $B_{4,1}$ (in general, by

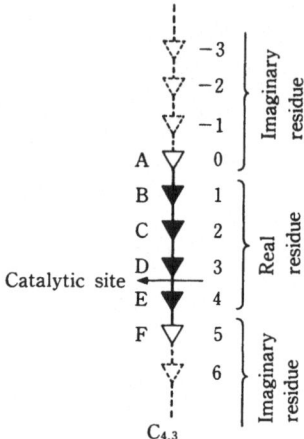

**Fig. 4. 14**   Rule for representation of substrate binding.

$B_{i,n-i}$). The complex $B_{4,1}$ releases $(GlcNAc)_1$ from subsite E and becomes a reactive intermediate complex $A_4$ (in general, denoted by $A_i$). In complex $A_i$, subsites E and F are empty.

After formation of $A_i$, the reaction branches off to either the hydrolysis (hydration) or the transglycosylation processes. In the hydration process, $A_4$ reacts directly with a water molecule to form the complex $C_{4,4}$, which releases the product $M_4$. In the transglycosylation process, $A_4$ binds with another free $M_5$ molecule as an acceptor, which occupies subsites E and F, then becoming the complex $B_{4,5}$. Lysozyme can generate new glycosidic linkage between subsites D and E for the intermediate and acceptor. That is, $B_{4,5}$ converts into $C_{9,4}$, from which $M_9$ is released. The products of lysozyme-catalyzed reaction from chitooligosaccharide serve again as substrate or acceptor.

**Rate Equation**

For formulation of a simple and practical rate equation [7-9], the following assumptions are made.   a) The binding processes are always in rapid equilibrium. For the substrate-binding process, the rate constants are known to be extremely (about $10^4$ times) larger than those in the catalytic process followed.   b) Complexes of the 1:2 (enzyme:substrate) to 1:6 types are ignored except for $B_{i,j}$ complex, since their amounts are found from the simulation to be very small and not to practically affect the time course.   c) The binding free energy of subsite is additive.   d) The rate constant $k_{-2}$ of the dehydration process is zero. This is confirmed experimentally.   e) The release of the intermediate from $A_i$ does not occur. In complex $A_i$, the terminal residue with $C(1)$ carbonium ion is thought to bind strongly to

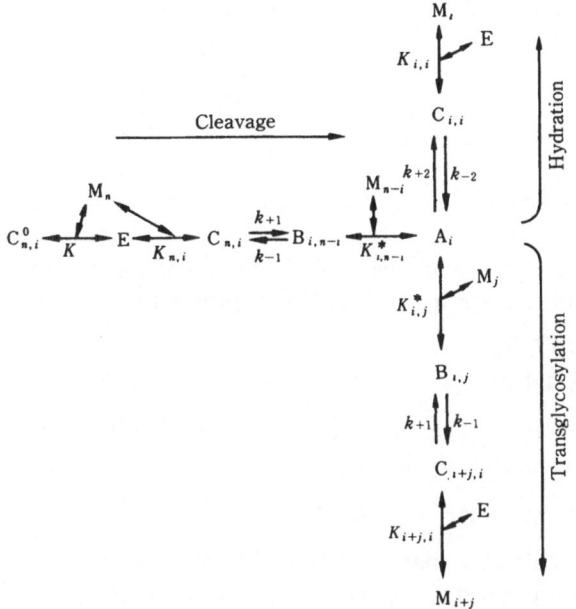

Scheme 4.11.   Lysozyme-catalyzed reaction

subsite D, because this residue takes the half-chair structure which fits well to subsite D.

With these assumptions, a reaction scheme for the computer simulation is constructed as Scheme 4.11. In the scheme, $C_{n,i}^0$ indicates the nonproductive complex. Let $S_n$, $M_n$ and $I_j$ be the total concentration of $(GlcNAc)_n$, concentration of free $(GlcNAc)_n$ and concentration of $(GlcNAc)_j^+$, respectively. These variables may be represented by

$$S_n = M_n + \sum_{i=1}^{n+3} C_{n,i} + \sum_{j=1}^{p-1} B_{j,n}, \quad n=1,2,\cdots,p \tag{16}$$

$$C_{n,i} = \begin{cases} \text{productive,} & i=1,2,\cdots,n-1 \\ \text{nonproductive,} & i=-1,0,n,n+1,n+2,n+3 \end{cases}$$

$$I_j = A_j + \sum_{n=1}^{p} B_{j,n}, \quad j=1,2,\cdots,p-1. \tag{17}$$

From Scheme 4.11 the differential equations for these variables may be written as

$$\frac{dS_1}{dt} = k_{+2}A_1 - k_{-1}\sum_{j=1}^{p-1} B_{j,1} + k_{+1}\sum_{l=2}^{p} C_{l,l-1}$$

$$\frac{dS_n}{dt} = k_{+2}A_n - k_{-1}\left(\sum_{m=1}^{n-1} B_{m,n-m} - \sum_{j=1}^{p-1} B_{j,n}\right)$$

$$+ k_{+1}\left(\sum_{l=n+1}^{p} C_{l,l-n} - \sum_{i=1}^{n-1} C_{n,i}\right), \quad n=2,3,\cdots,p \qquad (18)$$

$$\frac{dI_j}{dt} = -k_{+2}A_j + k_{+1}\sum_{l=j+1}^{p} C_{l,j} - k_{-1}\sum_{n=1}^{p} B_{j,n}, \quad j=1,2,\cdots,p-1.$$

By introducing the equilibrium (binding) constants, the amounts of complexes $C_{n,i}$ and $B_{j,n}$ are approximately expressed by

$$C_{n,i} = K_{n,i}EM_n$$
$$B_{j,n} = K_{j,n}^{*}A_jM_n. \qquad (19)$$

The binding constants $K$ and $K^{*}$ are calculated from the binding free energy $\Delta F^0$ according to the relationship,

$$K_{X\sim Y} = \exp(-\Delta F^0_{X\sim Y}/RT), \qquad (20)$$

where $X\sim Y$ indicates that the sequential subsites X to Y in an arbitrary region in the all subsites (A, B, C, D, E and F) are participating in the binding of reactants. The conservation equation for the total amounts of substrate and enzyme may be written as

$$\sum_{n=1}^{p} nM_{n,0} = \sum_{n=1}^{p} nS_n + \sum_{n=1}^{p-1} nI_n$$

$$E_0 = E + \sum_{n=1}^{p}\sum_{i=-1}^{n+3} C_{n,i} + \sum_{j=1}^{p-1} I_j, \qquad (21)$$

where the subscript 0 indicates the initial or total concentration of the respective species.

The variables $C_{n,i}$ and $B_{j,m}$ in equations (16), (17) and (21) can be eliminated with substitution of equation (19). Then, we have

$$S_n = M_n + \sum_{i=-1}^{n+3} K_{n,i}EM_n + \sum_{j=1}^{p-1} K_{j,n}^{*} A_jM_n, \quad n=1,2,\cdots,p \qquad (22)$$

$$A_j = I_j \Big/ \Big(1 + \sum_{n=1}^{p} K_{j,n}^{*} M_n\Big), \quad j=1,2,\cdots,p-1 \qquad (23)$$

$$E = \Big(E_0 - \sum_{j=1}^{p-1} I_j\Big) \Big/ \Big(1 + \sum_{n=1}^{p}\sum_{j=1}^{n+3} K_{n,j}M_n\Big). \qquad (24)$$

Furthermore, $A_j$ and $E$ in equation (22) can be eliminated using equations (23) and (24). Thus, $S_n$ ($n=1,2,\cdots,p$) can be calculated by

$$S_n = M_n + \frac{\sum\limits_{i=-1}^{n+3} K_{n,i}\left\{\left(E_0 - \sum\limits_{j=1}^{p-1} I_j\right)M_n\right\}}{1 + \sum\limits_{q=1}^{p}\sum\limits_{i=-1}^{n+3} K_{q,i}M_q} + \frac{\sum\limits_{j=1}^{p-1} K_{j,n}{}^* I_j M_n}{1 + \sum\limits_{q=1}^{p} K_{j,q}{}^* M_q} \qquad (25)$$

$$n = 1, 2, \cdots, p.$$

Finally, equations (18), (19), (21), (23), (24) and (25) are simultaneously and numerically solved to obtain the time course.

**Numerical Solution**

In the rate equation, the nonlinear differential equations are solved by the modified Fowler-Warten method [1] or the Gear method [2] and the nonlinear algebraic equations are solved by the modified Newton method. The standard values of parameters used for the simulation are given in Table 4.12. The computational error is defined as

$$Er = \frac{\left(\sum\limits_{n=1}^{p} nS_n + \sum\limits_{n=1}^{p-1} nI_n\right) - \sum\limits_{n=1}^{p} nM_{n,0}}{\sum\limits_{n=1}^{p} nM_{n,0}} \times 100. \qquad (26)$$

At first, the simulation was performed assuming that the initial substrate is $(GlcNAc)_5$ and the products up to $(GlcNAc)_{10}$ $(p=10)$ are to be included. The undefined $(GlcNAc)_n$ $(n>10)$ produced by transglycosylation was ignored and the computation yielded a large error up to 10%. This suggests that a much larger value should be used for $p$ and that the rate equation should be partly revised. Consequently, the transglycosylation process

$$B_{i,j} \underset{k_{+1}}{\overset{k_{-1}}{\rightleftarrows}} C_{i+j,i} \quad i+j = 2, 3, \cdots, p$$
$$\overset{k_{-1}}{\searrow} C_{i+j,i} \quad i+j = p+1, p+1, p+2, \cdots, 2p-1$$

Scheme 4.12.   Revised process of transglycosylation

shown in Scheme 4.11 was revised as Scheme 4.12. Then the differential equation in equation (18) is replaced by

Table 4.12   Kinetic parameters of lysozyme catalysis

| Subsite | Binding free energy $\Delta F^0$ (kcal/mol) | Rate constant (sec$^{-1}$) | |
|---------|:---:|:---:|:---:|
| A | $-2.0$ | $k_{+1}$ | 0.94 |
| B | $-3.0$ | $k_{-1}$ | 40.0 |
| C | $-5.0$ | $k_{+2}$ | 0.30 |
| D | $+4.5$ | | |
| E | $-2.5$ | | |
| F | $-1.5$ | | |

$$\frac{dS_n}{dt} = k_{+2}A_n + k_{-1}\left(\sum_{m=1}^{n-1} B_{m,n-m} - \sum_{j=1}^{p-3} B_{j,n}\right)$$

$$+ k_{+1}\left(\sum_{l=n+1}^{p} C_{l,l-n} - \sum_{i=1}^{n-1} C_{n,i}\right), \quad n = 2, 3, \cdots, p-1 \tag{27}$$

$$\frac{dS_p}{dt} = \left(\frac{dS_n}{dt}\right)_{n=p} = k_{-1} \sum_{n=p+1}^{2p-1} \sum_{m=n+p}^{p-1} B_{m,n-m}. \tag{28}$$

These equations imply that when $p$ is taken to be 10 the products of trans-glycosylation from $(GlcNAc)_{10}$ to $(GlcNAc)_{19}$ are included as $(GlcNAc)_{10}$. The error is reduced to 2% by this revision of the rate equation.

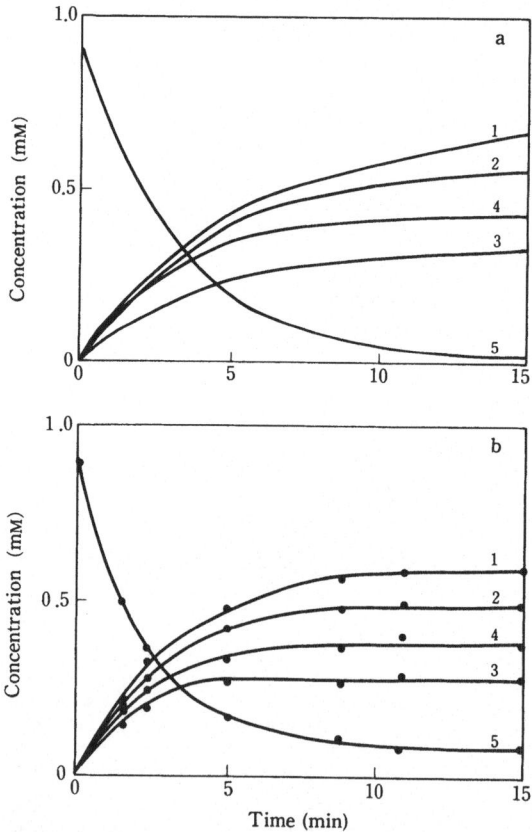

Fig. 4.15   Time courses of lysozyme-catalyzed reaction with $(GlcNAc)_5$. Initial condition: Lysozyme$=77.2\mu M$, $(GlcNAc)_5 = 894\mu M$. a) Computed time courses; b) Experimental time courses at 50°C and pH5.0. Numbers within the figure indicate $n$-value of $(GlcNAc)_n$.

### Evaluation of Rate Constants

Using the experimental values of binding free energy of each subsite, the computations are repeated with variation in the values of $k_{+1}$, $k_{-1}$ and $k_{+2}$ until the simulated time course fits well to that obtained experimentally with the substrate $(GlcNAc)_5$ (Fig.4.15). The evaluated values of rate constants are listed in Table 4.12 [9,10].

### Effect of Rate Constants on Time Course

As expected, the rate constant $k_{+1}$ for the cleavage of glycosidic linkage affects sensitively the overall rate of disappearance of substrate; a large $k_{+1}$ accelerates the reaction. On the other hand, contrary to expectation, the increase in $k_{-1}$ for the transglycosylation accelerates the overall rate and $k_{+2}$ for the hydration causes the reverse effect. Furthermore, the rate constants $k_{-1}$ and $k_{+2}$ govern the profile of the time course. If there is no transglycosylation, the substrate $(GlcNAc)_5$ decomposes to $(GlcNAc)_4 +$ $(GlcNAc)_1$ and $(GlcNAc)_2 + (GlcNAc)_3$, leading to an identical time course for both paired products. Thus, a larger $k_{-1}$ generates more different time courses in the paired products.

### Effect of Binding Free Energy on Time Course

The binding constants of each subsite may vary their values due to the change in milieu conditions and the chemical modification of subsite or its proximal groups.

A-site: The increase in negative value of binding free energy from the subsite E or F enhances the rate of transglycosylation and reduces that of the reaction (Fig. 4.16). On the other hand, the decrease in binding free

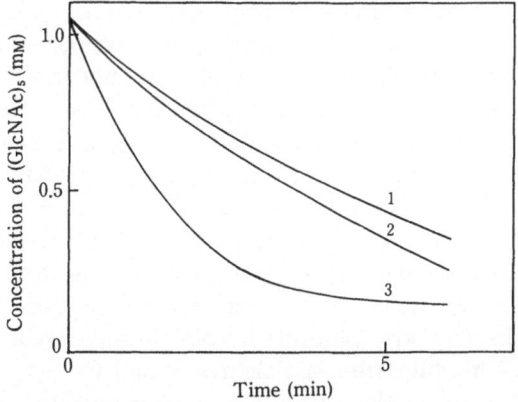

**Fig. 4. 16** Effect of binding constant of subsite A $(K_A)$ on rate of disappearance of substrate $(GlcNAc)_5$. 1: $K_A = 22.5 \times$ standard value, 2: $K_A =$ standard value, 3: $K_A = 1/22.5 \times$ standard value.

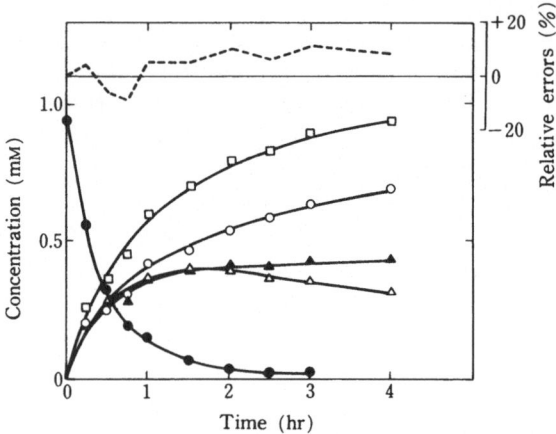

**Fig. 4. 17** Experimental time courses of NBS-lysozyme with substrate (GlcNAc)$_5$. Initial condition: NBS-lysozyme = 62.7$\mu$M, (GlcNAc)$_5$ = 942$\mu$M. Experimental time courses at pH 5.0 and 50°C; $\square$:(GlcNAc)$_1$, $\bigcirc$: (GlcNAc)$_2$, $\blacktriangle$: (GlcNAc)$_3$, $\triangle$: (GlcNAc)$_4$, $\bullet$: (GlcNAc)$_5$.

energy to $-1.0$ kcal/mol accelerates the reaction rate about two-fold. The finding that lowering of the substrate binding constant causes the increase in reaction rate seems unreasonable, but it can be explained by assuming that lowering of binding free energy of subsite A results in relative increase in the formation of productive complexes which span subsites D and E. Such phenomenon would only be observable in complicated enzymatic reactions and not in simple Michaelis-Menten-type enzymes.

B-, C- and D-sites: The change in binding free energy of subsites B, C and D results only in a change in the overall rate, not in the order of the amounts of the products. The time course is not sensitive to the binding free energy of subsites B and C; the change in binding free energy by $\pm 1.0$ kcal/mol of the two subsites does not shift the time course noticeably.

E- and F-sites: The binding free energy of subsites E and F governs the rate of transglycosylation. The increase in binding free energy of either subsite E or F enhances the rate of transglycosylation and reduces that of hydrolysis. This is easily understood because subsites E and F are binding sites for the acceptor molecule in the transglycosylation process.

**Effect of Chemical Modification of Subsites A and C**

The experimental data on the reaction of the chemically modified lysozymes with substrate (GlcNAc)$_5$ are analyzed by computer simulation [11] (for analytical technique, see Chapter 8). The oxidation or substitution of Trp 62 residue at subsite C causes a decrease in binding free energy of

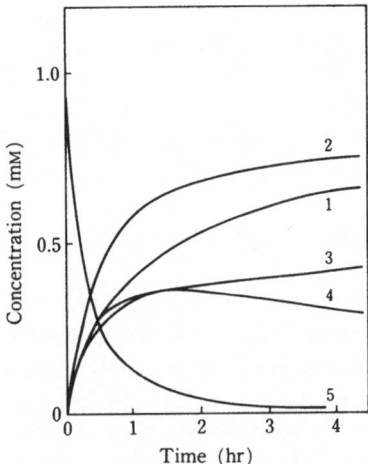

**Fig. 4.18** Simulated time courses of NBS - lysozyme. $\Delta F_A = \Delta F_C = -1.25$ kcal/mol. Numbers within the figure indicate $n$-value of (Glc-NAc)$_n$.

**Table 4.13**  Binding free energy of modified lysozymes

| Lysozyme | Binding free energy (kcal/mol) | | | | | |
| --- | --- | --- | --- | --- | --- | --- |
| | A | B | C | D | E | F |
| Native | −2.42 | −3.02 | −4.89 | +4.50 | −2.50 | −1.50 |
| NBS–oxidized | −1.50 | −3.00 | −1.25 | +4.50 | −2.50 | −1.50 |
| Asp 101–modified | −0.90 | −3.00 | −2.50 | +4.50 | −2.50 | −1.50 |

subsite A as well as that of subsite C, while the binding free energy of subsite B is not affected. The experimental and simulated time courses for the lysozyme oxidized by $N$-bromosuccinimide (NBS) are shown in Figs. 4.17 and 4.18, respectively. The values of binding free energy used for the simulation are listed in Table 4.13.

It is shown in the table that the substitution of $\beta$-carboxyl group of Asp 101 by nucleophile glucosamine yields the similar effect as that observed with the chemical modification of Trp 62. These findings indicate that there exists a strong connection between subsites A and C. This would reflect that in binding of the substrate the rod-like molecular structure of substrate chitooligosaccharide should be in contact with the binding site consisting of subsites A, B and C and that deviation of the position of sugar residue at subsite A or C would alter the position of sugar residue at those subsites with the fulcrum of subsite B.

In addition, as proposed by Pincus and Scheraga [12], the analysis of experimental data with the modified substrate $(GlcNAc)_6$, in which the reducing terminal is reduced to sugar alcohol, suggests that subsites E and F have rather wider spatial region than so far believed, and that, when the substrate binds on subsites D, E and F, the sugar residue at subsite D would bind strongly to subsite D with negative binding free energy.

## 4.2 Reactions of Allosteric Enzymes

It has been recognized that enzymes serve as a regulatory element as well as a catalyzer in enzyme systems. The dynamic behavior of allosteric enzymes has recently been regarded as an increasingly important aspect of enzymatic function in *in vivo* enzyme systems. Allosteric enzymes are characterized by a feature that the relationship between substrate concentration and initial velocity of reaction ($S$-$v$ curve) is sigmoidal. This characteristic is considered to play an important role in regulatory mechanisms for metabolic processes through allosteric interactions with various kinds of ligands: A small change in substrate concentration can yield a large variation in reaction rate, and regulation of enzymatic function is realized specifically by inhibition and activation. In fact, almost all enzymes in the first step are allosteric in metabolic pathways.

Indeed, the experimental and theoretical data on allosteric enzymes have accumulated extensively. The kinetic aspects of the enzymes, however, have been investigated on the quasi-steady-state and rapid-equilibrium assumptions. Their dynamic behavior in nonsteady states has been little known even for *in vitro* systems. We here analyze the dynamic behavior of allosteric enzymes of dimeric models in closed system ( *i. e.,* usual experimental system ) to provide some fundamental data for enzyme dynamics.

In this section the analysis is often concerned with a comparison of the results from simulation and experiment. "Results from experiment (empirical values)" refer to the data obtained as follows. A reaction process treated in an experiment is realized by computer simulation. The values at corresponding times in a time course are taken as observed values at the measuring times in an experiment. The results from the experiment are obtained by application of common analytical procedures like regression line for determination of initial velocity and Lineweaver-Burk plot to the observed values.

### 1. Allosteric Models
The allosteric enzyme is generally considered as an oligomeric molecule

consisting of more than two protomers [13]. The biochemical properties of many allosteric enzymes in addition to the most famous aspartate trans-carbamoylase have been studied with respect to the numbers of subunits and protomers, and the effects of substrate, product and other ligands on reaction rate and states of enzyme species. Many models have been proposed for the allosteric mechanism. The models proposed by Monod, Wyman and Changeux (MWC model) [14] and by Koshland, Némethy and Filmer (KNF model) [15] are often employed for the analysis.

The MWC model is also called a symmetrical model, in which protomers are allowed to be in two states with different conformations as long as all constituent protomers of an enzyme are in the same state and binding of the protomer with ligands gives rise to no transition between the two states. In contrast, the KNF model is called a sequential model, in which the state transition arises sequentially in the protomer by its binding with respective ligands. The change in state of a protomer causes other protomers in the enzyme to have different affinities to ligands. In general, the protomer may take any number of states specific to its various ligands.

The most basic MWC and KNF models for dimeric allosteric enzymes are adopted in the following simulation and analysis to derive the characteristics of dynamic behavior of allosteric enzymes in closed system. We are also concerned here with MWC tetrameric model and monomeric enzyme with cooperativity.

## 2. MWC Dimeric Model — Time Course and $S$-$v$ Relationship
We begin the dynamic analysis with an MWC dimeric model of Scheme 4.13. In this mechanism substrate S binds with the enzyme species in both R- and T-states, but product P is produced only from the enzyme–sub-

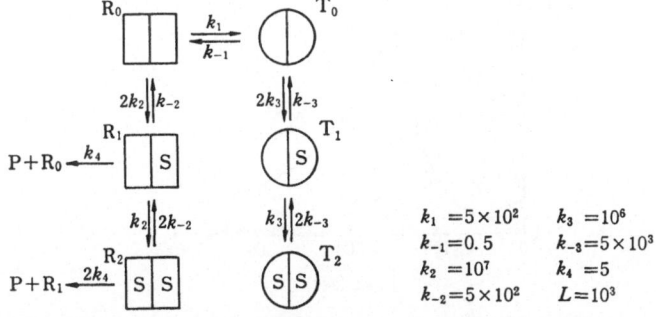

Scheme 4.13.   MWC dimeric model of allosteric enzyme (I)
The product is produced only from the R-form species.

strate complexes in R-state. The rectangle and half-circle in Scheme
4.13 designate the R- and T-forms of the protomer, respectively. $k_1$ and $k_{-1}$
are the rate constants for allosteric transition, the ratio of which is defined
as the allosteric constant $L$ $(=k_1/k_{-1})$. The dissociation constants for the
substrate-protomer complexes are denoted by $K_R = k_{-2}/k_2$ for R-state and
$K_T = k_{-3}/k_3$ for T-state. Their ratio $c$ $(=K_R/K_T)$ and the allosteric constant $L$ are often used to represent the characteristics of this allosteric
model in steady states [16]. In the scheme the units of kinetic parameters
($M^{-1}sec^{-1}$, $sec^{-1}$ and others) are omitted with no confusion foreseen. This
convention is used in the following schemes as well.

Phosphorylase b of rabbit muscle is a known example studied as an
allosteric dimeric enzyme. The results of the following analysis are not
only applied to such dimeric enzymes, but are extended to allosteric
enzymes in general. This is also inferred from the analysis of dynamic
behavior of the KNF dimeric model, MWC tetrameric model and monomeric enzyme with cooperativity.

**Time Course of Reaction**

Figures 4.19 and 4.20 illustrate the respective time courses of a reaction
with $L = 1000$ and $c = 0.01$ in Scheme 4.13 for 10 mM and 1.0 mM of $S_0$
(initial substrate concentration). The total enzyme concentration is 10 $\mu$M

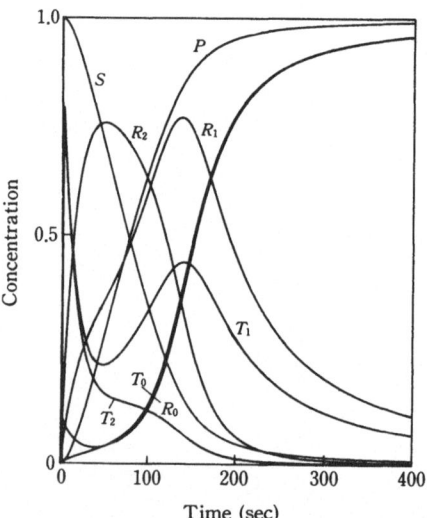

**Fig. 4.19**  Time course of Scheme 4.13 (a). Initial condition: $S = 10$ mM,
$R_0 = 10$ nM, $T_0 = 9.99\mu$M. Full scale (1.0) on the ordinate: $S = P = 10$ mM,
$R_0 = 10$ nM, $R_1 = 0.3\mu$M, $R_2 = T_0 = 10\mu$M, $T_1 = T_2 = 5\mu$M.

in both cases, ensuring very high concentration-ratios of substrate and enzyme of 1000 and 100, respectively.  In both cases the reaction gets through the very fast transient-phase for a few seconds after the start of reaction, never reaching the steady state thereafter in contrast to Michaelis-Menten-

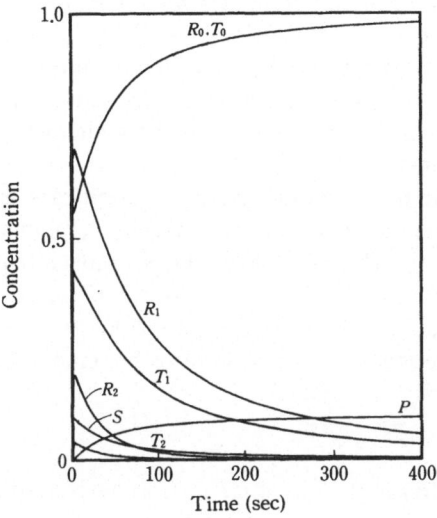

**Fig. 4. 20**  Time course of Scheme 4.13 (b).  Initial condition : $S=1.0$ mM.  Other conditions are the same as in Fig. 4. 19.

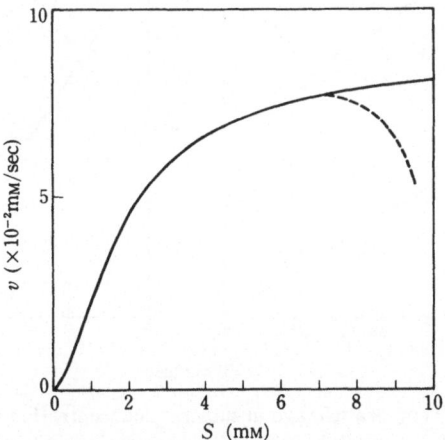

**Fig. 4. 21**  $S$-$v$ curve for Scheme 4.13.  Initial substrate concentration : $S_0=10$ mM.  Solid line : equation (29) ; broken line : simulation.

type reaction in Fig. 4.1. On the other hand, in the case of $S_0 = 10$ mM, each enzyme species has its extremals at about 45 sec when $R_2$ attains a peak, leading to a sort of steady state. Rapid equilibrium apparently arises after 45 sec for $S_0 = 10$ mM and 2 sec for $S_0 = 1.0$ mM, respectively, because the ratio of $T_0$ to $R_0$ is almost equal to $L$ ($= 1000$).

The time course for $S_0 = 1.0$ mM coincides with that for $S_0 = 10$ mM after 160 sec at which $S$ reaches 1.0 mM. Of course, $P$ in Fig. 4.20 is added by the amount produced until that time. In the following study, the simulation is performed for $S_0 = 10$ mM as standard, since the time courses for lower substrate concentration can be obtained from the standard time-course.

### $S$-$v$ Curve from Simulation

The simulation yields the temporal change in reaction rate as well as concentrations of every chemical species in the system. The evaluated values of $S$ and corresponding $v$ ($= dP/dt$) lead to the $S$-$v$ curve from the simulation.

On the other hand, application of rapid-equilibrium approximation gives the following $S$-$v$ relationship for the system of Scheme 4.13:

$$\frac{v}{V_{\max}} = \frac{\alpha(1+\alpha)}{L(1+c\alpha)^2+(1+\alpha)^2} \text{ ,} \tag{29}$$

where $\alpha = S/K_R$, $c = K_R/K_T$ and $V_{\max} = 2k_4E_0$. Comparison of the $S$-$v$ curve

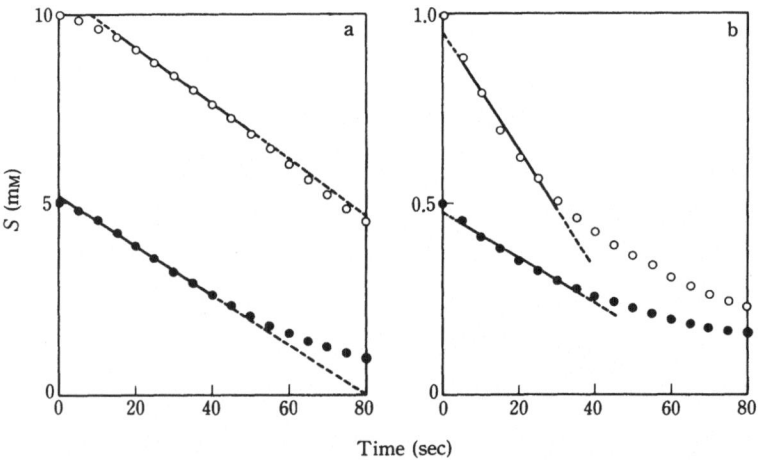

**Fig. 4.22** Temporal change in substrate concentration by simulation of Scheme 4.13. a) Initial substrate concentration : 10 mM($\bigcirc$) ; 5 mM($\bullet$). b) Initial substrate concentration : 1.0 mM($\bigcirc$) ; 0.5 mM($\bullet$). Initial velocities are obtained from the straight line drawn on the data points.

from a simulation for $S_0 = 10$ mM (Fig. 4.19) with that from equation (29) is demonstrated in Fig. 4.21, in which the two curves are virtually identical for $S$ below 7.0 mM. This agreement corresponds to the phase of rapid equilibrium established as seen in Fig. 4.19. At the start of reaction the enzyme species are distributed in 10 nM of $R_0$ and 9.99 $\mu$M of $T_0$, respectively. This is extremely umbalanced toward T-state, compared to the equilibrium with $S_0 = 10$ mM. The process in reducing this imbalance by allosteric transition causes the two curves to deviate for $S$ from 10 mM to 7.0 mM.

Experimental determination of an $S$-$v$ curve requires repetition of the procedure to obtain the initial velocity from extrapolation of the temporal change in $S$ or $P$ for various $S_0$. It is concluded from Fig. 4.21, however, that a single experiment can yield an $S$-$v$ curve with accurate measurements of substrate concentration and reaction rate.

**Remarks on $S$-$v$ Curves from Experiments**

The temporal changes in $S$ for 10, 5.0, 1.0 and 0.50 (in mM) of $S_0$ are plotted in Fig. 4.22. In all cases, $S$ apparently decreases at a constant rate beginning a few seconds after the start of reaction. As shown in Fig. 4.19, however, the progress of reaction has to go through the state in which the rapid-equilibrium assumption is invalid. It is thus expected that the initial velocity determined from the slope of $S$ in Fig. 4.22 might not be equal to $v$ calculated with equation (29). This situation is indicated in Table 4.14. The temporal change in $dP/dt$ from simulation is given in Fig. 4.23. The experimental values in Table 4.14 are obtained by taking the time-average

**Table 4.14** Relationship between substrate concentration and initial velocity

| $S$ (mM) | Initial velocity | |
| --- | --- | --- |
| | Experimental | Simulated |
| 0.2 | 0.16 | 0.18 |
| 0.4 | 0.46 | 0.57 |
| 0.6 | 0.87 | 1.07 |
| 0.8 | 1.27 | 1.65 |
| 1.0 | 1.56 | 2.22 |
| 2.0 | 3.53 | 4.58 |
| 3.0 | 4.68 | 5.93 |
| 4.0 | 5.52 | 6.68 |
| 6.0 | 6.72 | 7.52 |
| 8.0 | 7.17 | 7.94 |
| 10.0 | 7.40 | 8.21 |

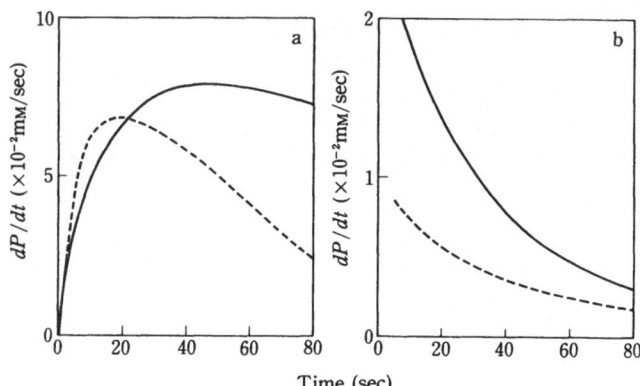

**Fig. 4. 23**  Temporal change in reaction rate by simulation of Scheme 4.13.  a) Initial substrate concentration: 10 mM (———) ; 5 mM (······).
b) Initial substrate concentration: 1.0 mM (———) ; 0.5 mM (······).

of $dP/dt$ during the period assigned for evaluating the slope of $S$. During this period $S$ changes so that the initial velocity obtained becomes an average reaction-rate in the process of reduction in $S$. Hence, the $S$-$v$ curve from these values shifts to the right side of that from the simulation or equation (29) as revealed in Fig. 4.24.

Another form of the $S$-$v$ relationship is represented by the Hill equation:

$$\log \frac{v}{V-v} = n \log S - \log K', \tag{30}$$

where the Hill coefficient $n$ correspond to the apparent number of protomers ($n_{app}$) and $K'$ is apparent dissociation constant. The value of $V$ in the equation is obtained by increasing $S$ to infinity in equation (29), resulting in a value of $9.09 \times 10^{-2}$ mM/sec and the Hill plot shown in Fig. 4. 25.  The Hill coefficient is evaluated in some regions of substrate concentration as given in Table 4.15.  The Hill plot from the data of simulation yields a Hill coefficient of 1.71, which cannot yet agree with the true value of 2.0.  Moreover, according to the above remark, the initial velocity from the experiment is estimated at a lower value, leading to smaller Hill coefficient.  The sigmoidicity is thus taken as weaker in the experiment than for the real enzyme.

**Effect of Allosteric Constant on Dynamic Behavior**
The effect of allosteric constant $L$ on the dynamic behavior of chemical species is revealed by the simulation of Scheme 4.13 for various values of $k_1$. The time course with $k_1 = 5000$ and $L = 10^4$ is shown in Fig. 4.26.  Its comparison with the time course in Fig. 4.19 indicates that increase in $L$

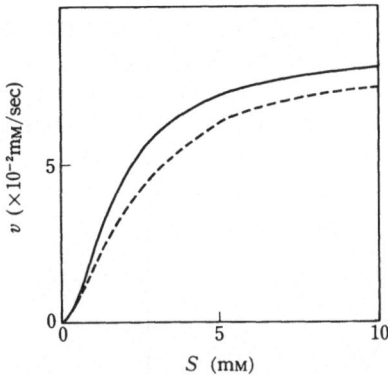

**Fig. 4. 24** Comparison of $S$-$v$ curves for Scheme 4.13. Initial substrate concentration: $S_0 = 15$ mM. Solid line: data from simulation, Broken line: data from the straight lines drawn as in Fig. 4.22.

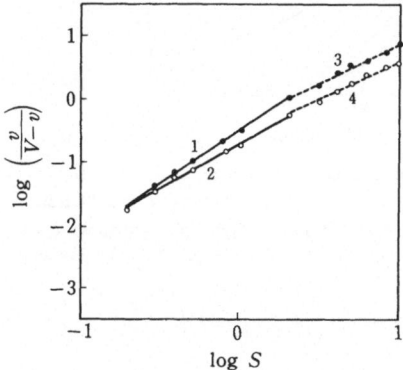

**Fig. 4. 25** Hill plot for Scheme 4.13. ○ : Simulated data ; ● : Experimental data. Solid line: 0.2 mM $< S <$ 2 mM ; Broken line: 2 mM $< S$ $<$ 10 mM. $n_{app}$ : 1: 1.71, 2: 1.54, 3: 1.30, 4: 1.24.

**Table 4. 15** Hill coefficients for Scheme 4. 13

| Range of $S$ (mM) | Hill coefficient ($n_{app}$) | |
|---|---|---|
| | Experimental | Simulated |
| 0.2~10 | 1. 44 | 1. 57 |
| 0.2~ 2 | 1. 54 | 1. 71 |
| 2~10 | 1. 24 | 1. 30 |

reduces the reaction rate and the time for establishing an equilibrium between $R_0$ and $T_0$. The $S-v$ curve and Hill plot from the experiment thus tend to be almost identical to those from the simulation, as demonstrated in Fig. 4.27 in contrast to Fig. 4.24. The Hill coefficient of 1.69 from the experiment also is in good agreement with 1.72 of that from the simulation.

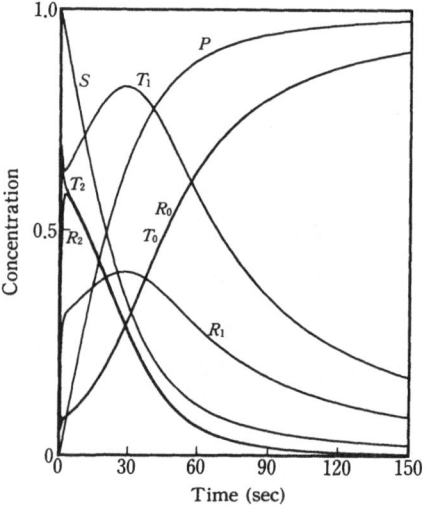

**Fig. 4.26** Time course of Scheme 4.13 with $L=10^4 (k_1=5\times10^3 \text{ sec}^{-1})$. Initial condition: $S=10$ mM, $R_0=1$ nM, $T_0=9.999\mu$M. Full scale (1.0) on the ordinate: $S=P=10$ mM, $R_2=T_1=T_2=5\mu$M, $R_0=1$ nM, $R_2=0.1\mu$M, $T_0=10\mu$M.

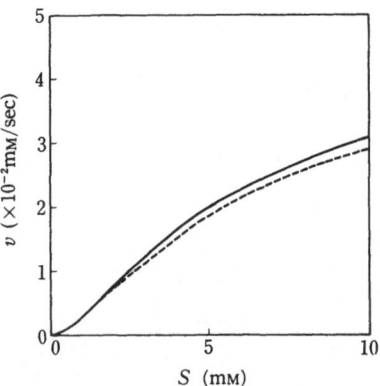

**Fig. 4.27** $S-v$ curve for Scheme 4.13 with $L=10^4 (k_1=5\times10^3 \text{ sec}^{-1})$. Solid line: simulated data; Broken line: experimental data.

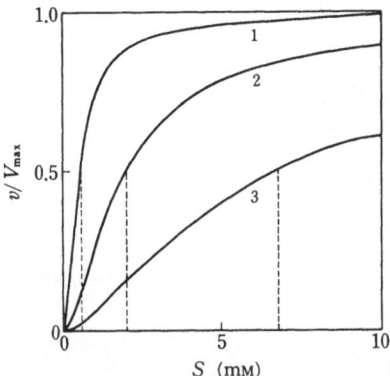

**Fig. 4.28**  Effect of $L$ on $S$-$v$ curve for Scheme 4.13.  1: $L=100$, $k_1=50 \text{ sec}^{-1}$; 2: $L=10^3$, $k_1=500 \text{ sec}^{-1}$; 3: $L=10^4$, $k_1=5\times10^3 \text{ sec}^{-1}$. The broken line indicates the substrate concentration for $V_{max}/2$.

The time course of a reaction with $L=10^4$ after 20 sec is similar to that with $L=1000$ after 50 sec, implying that the allosteric constant $L$ affects the dynamic behavior only before the establishment of equilibrium among the enzyme species. We can estimate the time course of reaction with an arbitrary $L$ from that with $L=1000$. Figure 4.28 illustrates the variation of $S$-$v$ curve with various values of $L$, indicating stronger sigmoidicity for larger $L$.

It should be mentioned for generalization of the effect of rate constant on the dynamic behavior that the $S$-$v$ curve from the simulation (*i.e.*, the actual reaction process) coincides with that from equation (29) only for a system having particular rate constants which possibly establish the rapid equilibrium among the enzyme species.

**Effect of Production from Enzyme Species in T-state**

We consider the system of Scheme 4.14, in which product P is produced from the enzyme species in T-state as well as the species in R-state. The effect of production from the T-forms on the dynamic behavior can be derived from the simulation for various values of $k_5$. Figure 4.29 presents the time course with $k_5=1$, which is quite similar to that with $k_5=0$ (shown in Fig. 4.19 for Scheme 4.13).

The effect of $k_5$ on $S$-$v$ curve is demonstrated in Fig. 4.30. It is inferred that addition of the V-system (*i.e.*, assignment of different rate constants for production from the R- and T-forms) to an allosteric mechanism strengthens the sigmoidicity of the system without much affecting the time course of reaction.

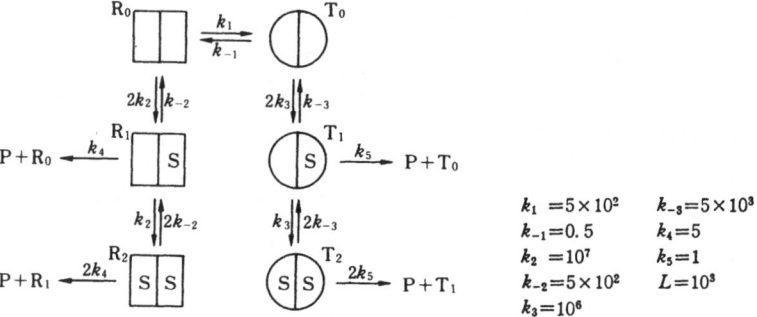

$$k_1 = 5 \times 10^2 \qquad k_{-3} = 5 \times 10^3$$
$$k_{-1} = 0.5 \qquad k_4 = 5$$
$$k_2 = 10^7 \qquad k_5 = 1$$
$$k_{-2} = 5 \times 10^2 \qquad L = 10^3$$
$$k_3 = 10^6$$

Scheme 4.14.  MWC dimeric model of allosteric enzyme (II)
The product is produced from both R- and T-form species.

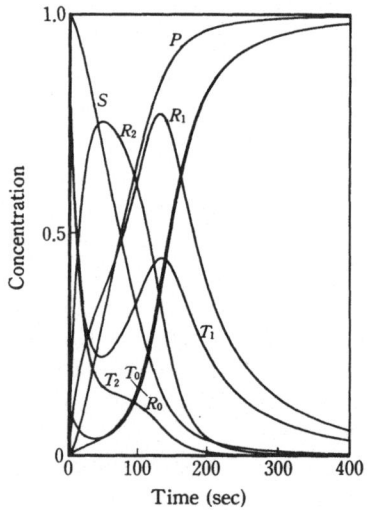

**Fig. 4. 29**  Time course of Scheme 4.14.  Initial condition: $S = 10$ mM, $R_0 = 10$ nM, $T_0 = 9.99 \mu$M.  Full scale (1.0) on the ordinate: $S = P = 10$ mM, $R_0 = 10$ nM, $R_1 = 0.3 \mu$M, $R_2 = T_0 = 10 \mu$M, $T_1 = T_2 = 5 \mu$M.

## System Including Allosteric Transition between Enzyme-Ligand Complexes

In principle, the MWC model allows two mechanisms for the allosteric transition: the first in which the transition takes place only between $R_0$ and $T_0$, enzyme species free of ligand, as in Scheme 4.13, and the second in

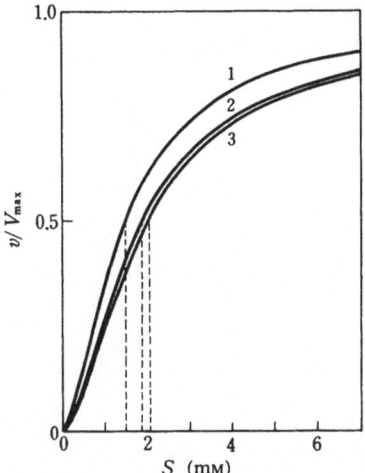

**Fig. 4. 30**　Effect of $k_5$ on $S$-$v$ curve for Scheme 4.14. $L=10^3$; $K_R/K_T$ $=0.01$; $k_5$: 1: 5 sec$^{-1}$, 2: 1 sec$^{-1}$, 3: 0. The broken line indicates the substrate concentration for $V_{max}/2$.

which the transition is possible for each pair of R- and T-form species, as in Scheme 4.15. We here compare the two mechanisms with respect to their dynamic behavior.

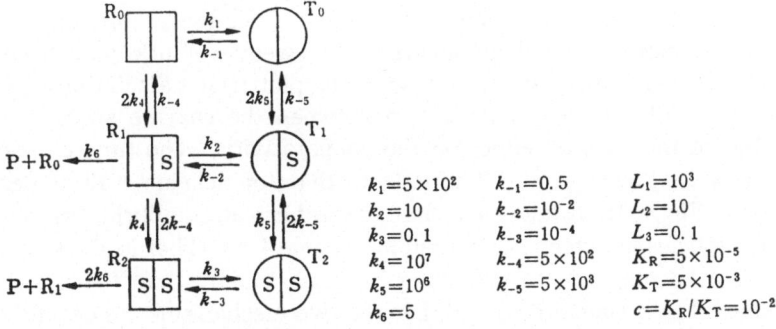

$$k_1=5\times10^2 \qquad k_{-1}=0.5 \qquad L_1=10^3$$
$$k_2=10 \qquad k_{-2}=10^{-2} \qquad L_2=10$$
$$k_3=0.1 \qquad k_{-3}=10^{-4} \qquad L_3=0.1$$
$$k_4=10^7 \qquad k_{-4}=5\times10^2 \qquad K_R=5\times10^{-5}$$
$$k_5=10^6 \qquad k_{-5}=5\times10^3 \qquad K_T=5\times10^{-3}$$
$$k_6=5 \qquad\qquad\qquad\qquad c=K_R/K_T=10^{-2}$$

Scheme 4. 15.　MWC dimeric model of allosteric enzyme (III)
The allosteric transitions take place between each pair of the
R- and T-form species.

In Scheme 4.15 the allosteric constants $L_1$, $L_2$ and $L_3$ are related to each other as follows:

$$L_2=L_1c \quad\text{and}\quad L_3=L_1c^2, \tag{31}$$

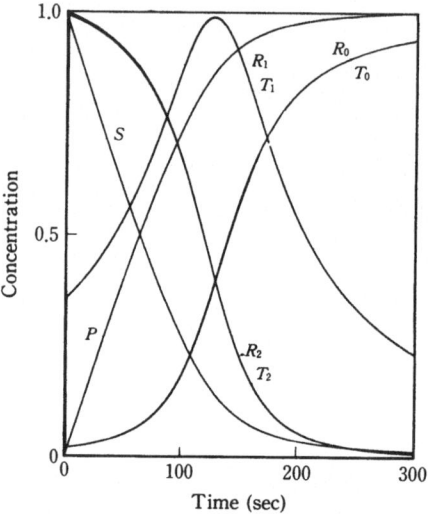

**Fig. 4.31** Time course of Scheme 4.15. Initial condition: $S = 10$ mM, $R_0 = 10$ nM, $T_0 = 9.99 \mu$M. Full scale (1.0) on the ordinate: $S = P = 10$ mM, $R_0 = 10$ nM, $R_1 = 0.13 \mu$M, $R_2 = 2.7 \mu$M, $T_0 = 10 \mu$M, $T_1 = T_2 = 3.9 \mu$M.

where $L_1 = k_1/k_{-1}$, $L_2 = k_2/k_{-2}$, $L_3 = k_3/k_{-3}$ and $c = K_R/K_T$ with $K_R = k_{-4}/k_4$ and $K_T = k_{-5}/k_5$. This relationship is based on the microscopic reversibility in closed loops in the scheme. The rate constants for the allosteric transitions are chosen as given in Scheme 4.15 for the simulation.

Both mechanisms yield an identical Lineweaver - Burk plot from the simulation, which also agrees well with the plot from calculation of equation (29). The allosteric transition between the enzyme species bound with ligand thus has no effect on the cooperativity. The time course for Scheme 4.15, however, is different from that for Scheme 4.13 as demonstrated in Fig. 4.31. It takes less time to establish an equilibrium among the enzyme species in Scheme 4.15 than in Scheme 4.13. That is, the system of Scheme 4.15 responds to disturbances more rapidly than the system of Scheme 4.13. It should be noted that the two mechanisms are not different from the aspect of cooperativity, while they are distinguished from other aspects of behavior in nonsteady states.

## 3. MWC Dimeric Model—Interaction with Effectors

Specific properties of allosteric enzymes are revealed from their interactions with effectors (inhibitors and activators) as well as with substrates. In enzyme systems of multi-step reactions the intermediate metabolites and end product often work as effectors for the first-step allosteric enzyme, regulating the action of the whole system. We here examine the effects of effectors on allosteric enzymes of the MWC dimeric model. In the analysis the concentrations of effectors in a system are assuemd to be constant during the progress of reaction.

**Effect of Inhibitor — Determination of Dissociation Constant for Enzyme-inhibitor Complex**

The mechanism in Scheme 4.16 is considered as the simplest model for inhibition, in which substrate S binds only with the R-form species and

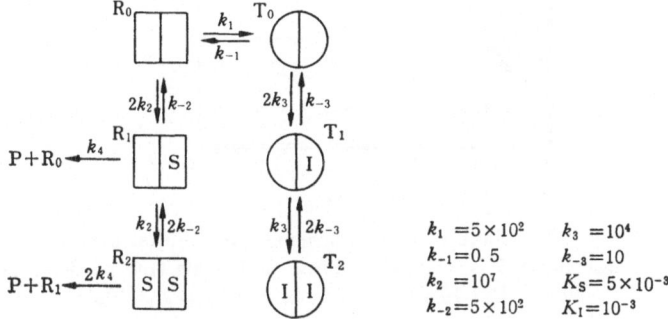

Scheme 4.16.  MWC dimeric model with an inhibitor

inhibitor I only with the T-form species. The time course of a reaction with $I=1.0$ mM is shown in Fig. 4.32. Enzyme species $T_1$ and $T_2$ do not reduce their concentrations with decrease in substrate concentration, increasing instead after 3 sec at which $R_2$ reaches a peak. The enzyme species in R-state behave similarly to those in a system without inhibition (Fig. 4.19). The $S$-$v$ curve from the simulation coincides well with that from calculation with the relationship,

$$\frac{v}{V_{max}}=\frac{\alpha(1+\alpha)}{L(1+\gamma)^2+(1+\alpha)^2}, \quad \alpha=S/K_S, \quad \gamma=I/K_I,$$
(32)

which is derived by the rapid-equilibrium approximation. Figure 4.33 demonstrates the change in $S$-$v$ curve with variation in $I$ or $K_I$ (dissociation constant for I) ; increase in $\gamma$ leads to stronger sigmoidicity.

We now evaluate a method for experimental determination of dissociation constant $K_I$ [5]. Measurement of the initial velocity $v$ with constant concentration in substrate and various concentrations in inhibitor yields a relationship between $v$ and $\gamma$ in equation (32) to determine $K_I$. Equation (32) is rearranged as

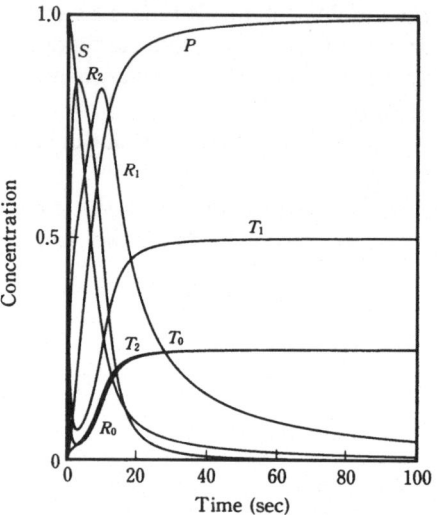

**Fig. 4. 32**  Time course of Scheme 4.16.  Initial condition: $S=10$ mM, $R_0=10$ nM, $T_0=9.99\mu$M, $I=1$ mM.  Full scale (1.0) on the ordinate: $S=P=10$ mM, $R_0=10$ nM, $R_1=0.2\mu$M, $R_2=T_0=T_1=10\mu$M.

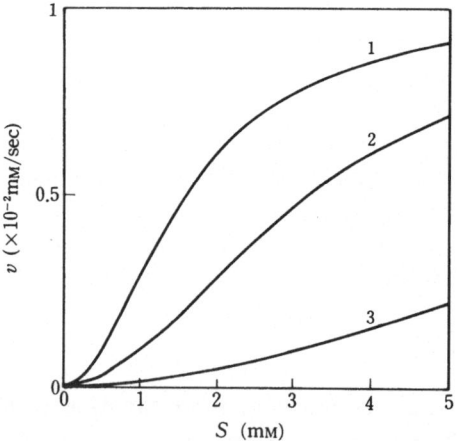

**Fig. 4. 33**  Effect of inhibitor on $S$-$v$ curve for Scheme 4.16.  Concentration in inhibitor: 1: 0, 2: 1 mM, 3: 5 mM.

$$\sqrt{\frac{V'-v}{v}}=\frac{\sqrt{L}}{1+\alpha}\left(1+\frac{I}{K_I}\right),$$

$$(33)$$

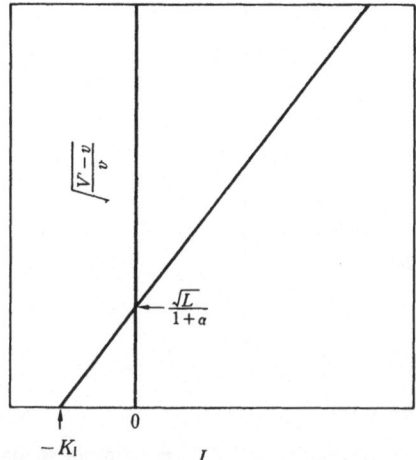

**Fig. 4. 34**  Plot of equation (33).

where $V' = \alpha V_{max}/(1+\alpha)$. This relationship is plotted as in Fig. 4.34, presenting the procedure for determination of $K_1$: evaluation of $V'$ followed by plotting $((V'-v)/v)^{1/2}$ against $I$ to read $-K_1$ on the $I$-axis. $V'$ can be obtained from the $S$-$v$ curve for a reaction without inhibition (*i. e.*, $I=0$ or $\gamma=0$). If the substrate concentration is set high enough to satisfy the condition,

$$L \ll (1+\alpha)^2, \tag{34}$$

equation (32) is reduced to

$$\frac{v}{V_{max}} = \frac{\alpha}{1+\alpha}. \tag{35}$$

The Lineweaver-Burk plot thus becomes linear in this range of substrate concentration, assigning $V'$ to $v$ at an appropriate $\alpha$. The condition of inequality (34) corresponds to

$$S \gg 1.5\,\text{mM}$$

for a reaction of Scheme 4.16 with $L=1000$ and $K_s=0.05$ mM. The plot for $I=0$ is displayed in Fig. 4.35, in which neither simulation nor calculation with equation (32) yield a linear relationship, while the data from the experiment are on a straight line without guarantee of true $V'$. Besides the evaluation of the condition (34), the values of $V'$ from $v$ at $S_0=10$ mM are determined as 97.1 and 96.2 (in $\mu$Msec$^{-1}$) from the simulation and

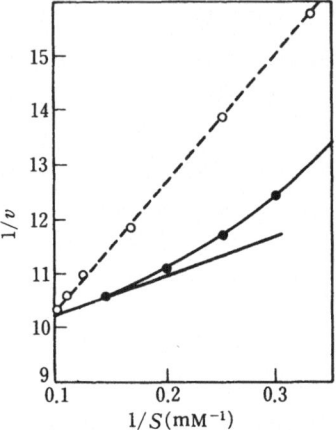

**Fig. 4. 35** Plot for evaluation of $V'$.  ●: Simulated data, $V'=97.1$ $\mu$M sec$^{-1}$; ○: experimental data, $V'=96.2$ $\mu$M sec$^{-1}$.

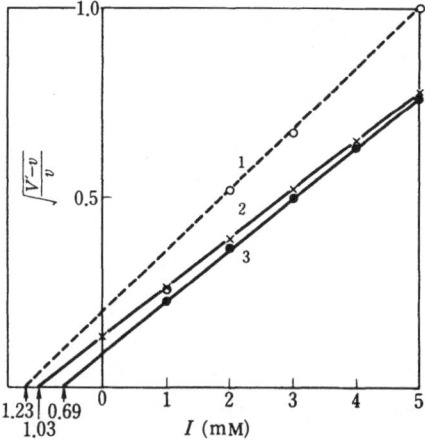

**Fig. 4. 36** Determination of $K_1$. 1: from $V'$ by experiment, yielding $K_1=1.23$ mM; 2: from true $V'$, yielding $K_1=1.03$ mM; 3: from $V'$ by simulation, yielding $K_1=0.69$ mM.

experiment, respectively, which are not in agreement with the true value of 99.5 $\mu$Msec$^{-1}$ calculated with the parameters in Scheme 4.16.

    Figure 4.36 illustrates plotting of $((V'-v)/v)^{1/2}$ against $I$ for the various $V'$ evaluated. With the true value of 99.5 $\mu$M sec$^{-1}$ of $V'$ we obtain the accurate dissociation constant $K_1$ of 1.03 mM. On the other hand, the use of $V'$ from Fig. 4.35 yields $K_1 = 0.690$ mM from the simulation and $K_1 = 1.23$ mM from the experiment. It should be noted that evaluation of $V'$ has a great effect on determination of $K_1$, and that a considerably high concentration of inhibitor must be used for experiments because $v$ at $I = 0$ is set as $V'$.

**Effect of Activator — Determination of Dissociation Constant for Enzyme-Activator Complex**

A model for a system with an activator is given in Scheme 4.17. Sub-

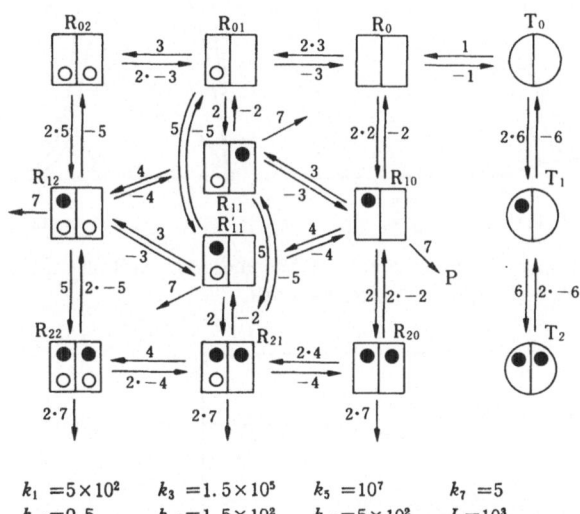

| | | | |
|---|---|---|---|
| $k_1 = 5 \times 10^2$ | $k_3 = 1.5 \times 10^5$ | $k_5 = 10^7$ | $k_7 = 5$ |
| $k_{-1} = 0.5$ | $k_{-3} = 1.5 \times 10^2$ | $k_{-5} = 5 \times 10^2$ | $L = 10^3$ |
| $k_2 = 10^7$ | $k_4 = 1.5 \times 10^5$ | $k_6 = 10^6$ | $K_S = 5 \times 10^{-3}$ |
| $k_{-2} = 5 \times 10^2$ | $k_{-4} = 1.5 \times 10^2$ | $k_{-6} = 5 \times 10^3$ | $K_A = 10^{-3}$ |

Scheme 4.17. MWC dimeric model with an activator
● : substrate ; ○ : activator. Rate constants are denoted by numerals. For example, $k_7$ and $2k_{-2}$ are indicated by 7 and 2 · −2, respectively. This convention is used in the following schemes as well.

strate S binds with both R- and T-form species, while activator A binds only with the R-form species. The time course of a reaction with $A = 1$ mM is shown in Fig. 4.37. Until about 60 sec after the start of reaction, equilibrium is not attained among the enzyme species, and then the rapid-equilibrium assumption becomes valid.

The simulation with various concentrations of activator yields $S-v$ curves in Fig. 4.38, indicating the dependency of sigmoidicity on the dissociation constant $K_A$. As with inhibition, we examine a method for de-

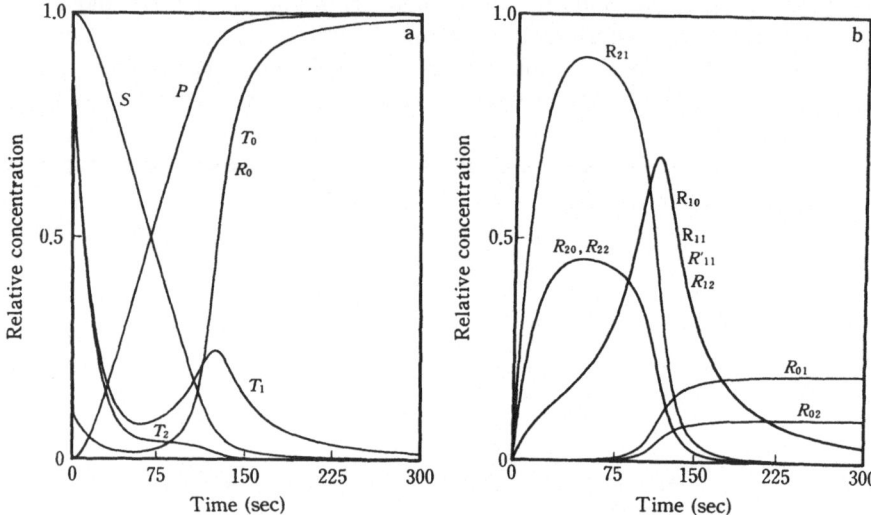

**Fig. 4. 37** Time course of Scheme 4.17. Initial condition: $S=10$ mM, $R_0=10$ nM, $T_0=9.99\mu$M, $A=1$ mM.

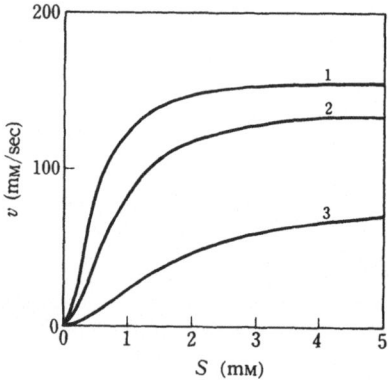

**Fig. 4. 38** Effect of activator on $S$-$v$ curve for Scheme 4.17. Concentration in activator: 1: 2 mM, 2: 1 mM, 3: 0.

termination of $K_A$ [5]. Based on the rapid-equilibrium approximation the velocity $v$ is given by

$$\frac{v}{V_{\max}}=\frac{\alpha(1+\alpha)(1+\beta)^2}{L(1+c\alpha)^2+(1+\alpha)^2(1+\beta)^2},$$
(36)

where

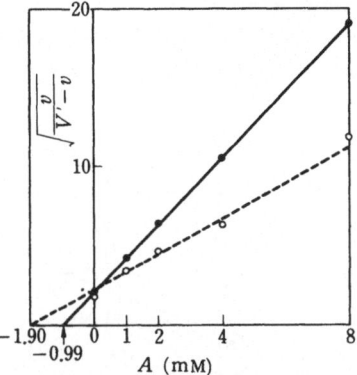

**Fig. 4. 39** Determination of $K_A$. Solid line: from $V'$ and $v$ by simulation, yielding $K_A = 0.99$ mM. Broken line: from $V'$ and $v$ by experiment, yielding $K_A = 1.90$ mM.

$$\alpha = S/K_R, \quad \beta = A/K_A, \quad c = K_R/K_T.$$

When a system is saturated with activator, that is, $\beta$ is very large, equation (36) is reduced to

$$V' = \frac{\alpha V_{max}}{1+\alpha}, \tag{37}$$

which is of the same form as $V'$ for determination of $K_1$. From equations (36) and (37) a linear relationship between $(v/(V'-v))^{1/2}$ and $A$,

$$\sqrt{\frac{v}{V'-v}} = \frac{1+\alpha}{(1+c\alpha)\sqrt{L}}\left(1+\frac{A}{K_A}\right), \tag{38}$$

is derived to determine $K_A$.

We adopt the initial velocity for $S_0 = 10$ mM and $A = 100$ mM as $V'$. The simulation yields a value of $99.5\ \mu$M sec$^{-1}$ of $V'$, which agrees with the true value from equation (37). The experimental data gives $V' = 98.4\ \mu$M sec$^{-1}$. The procedure of determination is applied to the data from the simulation and experiment, resulting in the plots in Fig. 4.39. A value of $0.99$ mM is obtained from the simulation for $K_A$, which is in good agreement with the true value of $1.0$ mM. This agreement is due to the accurate evaluation of $V'$ in activation, in contrast to the case of inhibition. On the other hand, the experimental data yield $K_A = 1.90$ mM, which is much different from the true value, because the reaction consumes the substrate rapidly and the initial velocity $v$ is measured at a substrate concentration much lower than $S_0$.

## Cumulative Effect of Inhibitor and Activator

Scheme 4.18 represents a generalized form of MWC dimeric model, in which substrate S binds with both R- and T-form species, while inhibitor I

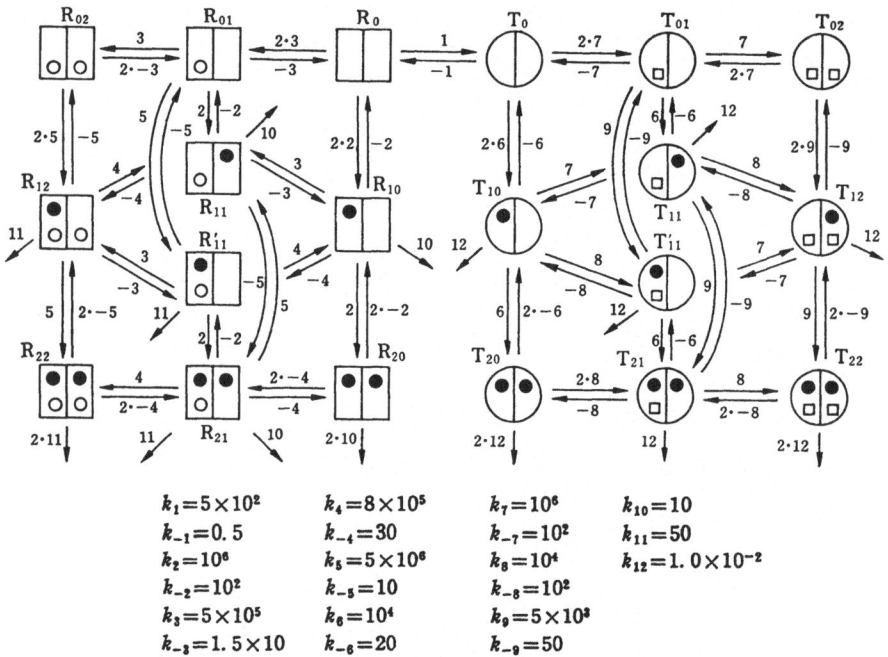

$$k_1 = 5 \times 10^2 \qquad k_4 = 8 \times 10^5 \qquad k_7 = 10^6 \qquad k_{10} = 10$$
$$k_{-1} = 0.5 \qquad k_{-4} = 30 \qquad k_{-7} = 10^2 \qquad k_{11} = 50$$
$$k_2 = 10^6 \qquad k_5 = 5 \times 10^6 \qquad k_8 = 10^4 \qquad k_{12} = 1.0 \times 10^{-2}$$
$$k_{-2} = 10^2 \qquad k_{-5} = 10 \qquad k_{-8} = 10^2$$
$$k_3 = 5 \times 10^5 \qquad k_6 = 10^4 \qquad k_9 = 5 \times 10^3$$
$$k_{-3} = 1.5 \times 10 \qquad k_{-6} = 20 \qquad k_{-9} = 50$$

Scheme 4.18. MWC dimeric model with both inhibitor and activator. ● : substrate ; ○ : activator ; □ : inhibitor.

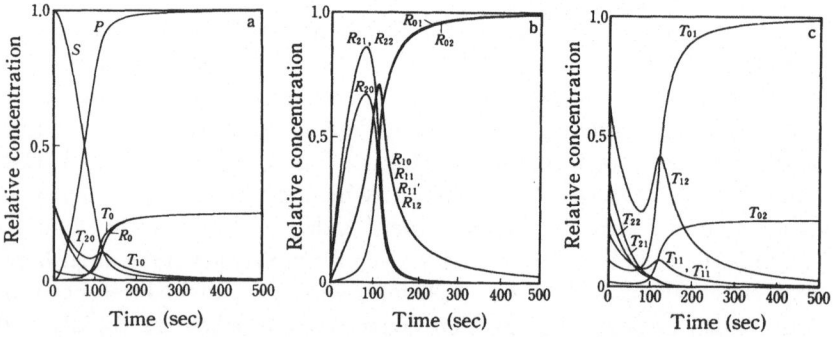

**Fig. 4.40** Time course of Scheme 4.18. Initial condition : $S = 10$ mM, $R_0 = 10$ nM, $T_0 = 9.99 \mu$M, $I = 1$ mM, $A = 1$ mM.

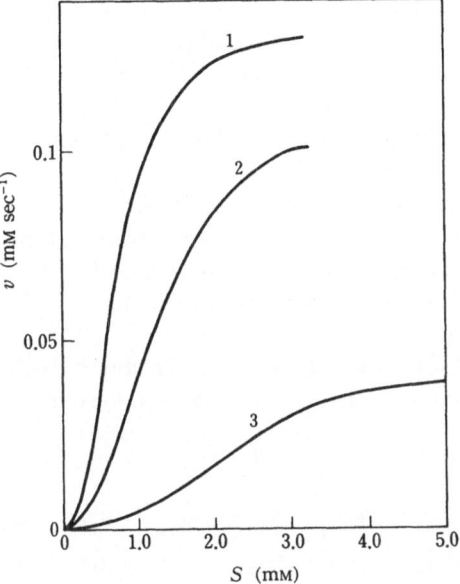

**Fig. 4.41**  $S$-$v$ curve for Scheme 4.18.  1: $I = 1$ mM, $A = 2$ mM; 2: $I = 1$ mM, $A = 1$ mM; 3: $I = 5$ mM, $A = 1$ mM.

and activator A bind only with the T- and R-form species, respectively. The time course of a reaction with $A = 1$ mM and $I = 1$ mM is shown in Fig. 4.40. The simulation with various concentrations of A and I yields the $S$-$v$ curves in Fig. 4.41. Similarity is noticed in the corresponding behavior of each chemical species in the time courses of various schemes of the MWC dimeric model.

In summary, the dynamic behavior of every chemical species in allosteric reactions of the MWC dimeric model in closed system, in which the enzyme species are distributed in $R_0$ and $T_0$ by a ratio of $L$ and addition of excessive substrate starts the reaction, are characterized as follows:

1) After a short lag, substrate S decreases at a nearly constant rate, which then gets smaller with lower $S$.
2) The time course of product P is symmetrical to that of substrate S.
3) With high concentration of S the ratio of $R_0$ to $T_0$ is larger than that at equilibrium, which is attained with decrease of $S$.
4) The R-form species bound with two molecules of substrate S ($R_{20}$, $R_{21}$ and $R_{22}$) increase rapidly after the start of reaction, reaching peaks at which an equilibrium is attained among the enzyme species, and then decreasing.

5) The T-form species bound with two molecules of substrate S ($T_{20}$, $T_{21}$ and $T_{22}$) increase through a very rapid transient-phase, then decreasing monotonously with shoulders in some cases.

6) The R-form species bound with one molecule of substrate S ($R_{10}$, $R_{11}$, $R_{11}'$ and $R_{12}$) reach their peaks at the base of the peaks of those species bound with two molecules of S.

7) The T-form species bound with one molecule of S ($T_{10}$, $T_{11}$, $T_{11}'$ and $T_{12}$) also increase through a very rapid transient-phase, and then decrease to the minimum when an equilibrium is attained among the enzyme species. They again increase to peaks at the almost same time as $R_{10}$, and then decrease.

### Dynamic Behavior of Two Substrates-Two Products System

Scheme 4.19 illustrates the enzymatic reaction system of an MWC di-

$$A + B \underset{}{\overset{E}{\rightleftharpoons}} P + Q$$

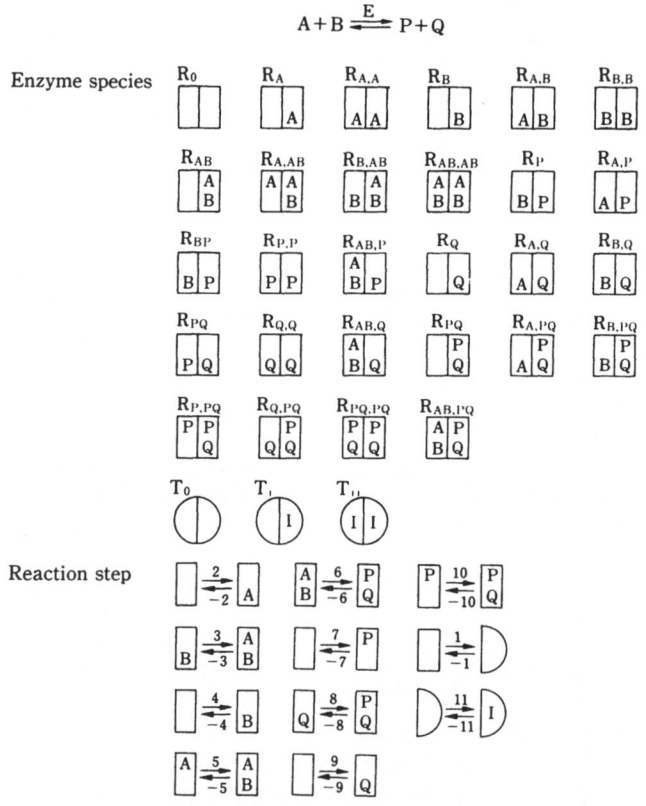

Scheme 4.19. MWC dimeric model of two substrate–two product allosteric reaction with an inhibitor

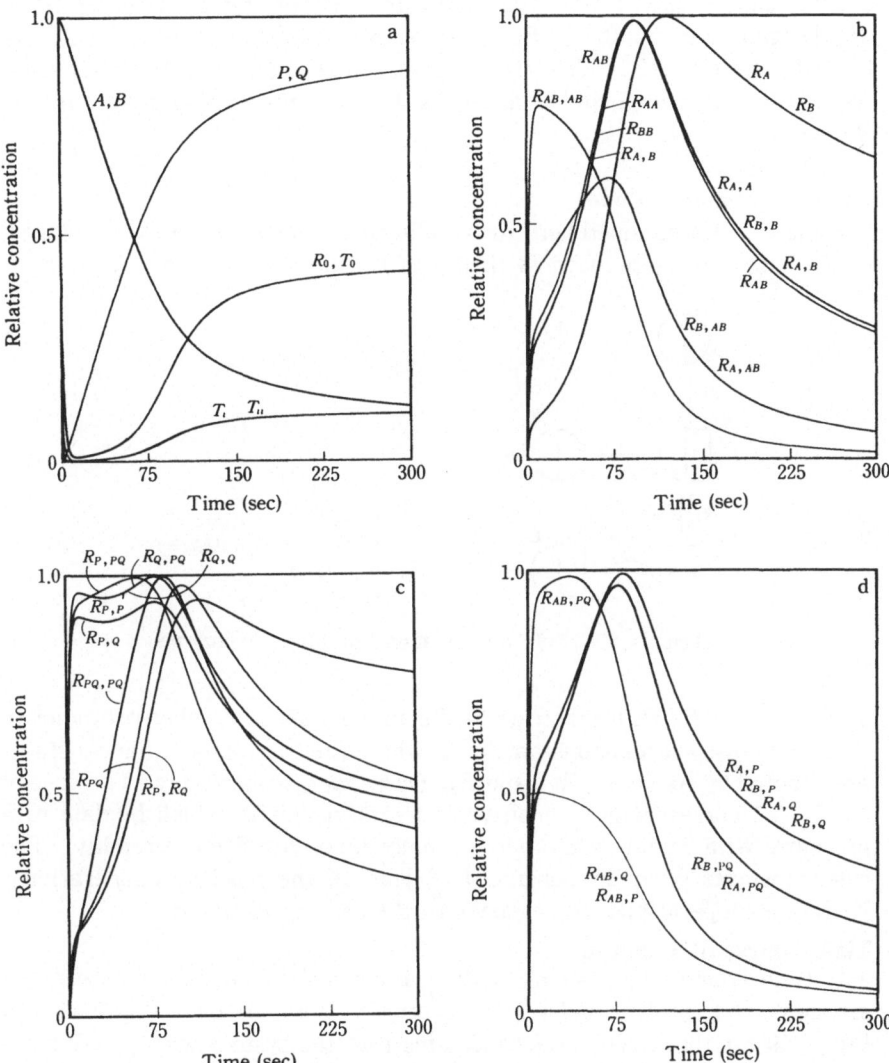

**Fig. 4. 42** Time course of Scheme 4.19.   Initial condition: $A = B =$ 10 mM, $R_0 = 10$ nM, $T_0 = 9.99\mu$M, $I = 10$ mM, $P = Q = 0$.

meric model producing two products P and Q from two substrates A and B. The reaction is reversible and responsive to inhibition, in which the substrates and products interact only with the R-form species and an inhibitor I binds only with the T-form species. As given in Scheme 4.19, the enzyme can occupy 31 states and the rate constants for the reaction steps are designated by numbers. For example, the transition from $R_0$ to $R_A$ has the rate constant $2k_2$. The time course of a reaction with initial concentration of 10 mM in A and B and constant concentration of 10 mM in I is shown in Fig. 4.42.

## 4. KNF Dimeric Model

Scheme 4.20 is examined here for an allosteric enzyme of the KNF dimeric model based on the induced-fit theory of Koshland *et al.* [15]. Binding of

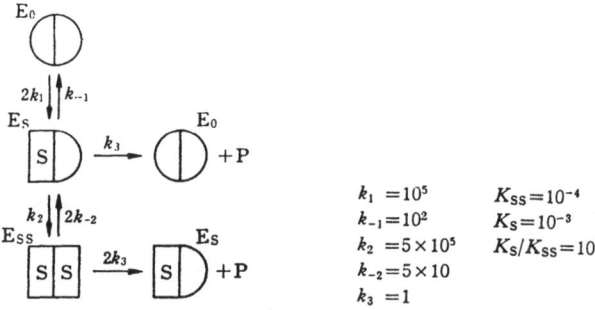

$$k_1 = 10^5 \qquad K_{ss} = 10^{-4}$$
$$k_{-1} = 10^2 \qquad K_s = 10^{-3}$$
$$k_2 = 5 \times 10^5 \qquad K_s/K_{ss} = 10$$
$$k_{-2} = 5 \times 10$$
$$k_3 = 1$$

Scheme 4.20.   KNF dimeric model of allosteric enzyme

a protomer in $E_0$ with substrate S affects the state of another protomer to change its dissociation constant for S, where the two dissociation constants are denoted by $K_s$ $(=k_{-1}/k_1)$ and $K_{ss}$ $(=k_{-2}/k_2)$, respectively. The case of $K_s/K_{ss} > 1$ corresponds to positive cooperativity, with which binding of a protomer with S makes another protomer bind with S more readily. The rate constants given in Scheme 4.20 lead to the positive cooperativity. Negative cooperativity corresponds to the case of $K_s/K_{ss} < 1$.

**Time Course of Reaction**
The time courses in systems with positive and negative cooperativities are shown in Figs. 4.43 and 4.44, respectively. Comparison of Fig. 4.43 with Fig. 4.19 for the MWC model indicates that the system with positive cooperativity follows the time course in which $E_s$ and $E_{ss}$ correspond to $R_1$ and $R_2$, and that the latter species is replaced by the former in the main role in production of P as the concentration of S decreases. On the other hand, P is always produced mainly from $E_s$ in the system with negative cooperativity.

*S-v* **Relationship**
The rapid-equilibrium approximation for the system of Scheme 4.20 leads to an *S-v* relationship,

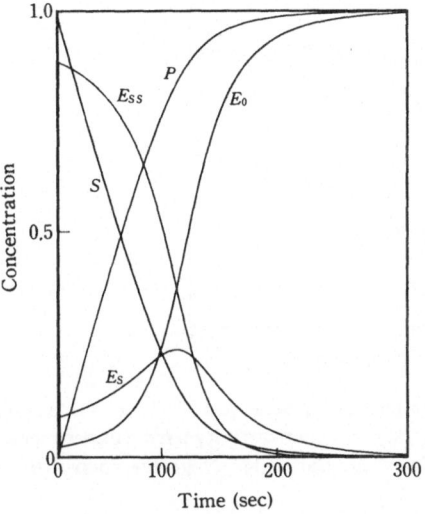

**Fig. 4. 43** Time course of Scheme 4.20 with positive cooperativity. Initial condition: $S=2$ mM, $E_0=10\mu$M. Full scale (1.0) on the ordinate: $S=P=2$ mM, $E_0=E_s=E_{ss}=10\mu$M.

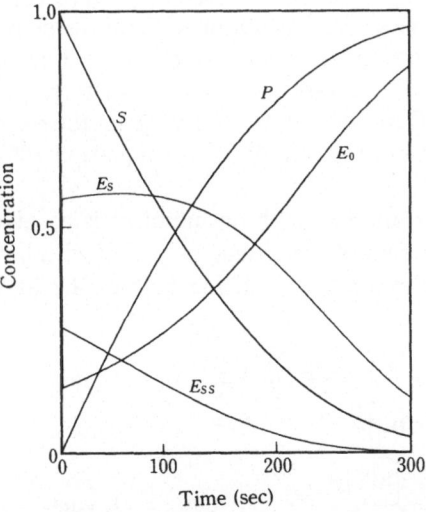

**Fig. 4. 44** Time course of Scheme 4.20 with negative cooperativity. $k_2=5\times10^4$M$^{-1}$ sec$^{-1}$, $k_{-2}=10^2$ sec$^{-1}$; other rate constants are equal to those in Scheme 4.20. The initial condition and full scale (1.0) on the ordinate are the same as in Fig. 4.43.

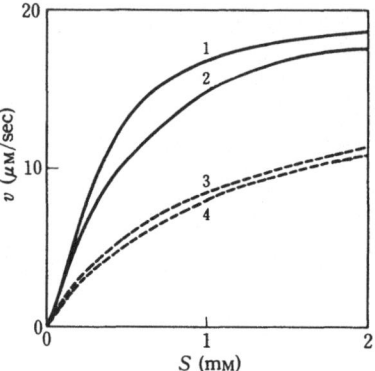

Fig. 4.45  $S-v$ curves for Scheme 4.20. 1: Positive cooperativity; from simulation, 2: Positive cooperativity; from experiment, 3: Negative cooperativity; from simulation, 4: Negative cooperativity; from experiment.

$$\frac{v}{V_{max}} = \frac{S(K_{SS}+S)}{K_S K_{SS} + K_{SS} S + S^2} ,$$

(39)

where $V_{max} = 2k_3 E_T$ and $E_T$ is the total concentration of the enzyme. The initial velocity from the experiment is obtained from the slope at which the concentration of S decreases in apparent linearity as in Figs. 4.43 and 4.44. Figure 4.45 presents the $S-v$ curves from the experiment and simulation. As in the case of the MWC model, the $S-v$ curve from the experiment shifts to the right side of that from the simulation or equation (39).

**Effect of Inhibitor**

One of the most simplified models with an inhibitor is as given in Scheme 4.21, in which a protomer cannot simultaneously bind with both substrate S and inhibitor I. This corresponds to Scheme 4.16 of the MWC model. In

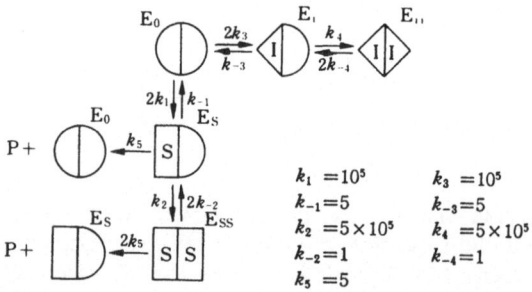

Scheme 4.21.  KNF dimeric model with an inhibitor

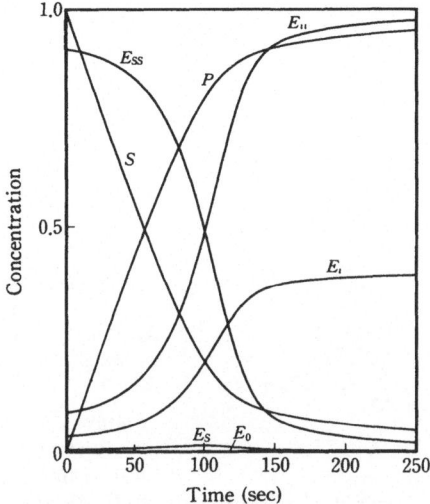

**Fig. 4. 46** Time course of Scheme 4.21. Initial condition: $S = 10$ mM, $I = 1$ mM, $E_0 = 10\mu$M. Full scale(1.0) on the ordinate: $S = P = 10$ mM, $E_0 = E_{ss} = E_{ii} = 10\mu$M, $E_s = 4\mu$M, $E_i = 10$ nM.

fact, correspondence between $E_s$ and $R_1$, $E_{ss}$ and $R_2$, $E_i$ and $T_1$, and $E_{ii}$ and $T_2$ is valid in their time courses. Figure 4.46 indicates the time course of a reaction in which the production rate of P (*i. e.*, $k_5$) is not rate-limiting nor is rapid equilibrium attained. In this case $E_s$ has no distinct peak, in contrast to the cases of Figs. 4.43 and 4.32.

## 5. MWC Tetrameric Model

We have so far been concerned with the cooperative behavior of dimeric allosteric enzymes. The MWC model is now extended to the tetrameric enzyme to examine its dynamic behavior. Phosphofructokinase of *Escherichia coli* and yeast glycelaldehyde-3-phosphate dehydrogenase are well-known among the many examples of tetrameric enzymes.

Scheme 4.22 represents an MWC tetrameric model in which substrate S binds with both R- and T-form species. As demonstrated in the following, the behavior of the tetrameric model is almost similar to that of dimeric models, revealing no features specific to tetrameric enzymes.

### Time Course of Reaction

The time course of a reaction with the rate constants given in Scheme 4.22 is shown in Fig. 4.47. The characteristics in the time course are not more than what are readily suggested from the time course in dimeric enzyme (Fig. 4.29). The concentration in $R_4$, which contributes most to production

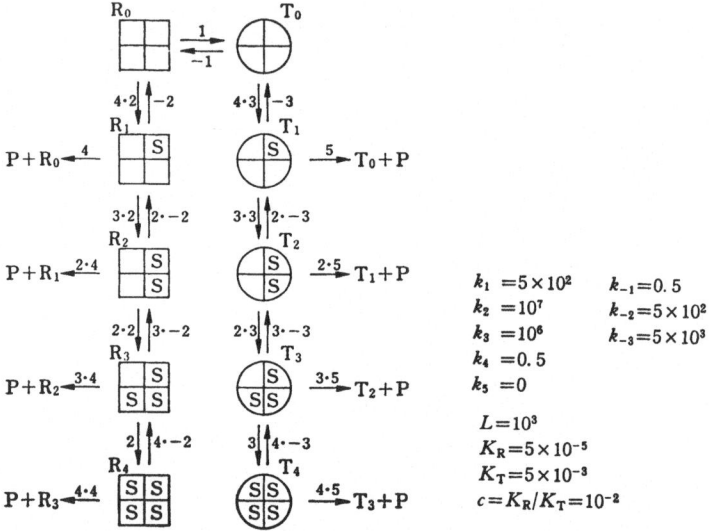

Scheme 4.22.   MWC tetrameric model of allosteric enzyme

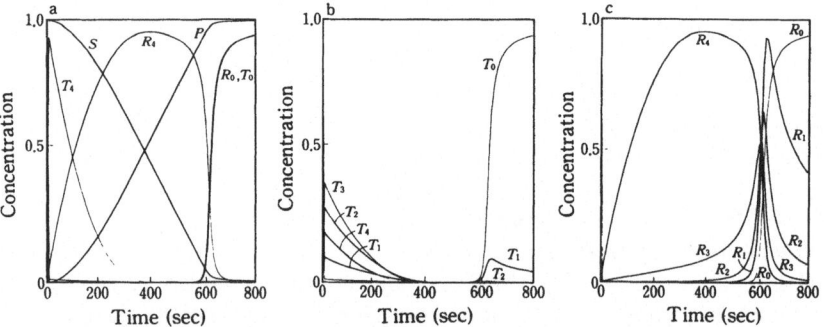

**Fig. 4.47**  Time course of Scheme 4.22. Initial condition: $S=10$ mM, $R_0=10$ nM, $T_0=9.99\mu$M. Full scale (1.0) on the ordinate: $S=P=10$ mM, $T_0=9.99\mu$M, $T_1=T_2=T_3=T_4=10\mu$M, $R_0=10$ nM, $R_1=0.1\mu$M, $R_2=1\mu$M, $R_3=5\mu$M, $R_4=10\mu$M.

of P, increases after the start of reaction, remaining almost constant from 250 sec to 600 sec. During this period $S$ and $P$ vary linearly, indicating the establishment of quasi-steady state in the system. At about 600 sec the concentration in S is reduced to less than 1.0 mM and $R_4$ decreases rapidly to be followed by peaks of $R_3$, $R_2$ and $R_1$ in sequence. The T-form species bound with S increase the concentration very sharply within 10 sec of the

transient phase, and then decrease monotonously as S decreases its concentration. At about 600 sec peaks appear in $T_3$, $T_2$ and $T_1$ in sequence.
Similar to $T_2$ in the dimeric model, $T_4$ sometimes has a shoulder in the
course of decreasing.

### $S$-$v$ Relationship

The rapid-equilibrium approximation for the system of Scheme 4.22
leads to an $S$-$v$ relationship,

$$v = \frac{k_4 \alpha (1+\alpha)^3 + k_5 Lc\alpha (1+c\alpha)^3}{L(1+c\alpha)^4 + (1+\alpha)^4} E_0, \tag{40}$$

where $\alpha = S/K_R$ and $c = K_R/K_T$, and $E_0$ expresses the total concentration of
enzyme. In Fig. 4.48 the $S$-$v$ curve from equation (40) with $k_5 = 0$ is plotted and compared with the curve from the time course in Fig. 4.47. At 400
sec after the start of reaction the rapid-equilibrium assumption becomes
valid, where $S$ is about 5.0 mM for the tetramer rather behind the dimer
with about 7.0 mM (Fig. 4.21). This corresponds to the time when $R_4$
reaches a peak, implying that the peak of the R-form species saturated
with the substrate points the phase valid of the rapid-equilibrium assumption regardless of the number of protomers in an enzyme. Time courses
after 400 sec are common in cases with various values of $L$ and $k_5$.

### Effect of Allosteric Constant

The effect of allosteric constant on behavior is revealed by the simulation
of reaction in Scheme 4.22 for the values of $10^4$ and 100 in $L$ (*i. e.*, 5000 and
50 [in sec$^{-1}$] in $k_1$). The time courses are quite similar in the two cases
because the initial velocity $v$ is almost equal to $V_{max}$ due to the high initial

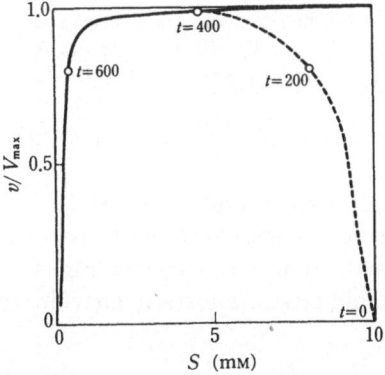

**Fig. 4.48** $S$-$v$ curve for Scheme 4.22. Solid line: equation (40);
Broken line: simulation. $t$ within the figure indicates the time after the
start of reaction.

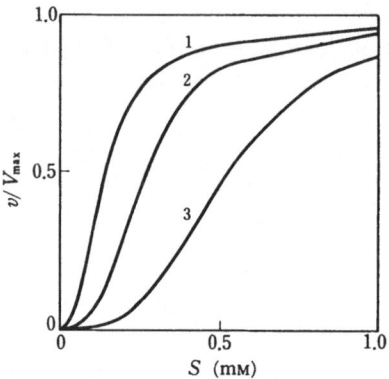

**Fig. 4. 49**  Effect of $L$ on $S$-$v$ curve for Scheme 4.22.  1 : $L=100$, $k_1=$ 50 sec$^{-1}$ ; 2 : $L=10^3$, $k_1=500$ sec$^{-1}$ ; 3 : $L=10^4$, $k_1=5\times10^3$ sec$^{-1}$.

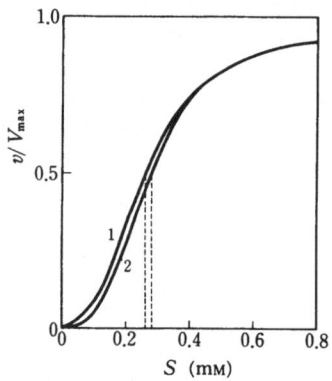

**Fig. 4. 50**  Effect of production from the T-form enzyme species on $S$-$v$ curve for Scheme 4.22.  1: $k_5=10$ sec$^{-1}$; 2: $k_5=0$. The broken line indicates the subutrate concentration for $V_{max}/2$.

concentration of 10 mM in the substrate.    Figure 4.49 displays the $S$-$v$ curves for various values of $L$.

**Effect of Product Production from T-Form Species**

The time courses of reactions of Scheme 4.22 with the values of 3, 5 and 10 (in sec$^{-1}$) in $k_5$ are similar to that with $k_5=0$ in Fig. 4.47.  Increase in $k_5$, however, reduces the sigmoidicity of allosteric enzyme, as shown in the $S$-$v$ curve in Fig. 4.50.

## 6.  Cooperativity and Transient State of Monomeric Enzyme

In addition to the allosteric enzymes with subunit structures, monomeric enzymes have recently been known to yield positive and negative cooper-

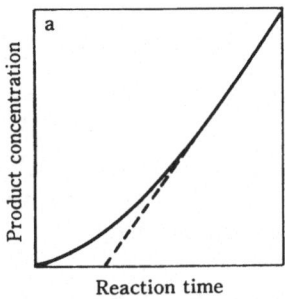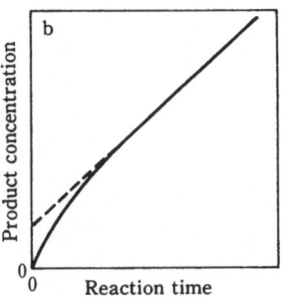

**Fig. 4. 51** Transient state of enzymatic reaction. a) Time-lag ; b) Burst.

ativities. The monomeric enzyme has no subunit structure, but is assumed to take possible multiple-states differing in three-dimensional structures due to interactions with ligands. The cooperativity is now defined as a phenomenon wherein the $S-v$ relationship, that is, the relationship between substrate concentration and initial velocity for enzymatic reaction, deviates from the hyperbolic curve (*i. e.*, the relationship for reaction of Michaelis-Menten-type enzyme).

The cooperativity is also known to be related with slow transient state. Michaelis-Menten-type and allosteric reactions generally display the transient states such as shown in Fig. 4.51 from the start of reaction until the quasi-steady state, that is, linear temporal change in product concentration is attained in the system. The time lag in the transient state in Fig. 4.51a is usually short under ordinary experimental condition of excess substrate. In some cases, however, time lags on the order of seconds or minutes are observed as slow transient states. On the other hand, the burst in Fig. 4.51b occurs in many reactions, for example, in bakery yeast hexokinase and homoserine dehydrogenase of *Escherichia coli*. A general analysis of relationships between reaction mechanisms and characteristics at the transient state will be described in detail in Section 6.2. We here discuss hysteretic and mnemonic enzymes as representative models of monomeric enzymes with cooperativity.

### Hysteretic Enzyme

Rabin [17] and Frieden [18] have proposed the mechanisms of Scheme 4.23 for cooperativity in a monomeric enzyme, which is induced by substrate binding to change slowly into a state of higher activity. It is assumed in the mechanisms that the elementary processes, $E \rightarrow E'$ and $ES \rightarrow E'S$, are slower than all other elementary processes, and that $E'$ is more active than E. The change in enzyme state induced by substrate binding is the basic process for the KNF model, which is assumed to be very slow in the models

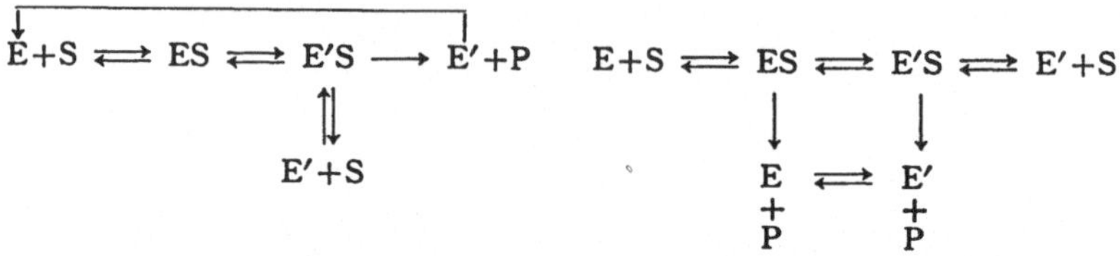

$$\text{E+S} \rightleftharpoons \text{ES} \rightleftharpoons \text{E'S} \longrightarrow \text{E'+P} \qquad \text{E+S} \rightleftharpoons \text{ES} \rightleftharpoons \text{E'S} \rightleftharpoons \text{E'+S}$$

$$\updownarrow \qquad\qquad\qquad\qquad\qquad \downarrow \qquad\qquad \downarrow$$

$$\text{E'+S} \qquad\qquad\qquad\qquad \text{E} \rightleftharpoons \text{E'}$$
$$+ \qquad\quad +$$
$$\text{P} \qquad\quad \text{P}$$

$$\text{(a)}\quad \text{Rabin's model} \qquad\qquad \text{(b)}\quad \text{Frieden's model}$$

Scheme 4.23.  Hysteretic enzyme

of Rabin and Frieden.  The enzyme responding slowly to abrupt change in a substrate concentration (for example, due to addition of substrate at the start of reaction) is called a hysteretic enzyme by Frieden, who emphasizes the physiological significance of such enzyme in acting as temporal buffering to variations in metabolite concentrations [18,19].

Ainslie *et al.* [20] have subsequently proposed Scheme 4.24 which is an extended model of Rabin's and Frieden's to analyze the relationship between cooperativity and transient state in hysteretic enzyme employing the rapid-equilibrium approximation.  In this mechanism E and E' are different

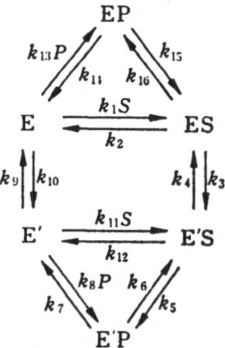

Scheme 4.24.  Ainslie's model of hysteretic enzyme

states of the enzyme, transforming each other through the very slow elementary processes represented by the rate constants $k_3$, $k_4$, $k_9$ and $k_{10}$. The $S$-$v$ relationship for Scheme 4.24 is given by

$$v = \frac{dS^2 + eS}{aS^2 + bS + c} , \tag{41}$$

where the coefficients $a$, $b$, $c$, $d$ and $e$ are the functions of 16 rate constants as specified in Table 4.16 by Ainslie *et al.* [20].  Their analysis with equation (41) for the various cases of rate constants given in Table 4.17

indicates the appearance of a time lag or burst in the time course of product production, and that the positive and negative cooperativities are observed as summarized in Table 4.16.

For the cases in Table 4.17, we here examine the dynamic behavior of hysteretic enzyme using computer simulation rather than rapid-equilibrium

**Table 4.16**  Coefficients in equation (41)

|   | Case 1 (N–B) | Case 2 (N–L) | Case 3 (P–B) | Case 4 (P–L) |
|---|---|---|---|---|
| $a$ | 1.00 | 1.00 | 1.00 | 1.00 |
| $b$ | 1.96 | 4.42 | 0.0313 | 0.0135 |
| $c$ | 0.0216 | 0.404 | $1.19 \times 10^{-4}$ | $1.09 \times 10^{-4}$ |
| $d$ | 106 | 333 | 106 | 82.1 |
| $e$ | 116 | 636 | 0.195 | 0.838 |

N–B: negative cooperativity and burst, N–L: negative cooperativity and time lag, P–B: positive cooperativity and burst, P–L: positive cooperativity and time lag

**Table 4.17**  Rate constants of Ainslie's model

|   | Case 1 (N–B) | Case 2 (N–L) | Case 3 (P–B) | Case 4 (P–L) |
|---|---|---|---|---|
| $k_1$ | 5 | 4 | 5 | 4 |
| $k_2$ | 3 | 3 | 3 | 3 |
| $k_3$ | −4 | −2 | −3 | −2 |
| $k_4$ | −2 | −2 | −1 | −3 |
| $k_5$ | 6 | 4 | 6 | 3 |
| $k_6$ | 6 | 4 | 6 | 2 |
| $k_7$ | 4 | 3 | 4 | 2 |
| $k_8$ | 5 | 4 | 5 | 4 |
| $k_9$ | −1 | 0 | −3 | −2 |
| $k_{10}$ | −2 | −1 | −4 | −3 |
| $k_{11}$ | 5 | 4 | 4 | 5 |
| $k_{12}$ | 4 | 2 | 3 | 2 |
| $k_{13}$ | 4 | 3 | 4 | 3 |
| $k_{14}$ | 4 | 3 | 4 | 3 |
| $k_{15}$ | 3 | 2 | 3 | 0 |
| $k_{16}$ | 1 | 2 | 1 | 1 |

The values represent $\log k_i$.

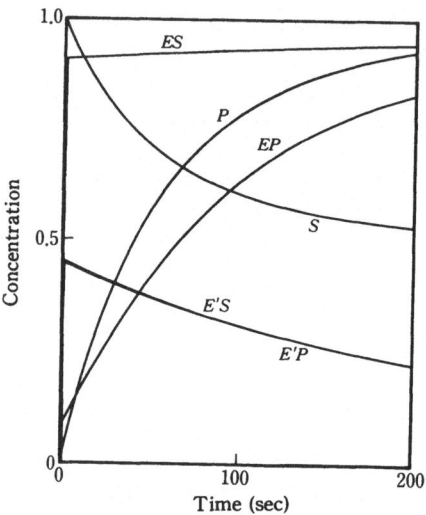

**Fig. 4.52** Time course of Scheme 4.24 (case 1). Initial condition: $S =$ 500 mM, $E = 9.1\mu$M, $E' = 0.9\mu$M. Full scale (1.0) on the ordinate: $S =$ 500 mM, $P = 250$ mM, $ES = 10\mu$M, $EP = 0.1\mu$M, $E'S = E'P = 1\mu$M.

approximation. Figure 4.52 illustrates the time course of case 1 (negative cooperativity and burst). The burst of product concentration is detected immediately after the start of reaction. The production rate of P is very high ($4.6$ mMsec$^{-1}$) at $0.04$ sec and still high ($4.1$ mMsec$^{-1}$) at 5 sec. It then decreases rapidly. During the period of burst the production results mostly from the cycle $E' \rightarrow E'S \rightarrow E'P \rightarrow E'$, because the concentration of E'P becomes several tens of times higher than that of EP. After 100 sec the reverse reactions affect the production strongly, causing more production from EP.

The time course of case 2 (positive cooperativity and time lag) is shown in Fig. 4.53. During 5 sec after the start of reaction the production rate is $1.2$ mMsec$^{-1}$, 60% of which is due to EP. The concentration of E'P then increases to produce most of P. The time lag thus arises from the transition of production of P from the cycle $E \rightarrow ES \rightarrow EP \rightarrow E$ to the cycle $E' \rightarrow E'S$ $\rightarrow E'P \rightarrow E'$.

The time course of case 3 (positive cooperativity and burst) is given in Fig. 4.54. During 5 sec of the burst the concentrations of EP and E'P remain almost constant with their ratio of 1 to 20~50. The production rate of P is $4.46$ mMsec$^{-1}$ at $0.05$ sec after the start of reaction, then reduces rapidly to $1.45$ mMsec$^{-1}$ at 5 sec mainly due to fast reverse-reactions.

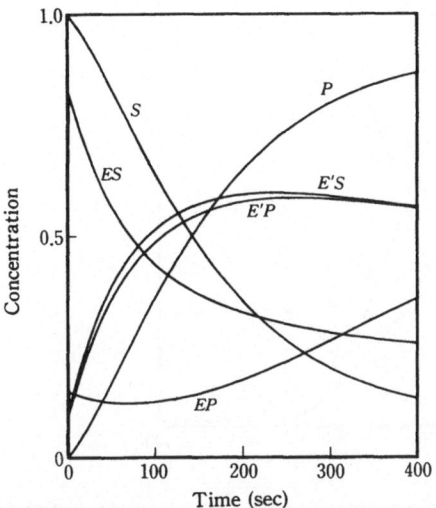

**Fig. 4.53**  Time course of Scheme 4.24 (case 2). Initial condition: same as in Fig. 4.52.  Full scale (1.0) on the ordinate: $S = P = 500$ mM, $ES = 10\mu$M, $EP = E'S = E'P = 5\mu$M.

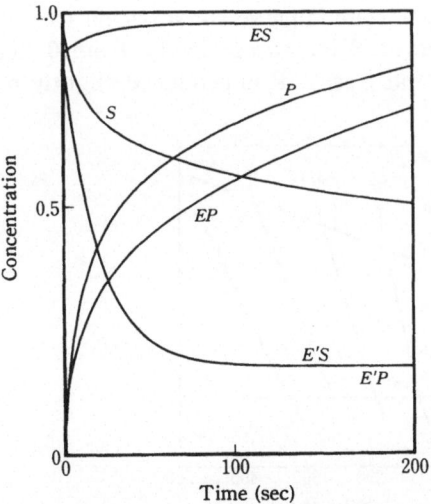

**Fig. 4.54**  Time course of Scheme 4.24 (case 3). Initial condition: same as in Fig. 4.52.  Full scale (1.0) on the ordinate: $S = 100$ mM, $P = 50$ mM, $ES = 10\mu$M, $EP = 1\mu$M, $E'S = E'P = 0.5\mu$M.

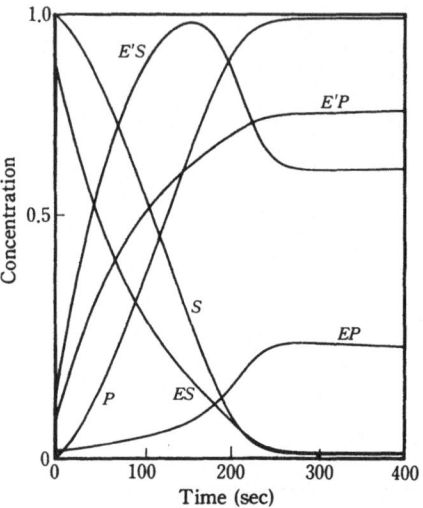

**Fig. 4.55** Time course of Scheme 4.24 (case 4). Initial condition: same as in Fig. 4.52. Full scale (1.0) on the ordinate: $S = P = 100$ mM, $ES = E'P = 10\mu$M, $EP = 5\mu$M, $E'S = 1\mu$M.

Figure 4.55 displays the time course of case 4 (positive cooperativity and time lag). With a long time-lag of about 20 sec, the production of P becomes linear very slowly, remaining so for a period from 100 sec to 180 sec. The production rate of P increases slowly from 0.166 mMsec$^{-1}$ at 0.04 sec to 0.197 mMsec$^{-1}$ at 5 sec. P is produced slightly more from EP

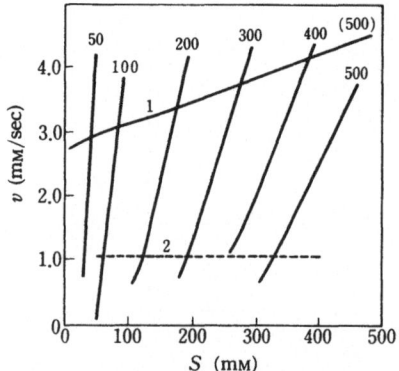

**Fig. 4.56** $S$–$v$ curve for Scheme 4.24 (case 1). Numeral on a curve indicates the initial substrate concentration. 1: without reverse reaction; 2: from equation (41).

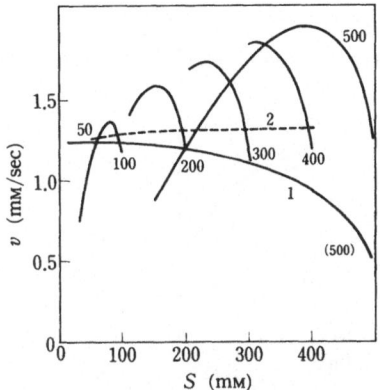

**Fig. 4. 57**  *S–v* curve for Scheme 4.24 (case 2). Notations are the same as in Fig. 4.56.

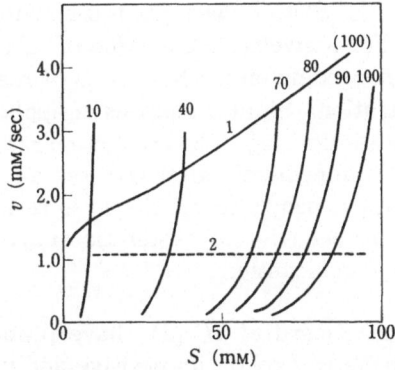

**Fig. 4. 58**  *S–v* curve for Scheme 4.24 (case 3). Notations are the same as in Fig. 4.56.

immediately after the start of reaction, and then gradually more from E′P.

The model of Ainslie *et al.* for hysteretic enzyme thus demonstrates that the process of slow change in conformation of an enzyme molecule could result in time lag and burst. From another point of view, these phenomena appear to be due to the cooperative effects of E- and E′-cycles.

We now consider the hysteresis of Scheme 4.24. The simulation performed with various initial concentrations of substrate S yields the relationship between substrate concentration and production rate of P through the progress of reaction for each case in Table 4.17. Figures. 4.56~4.59 indicate these relationships. Similar relationships are observed for cases 1

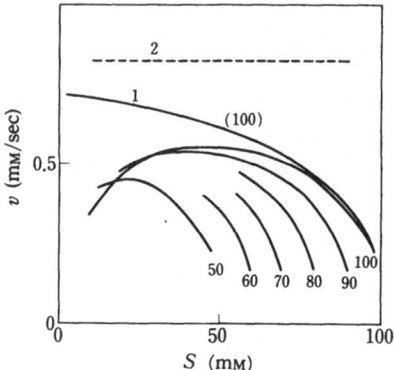

**Fig. 4.59** $S-v$ curve for Scheme 4.24 (case 4). Notations are the same as in Fig. 4.56.

and 3, which both give rise to burst. The production rate of P decreases rapidly as the reaction proceeds. Sharper decrease is seen with lower initial concentration of substrate. For a given concentration of substrate the production rates are different, depending on the hysteresis of reaction, that is, the initial substrate concentration. Similar analysis is applicable to cases 2 and 4, which both demonstrate time lag. The production rate of P is lower than that from equation (41) immediately after the start of reaction, then increasing rapidly again with dependence on the hysteresis of reaction. Hence, we could say that the behavior of hysteretic enzyme affects the regulation of metabolic flow significantly.

**Mnemonic Enzyme**

Instead of hysteretic enzyme, Ricard *et al.* [21] have proposed a mechanism of the mnemonic enzyme to explain its positive and negative cooperativities. The model of mnemonic enzyme is based on the following three postulates, part of which take the induced-fit theory of Koshland [15] into consideration:

1) Two enzyme-states with different conformations are in equilibrium.
2) Interaction of substrate with the two states induces a new conformation suitable for enzyme-substrate complex.
3) Product P binds only with one of the two states, whose conformation is stabilized with complex formation.

This model is applied to the analysis of reaction of wheat germ hexokinase.

We here consider a model for one substrate-one product reaction, which is given in Scheme 4.25. The most important feature of the model is the memory of conformation E' stabilized by product P in a certain period in the system, which results in positive and negative cooperativities. The

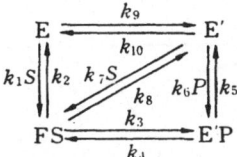

Scheme 4. 25.   Mnemonic enzyme

dynamic behavior of a mnemonic enzyme is obtained from the simulation for cases with the rate constants specified in Table 4.18 by Ricard *et al.* [21]. The time course of each case is respectively shown in Figs. 4.60∼ 4.62.

In each case the rapid-equilibrium assumption is apparently valid until 200 sec from the start of reaction. During that period the consumption of substrate S turns out to be through flow 2 (E′ + S → FS) by more than 99.9%, compared to flow 1 (E+S → FS). The reaction system thus operates on a cycle E′→FS→E′P. On the other hand, in cases 2 and 3 the consumption depends on the substrate concentration. More than 90% of the consumption is through flow 2 at the early stage of reaction, that is, at high substrate concentration, while flow 1 starts working quickly after 150 sec when the substrate concentration becomes lower than 5 mM, then contributing to more than 66% of the consumption at about 200 sec. This shift corresponds to the temporary memory of E′ in the system by stabilization of the conformation with product P.

The positive and negative cooperativities realized in mnemonic enzyme are seen in the Lineweaver-Burk plots from the simulation data in Fig. 4.63. The system of case 1 corresponds to the Michaelis-Menten mechanism. The systems of cases 2 and 3 give rise to positive and negative co-

**Table 4.18**  Rate constants for Scheme 4.25

|          | Case 1    | Case 2     | Case 3     |
|----------|-----------|------------|------------|
| $k_1$    | 2         | 1          | $10^3$     |
| $k_2$    | 0. 2      | $10^{-3}$  | $10^{-3}$  |
| $k_3$    | $10^3$    | $10^3$     | $10^3$     |
| $k_4$    | 1         | 1          | 1          |
| $k_5$    | $10^6$    | $10^6$     | $10^6$     |
| $k_6$    | 0         | 0          | 0          |
| $k_7$    | 2         | $10^3$     | 1          |
| $k_8$    | 0. 2      | $10^{-6}$  | $10^{-6}$  |
| $k_9$    | 1         | 1          | $10^3$     |
| $k_{10}$ | 1         | $10^6$     | $10^3$     |

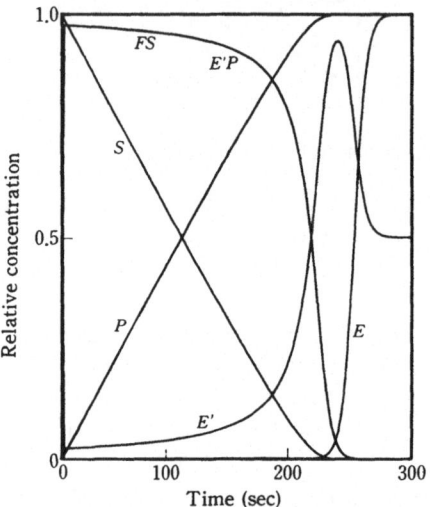

**Fig. 4. 60** Time course of Scheme 4.25 (case 1). Initial condition: $S = 20$ M, $E = E' = 50 \mu$ M.

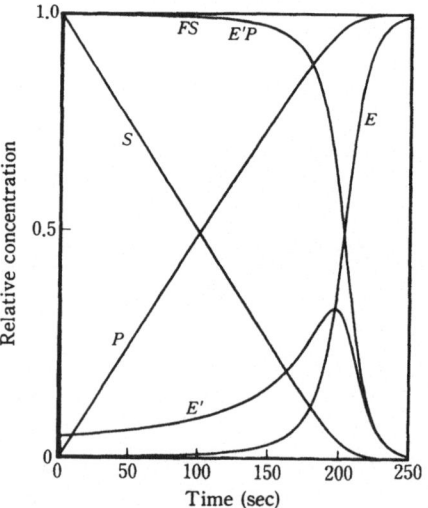

**Fig. 4. 61** Time course of Scheme 4.25 (case 2). Initial condition: $S = 20$ M, $E = E' = 0.1$ mM.

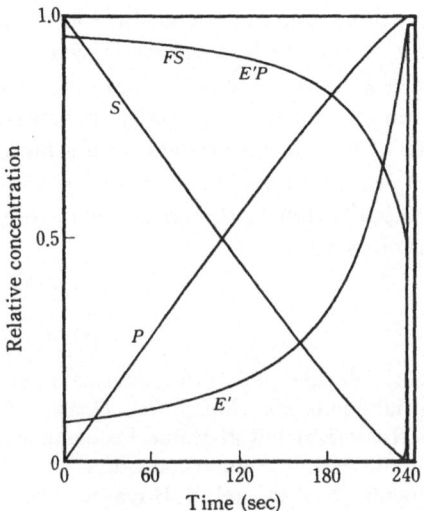

**Fig. 4. 62** Time course of Scheme 4.25 (case 3). Initial condition: $S = 20$ M, $E = E' = 50\mu$M.

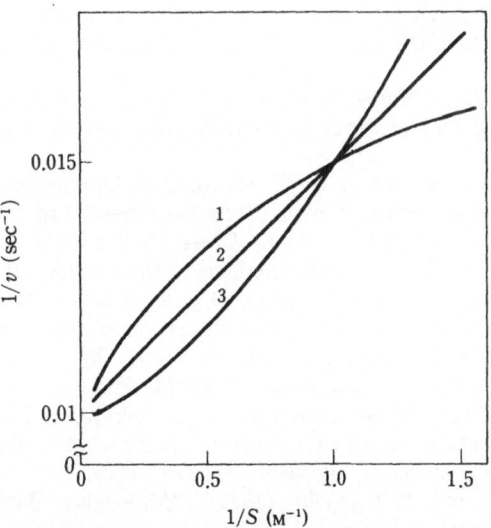

**Fig. 4. 63** Lineweaver-Burk plot for Scheme 4.25. 1: case 3; 2: case 1; 3: case 2.

operativities, respectively.

The analysis of cooperativity and transient state in monomeric enzyme has so far been done with respect to homogeneous system. Of course, the idea of state transition $E \rightarrow E'$ in the hysteretic and mnemonic enzymes can be readily extended to inhomogeneous (spatially structured) system. $E'$ would represent the state of the enzyme bound with a biological structure such as membrane, that is, the membrane-bound enzyme can have a different conformation. Application of this mechanism to enzyme systems remains to be done in future studies.

## References

[ 1 ]  Fowler, M.E. and R.M. Warten (1967). A numerical integration technique for ordinary differential equations. *IBM J. Res. Develop.*, 11, 537-543.

[ 2 ]  Gear, C.W. (1971). "Numerical Initial Value Problems in Ordnary Differential Equations," Prentice-Hall, Englewood Cliffs.

[ 3 ]  Okamoto, M., Y.Takeda, Y.Aso and K.Hayashi (1983). Steady-state approximation of enzyme activation and inhibition. *Biotech. Bioeng.*, 25, 1453-1463.
       Okamoto, M., Y.Aso, D.Koga and K.Hayashi (1975). Note on steady-state approximation in enzyme kinetics. *J. Fac. Agr., Kyushu Univ.*, 19, 125-138.

[ 4 ]  Frere J.-M., B.Leyh and A.Renard (1983). Lineweaver-Burk, Hanes, Eadie-Hofstee and Dixon plots in non-steady-state situations. *J. Theor. Biol.*, 101, 387-400.

[ 5 ]  Segel, I.H. (1975). "Enzyme Kinetics. Behavior and Analysis of Rapid Equilibrium and Steady-State Enzyme Systems," Wiley-Interscience, New York.

[ 6 ]  Imoto, T., L.N. Johnson, A.C.T. North, D.C. Phillips and J.A. Rupley (1972). Vertebrate lysozymes. *In* "The Enzymes," 3 rd ed., ed. by P.D. Boyer, Vol. 7, pp. 665-868, Academic Press, New York.

[ 7 ]  Chipman, D.M. (1971). A kinetic analysis of the reaction of lysozyme with oligosaccharides from bacterial cell walls. *Biochemistry,* 10, 1714-1722.

[ 8 ]  Tada, H. and T.Kakitani (1973). Determination of catalytic rate constants in the reaction of lysozyme and oligosaccharide by computer simulation analysis. *Bull. Chem. Soc. Japan*, 46, 1226-1232.

[ 9 ]  Masaki, A., T.Fukamizo, A.Otakara, T.Torikata, K.Hayashi and T. Imoto (1981). Estimation of rate constants in lysozyme-catalyzed reaction of chitooligosaccharides. *J. Biochem.*, 90, 1167-1175.

[10]  Hamaguchi, K. and K.Hayashi (1978). "Molecular Basis of Enzyme Function," Kodansha, Tokyo.

[11]  Fukamizo, T., S.Kuhara and K.Hayashi (1982). Enzymatic activity of Trp 62-modified lysozyme. *J. Biochem.*, 92, 717-724.

[12]  Pincus, M.R. and H.A. Scheraga (1981). Prediction of the three-dimensional structures of complexes of lysozyme with cell wall substrates.

*Biochemistry,* **20**, 3960-3965.

[13]   Klotz, I.M., N.R. Langerman and D.W. Darnall (1970). Quaternary structure of proteins. *Annu. Rev. Biochem.,* **39**, 25-62.

[14]   Monod, J., J.Wyman and J.-P.Changeux (1965). On the nature of allosteric transitions: a plausible model. *J. Mol. Biol.,* **12**, 88-118.

[15]   Koshland, D.E., Jr., G.Némethy and D.Filmer (1966). Comparison of experimental binding data and theoretical models in proteins containing subunits. *Biochemistry,* **5**, 365-385.

[16]   Hammes, G.H.and C.-W. Wu (1974). Kinetics of allosteric enzymes. *Annu. Rev. Biophys. Bioeng.,* **3**, 1-33.

[17]   Rabin, B.R. (1967). Co-operative effects in enzyme catalysis: a possible kinetic model based on substrate-induced conformation isomerization. *Biochem. J.,* **102**, 22 c-23 c.

[18]   Frieden, C.(1970). Kinetic aspects of regulation of metabolic processes: the hysteretic enzyme concept. *J. Biol. Chem.,* **245**, 5788-5799.

[19]   Frieden, C.(1979). Slow transitions and hysteretic behavior in enzymes. *Annu. Rev. Biochem.,* **48**, 471-489.

[20]   Ainslie, G.R., Jr., J.P. Shill and K.E. Neet (1972). Transients and cooperativity. A slow transitioon model for relating transients and cooperative kinetics of enzymes. *J. Biol. Chem.,* **247**, 7088-7096.

[21]   Ricard, J., J.C. Meunier and J.Buc (1974). Regulatory behavior of monomeric enzymes. 1. The mnemonical enzyme concept. *Eur. J. Biochem.,* **49**, 195-208.

# CHAPTER 5

# Microscopic Analysis of Enzyme Systems

The microscopic analysis of enzyme systems aims at the clarification of the dynamic behavior of each chemical species in an objective system using knowledge of the molecular mechanisms of enzymatic reactions. This would be the first step in a detailed observation of the metabolic processes (systems of enzymatic reactions) functioning in actual *in vivo* environments. The systems under observation hence include the reactions of allosteric enzymes exhibiting complicated behavior as well as the Michaelis-Menten-type reactions. An increase in the number of enzymes in such a system, which operates in both open and closed systems, would introduce further difficulty in accurately observing the behavior in biochemical and physiological experiments. Fortunately, in addition to efforts toward the technical development of experimental analysis, microscopic analysis now is possible to a certain extent by computer simulation employing such numerical methods as described in Chapter 3.

In this chapter, the simulation of the dynamic behavior which enzymes display in functioning as the constituent elements of various enzyme systems is described for the Michaelis-Menten-type and allosteric enzymes; their behavior in closed system is presented individually in Chapter 4. In the following, the reaction systems will be treated as open system to observe the enzymatic reactions behaving in actual *in vivo* environments. We are naturally concerned with the behavior of allosteric enzymes in open system. First, the time course of reaction of a single enzyme in open system is obtained and analyzed for enzymes of the Michaelis-Menten type and allosteric dimeric models of MWC and KNF. Reaction systems consisting of two enzymes, the Michaelis-Menten-type enzyme and the allosteric

---

This chapter was written by Naoto Sakamoto, Kiyokazu Nemoto and Yukihiro Eguchi.

enzyme of the MWC dimeric model, are studied with respect to their transient responses to disturbances in systems undergoing negative and positive feedback controls and sustained oscillations.

The chapter concludes with the simulation of more complex systems such as branched and coupled systems. Branched systems, which are considered important in metabolic regulation, are analyzed with their time courses in order to derive the relationship between the feedback mechanisms and the regulatory features in the systems. The coupled systems consisting of two Michaelis-Menten-type reactions of two substrates-two products are analyzed for cases of product flowing and recycling. The analysis is preceded by simulation of time courses of two substrates-two products reactions for the representative mechanisms.

This chapter mainly deals with the simplest models for enzyme systems, where the allosteric enzyme is dimeric and the system consists of only two enzymes. These systems, if simplified, still retain the fundamental characteristics of dynamics in enzymatic reaction systems functioning in actual *in vivo* environments. Simulation of these systems provides fundamental findings on the dynamic behavior of *in vivo* enzyme systems.

## 5.1 Allosteric Enzymes in Open System

The systems we define in the biological phenomena operate mostly in open system. Naturally, enzymatic reaction systems in *in vivo* environments work in open system, in which the substrate is supplied to the system from the outside and the product is taken out of the system to be consumed as the substrate for the following enzyme system. In a more general open system (Fig. 5.1), the intermediates, enzymes and effectors as well as the substrate and product are interchangeable between the system and the environment. In contrast, the closed system is defined as a common *in vitro* (or laboratory) system, in which chemical species such as the substrate and product in an enzyme system cannot flow through the system boundary. Closed

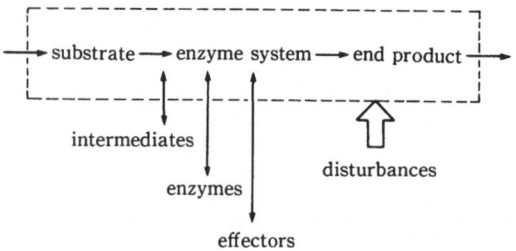

**Fig. 5.1** Enzyme system operating in open system.

and open systems are discussed in relation to thermodynamics in the Introduction and Chapter 9.

The dynamic behavior in individual enzymatic reactions has long been
under investigation in closed system to assay their activity and estimate the
reaction mechanism. This has led to an understanding of the fundamental
properties and functions of enzymes. It is still probable that the behavior
in open system is quite different from that in closed system. Individual
behavior would again be different from that in multi-enzyme systems. In
this section, the dynamic behavior in open system is examined for the one-
step and two-step systems of one substrate-one product enzymatic reactions, in contrast to the behavior of allosteric enzymes in closed system as
described in Section 4.2.

The open system is considered to maintain a constant substrate concentration through its appropriate supply from the outside. The end product
flows out of the system at a certain efflux rate, and other chemical species
are not exchanged between the system and the environment. The dynamic
behavior in open system is best observed by an analysis of the time courses
from simulation of transient responses of chemical species in a system to
disturbances in the concentrations of substrate and end product and the
efflux rate of the end product.

The dynamic behavior of enzymatic reactions in open system has been
examined in relation to the oscillatory phenomena [1]. Individual enzymatic reactions in oscillatory systems, however, are treated mostly by the
quasi-steady-state or rapid-equilibrium approximation. Instead of such treatment, molecular models of reaction mechanisms for individual enzymes
are employed here for the analysis of transient responses, which yields in-
tegrally the behavior of the whole system at the molecular level.

Future implementation of corresponding experiments in open system is
also desirable to prove the simulation results. Once certain features on the
dynamic behavior of enzymatic reactions in open system are deduced from
the simulation and experimental data, an exact evaluation of the behavior
of that system will be possible based on the large amount of data already
on hand for closed system. Furthermore, an open system always contains
some control processes such as the feedback inhibition of enzymatic
reaction, supply of substrate and efflux of end product. The observation of
behavior thus provides more accurate data on the *in vivo* behavior of
enzyme systems.

## 1. One-Step Systems

The time course of an *in vitro* reaction of a single enzyme is always observed to assay enzymatic activity. As discussed for the Michaelis-

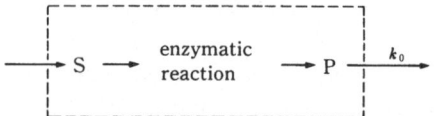

**Fig. 5. 2**   One-step system in an open system. The concentration in S is constant ; efflux rate of $P = k_0 P$.

Menten-type and allosteric reactions in the preceding chapters, however, the experiments for activity assay reveal only restricted aspects of the time course of enzymatic reaction.   Moreover, a careless application of quasi-steady-state approximation often leads to incorrect interpretation on the characteristics of enzymatic reaction.   In addition, the activity assay is commonly employed only to detect and verify the entity of an enzyme, or at most to evaluate the Michaelis constant and maximum velocity of enzyme. Hence, the time course of reaction in closed system is rarely reported from experiment in a form comparable to the entire progress of reaction from computer simulation.   Few time courses in open system have been examined experimentally.

We deal here with the time courses of one-step systems (*i.e.*, reactions of single enzymes) in open system obtained from computer simulation. One-step systems operate in an open system such as shown in Fig. 5.2, in which substrate S is supplied into the system so as to maintain a constant concentration and product P is taken out of the system at a rate of $k_0 P$. The dynamic behavior is compared with that in closed system to derive the catalytic and regulatory characteristics of individual enzymes as constituents in metabolic systems.   The enzymatic reactions considered are the allosteric ones of MWC and KNF dimeric models as well as a Michaelis-Menten-type reaction as a standard for comparison.

The treatment of open system attempts to explain the actual *in vivo* behavior of enzymatic reaction systems.   This analysis might lead to generation of a method to deduce the behavior in open system from data of enzymatic reactions in closed system.   It should also be noted that the existence and stability of steady states in the systems are intrinsically implied in the behavior in open system.   It is thus reasonable to assume that even one-step systems undergo some regulatory mechanisms.  In the system of Fig. 5.2, product P flows out at a rate proportional to its concentration to control the system.

**Michaelis-Menten-Type Reaction**

We first consider a case wherein the enzymatic reaction in the open system of Fig. 5.2 is of the Michaelis-Menten type.  Scheme 5.1 specifies the reaction scheme and rate constants.  The simulation of the system for a

$$E + S \underset{k_{-1}}{\overset{k_1}{\rightleftharpoons}} X \overset{k_2}{\longrightarrow} E + P$$

$$k_1 = 1.0 \times 10^5, \quad k_{-1} = 10.0, \quad k_2 = 1.0$$

Scheme 5.1.  Michaelis-Menten-type reaction
X represents the enzyme-substrate complex ES.

constant concentration of $0.1\,\mathrm{mM}$ in substrate S and $k_0 = 0.05\,\mathrm{sec}^{-1}$ and an initial condition with $E(0) = 0.01\,\mathrm{mM} = E_0$ and $P(0) = X(0) = 0$ demonstrates that the system reaches a steady state, in which the chemical species have the respective steady-state concentrations listed in Table 5.1. When a system in the steady state is disturbed by changes in substrate concentration from $0.1$ to $1.0$ or $0.01$ (in mM), it reaches new steady states as observed in Fig. 5.3.

It follows that the quasi-steady-state assumption is valid for the system

**Table 5.1**  Steady-state concentrations of Michaelis-Menten-type reaction (Scheme 5.1) in the open system

| Species | Concentration (mM) | | |
|---|---|---|---|
| S | $1.0 \times 10^{-2}$ | $1.0 \times 10^{-1}$ | $1.0$ |
| $\bar{P}$ | $1.67 \times 10^{-2}$ | $9.52 \times 10^{-2}$ | $1.80 \times 10^{-1}$ |
| $\bar{E}$ | $9.17 \times 10^{-3}$ | $5.24 \times 10^{-3}$ | $9.90 \times 10^{-4}$ |
| $\bar{X}$ | $8.30 \times 10^{-4}$ | $4.76 \times 10^{-3}$ | $9.01 \times 10^{-3}$ |

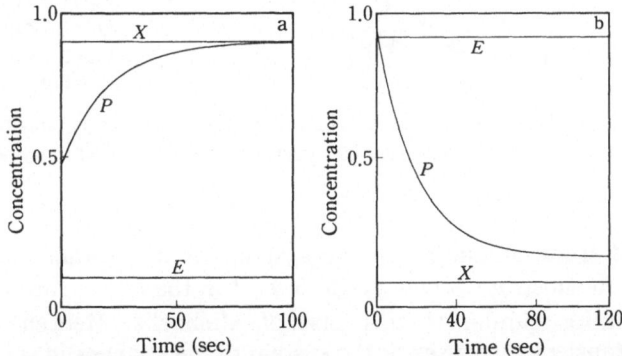

**Fig. 5.3**  Transient responses of Scheme 5.1 in the open system. a) Increase in S from $0.1\,\mathrm{mM}$ to $1.0\,\mathrm{mM}$. Full scale (1.0) on the ordinate: $P = 200\,\mu\mathrm{M}$, $E = X = 10\,\mu\mathrm{M}$; b) Decrease in S from $0.1\,\mathrm{mM}$ to $10\,\mu\mathrm{M}$. Full scale (1.0) on the ordinate: $P = 100\,\mu\mathrm{M}$. $E = X = 10\,\mu\mathrm{M}$.

since the enzyme-substrate complex X attains a new steady-state value $\bar{X}$ much more rapidly than product P. The variation in concentration of P is expressed by

$$\frac{dP}{dt}=k_2\bar{X}-k_0P, \tag{1}$$

indicating that P reaches a new steady state exponentially. The concentrations at new steady states also are given in Table 5.1.

We can thus conclude that the transient responses of a Michaelis-Menten-type reaction in open system are similar to the behavior in the initial phase of reaction in closed system.

**Allosteric Reaction (MWC Dimeric Model)**

We are now concerned with the dynamic behavior of allosteric enzymes of the following MWC dimeric models in open system.

( 1 ) *Substrate S binds only with the enzyme species in R-state*:

The reaction scheme and rate constants are as given in Scheme 5.2. The steady-state concentrations of the chemical species listed in Table 5.2 are

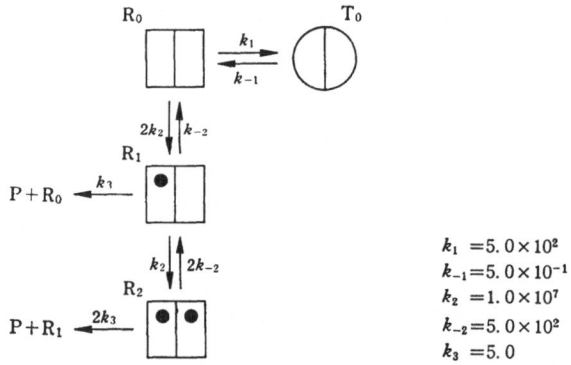

$$k_1 = 5.0 \times 10^2$$
$$k_{-1} = 5.0 \times 10^{-1}$$
$$k_2 = 1.0 \times 10^7$$
$$k_{-2} = 5.0 \times 10^2$$
$$k_3 = 5.0$$

Scheme 5.2. MWC dimeric model of allosteric enzyme Case (1) : substrate (●) binds only with the R-form enzyme-species.

obtained from simulation of the system for a constant concentration of 0.1 mM in substrate S and $k_0 = 0.05$ sec$^{-1}$ in the efflux rate of product P. A simulation similar to this case of Michaelis-Menten-type reaction yields transient responses of the system to the changes in substrate concentration from 0.1 to 1.0 or 0.01 (in mM), as shown in Fig. 5.4. Comparison with the behavior in closed system given in Fig. 3.9 reveals a characteristic in open system that the enzyme species reach a steady state much more rapidly than product P.

**Table 5.2**  Steady-state concentrations of allosteric reaction of MWC dimeric model (Scheme 5.2) in the open system

| Species | Concentration (mM) | | |
|---|---|---|---|
| $S$ | $1.0 \times 10^{-2}$ | $1.0 \times 10^{-1}$ | $1.0$ |
| $\bar{P}$ | $4.74 \times 10^{-4}$ | $1.17 \times 10^{-2}$ | $5.75 \times 10^{-1}$ |
| $\bar{R}_0$ | $9.99 \times 10^{-6}$ | $9.91 \times 10^{-6}$ | $6.98 \times 10^{-6}$ |
| $\bar{R}_1$ | $3.95 \times 10^{-6}$ | $3.93 \times 10^{-5}$ | $2.76 \times 10^{-4}$ |
| $\bar{R}_2$ | $3.92 \times 10^{-7}$ | $3.89 \times 10^{-5}$ | $2.74 \times 10^{-3}$ |
| $\bar{T}_0$ | $9.99 \times 10^{-3}$ | $9.91 \times 10^{-3}$ | $6.98 \times 10^{-3}$ |

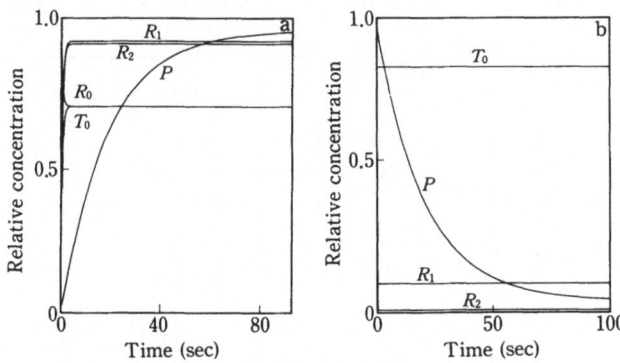

**Fig. 5.4**  Transient responses of Scheme 5.2 in the open system. a) Increase in $S$ from 0.1 mM to 1.0 mM; b) Decrease in $S$ from 0.1 mM to 10$\mu$M.

( 2 ) *Substrate S binds with both R- and T-form enzyme-species*:

The reaction scheme and rate constants are as given in Scheme 4.14. The steady-state concentrations of each chemical species are listed in Table 5.3 for the substrate concentrations of 0.01, 0.1 and 1.0 (in mM), respectively. Similar to case (1), the steady state for $S=0.1$ mM is disturbed by changing the substrate concentration to 1.0 or 0.01 (in mM). The system follows the time courses indicated in Fig. 5.5 to attain new steady states. The difference from the behavior in closed system (shown in Fig. 4.29) is similar to that for case (1). The concentration of the R-form species is lower, however, causing slower establishment of steady states in product P.

( 3 ) *Substrate S binds only with the R-form enzyme-species and a negative effector I binds only with the T-form enzyme-species*:

The reaction scheme and rate constants are as given in Scheme 4.16. The steady-state concentrations of each chemical species are listed in Table

**Table 5.3**  Steady-state concentrations of Scheme 4.14 in the open system

| Species | Concentration (mM) | | |
|---|---|---|---|
| $S$ | $1.0 \times 10^{-2}$ | $1.0 \times 10^{-1}$ | $1.0$ |
| $\bar{P}$ | $4.72 \times 10^{-4}$ | $1.12 \times 10^{-2}$ | $4.40 \times 10^{-1}$ |
| $\bar{R}_0$ | $9.95 \times 10^{-6}$ | $9.53 \times 10^{-6}$ | $5.34 \times 10^{-6}$ |
| $\bar{R}_1$ | $3.94 \times 10^{-6}$ | $3.77 \times 10^{-5}$ | $2.11 \times 10^{-4}$ |
| $\bar{R}_2$ | $3.90 \times 10^{-7}$ | $3.74 \times 10^{-5}$ | $2.09 \times 10^{-3}$ |
| $\bar{T}_0$ | $9.95 \times 10^{-3}$ | $9.53 \times 10^{-3}$ | $5.34 \times 10^{-3}$ |
| $\bar{T}_1$ | $3.98 \times 10^{-5}$ | $3.81 \times 10^{-4}$ | $2.14 \times 10^{-3}$ |
| $\bar{T}_2$ | $3.98 \times 10^{-8}$ | $3.81 \times 10^{-6}$ | $2.14 \times 10^{-4}$ |

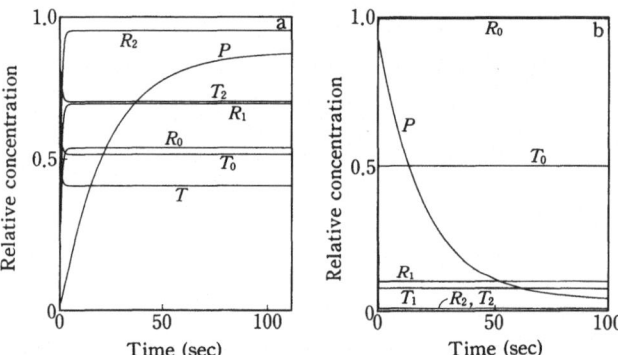

**Fig. 5.5**  Transient responses of Scheme 4.14 in the open system. a) Increase in $S$ from 0.1 mM to 1.0 mM; b) Decrease in $S$ from 0.1 mM to $10\mu$M.

5.4 for the substrate concentrations of 0.01, 0.1 and 1.0 (in mM), respectively, with 3.0 mM of negative effector I.  Figure 5.6 illustrates the time courses of the transition from a steady state to new steady states caused by similar disturbances in substrate concentration.  The negative effector makes the responses of the system slightly slower than in case (1).  The concentration of the R-form species is less than 10% of that in case (1).

( 4 )  *Substrate S binds with both R- and T-form enzyme-species and a negative effector I binds only with the T-form enzyme-species* :

The reaction scheme and rate constants are as given in Scheme 5.3.  The steady-state concentrations of each chemical species are listed in Table 5.5 for the substrate concentrations of 0.01, 0.1 and 1.0 (in mM), respectively, with 3.0 mM of negative effector I.  Figure 5.7 illustrates the time courses of the transition from a steady state to new steady states caused by simi-

**Table 5.4** Steady-state concentrations of Scheme 4.16 in the open system

| Species | Concentration (mM) | | |
|---|---|---|---|
| $S$ | $1.0 \times 10^{-2}$ | $1.0 \times 10^{-1}$ | $1.0$ |
| $\bar{P}$ | $2.98 \times 10^{-5}$ | $7.37 \times 10^{-4}$ | $5.01 \times 10^{-2}$ |
| $\bar{R}_0$ | $6.25 \times 10^{-7}$ | $6.25 \times 10^{-7}$ | $6.09 \times 10^{-7}$ |
| $\bar{R}_1$ | $2.48 \times 10^{-7}$ | $2.47 \times 10^{-6}$ | $2.41 \times 10^{-5}$ |
| $\bar{R}_2$ | $2.45 \times 10^{-8}$ | $2.45 \times 10^{-6}$ | $2.39 \times 10^{-4}$ |
| $\bar{T}_0$ | $6.25 \times 10^{-4}$ | $6.25 \times 10^{-4}$ | $6.09 \times 10^{-4}$ |
| $\bar{T}_1$ | $3.75 \times 10^{-3}$ | $3.75 \times 10^{-3}$ | $3.65 \times 10^{-3}$ |
| $\bar{T}_2$ | $5.62 \times 10^{-3}$ | $5.62 \times 10^{-3}$ | $5.48 \times 10^{-3}$ |
| $\bar{I}$ | $3.0$ | $3.0$ | $3.0$ |

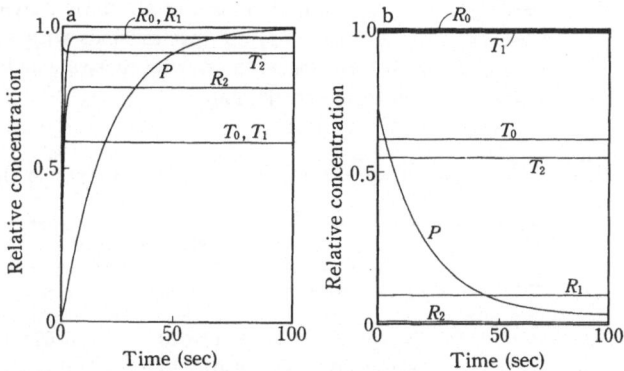

**Fig. 5.6** Transient responses of Scheme 4.16 in the open system. a) Increase in $S$ from 0.1 mM to 1.0 mM ; b) Decrease in $S$ from 0.1 mM to 10 $\mu$M.

lar disturbances in substrate concentration. The behavior is almost identical to that in case (3), except that product P reaches new steady states more rapidly than in case (2). The concentration of the R-form species is less than 10% of that in case (2).

( 5 ) *The R-form enzyme-species bind with substrate S and a positive effector A, and the T-form enzyme-species bind only with a negative effector I* :

The reaction scheme and rate constants are as given in Scheme 5.4. The steady-state concentrations of each chemical species are listed in Table 5.6 for the substrate concentrations of 0.01, 0.1 and 1.0 (in mM), respectively, with 3.0mM of negative effector I and 1.0mM of positive effector A. Fig-

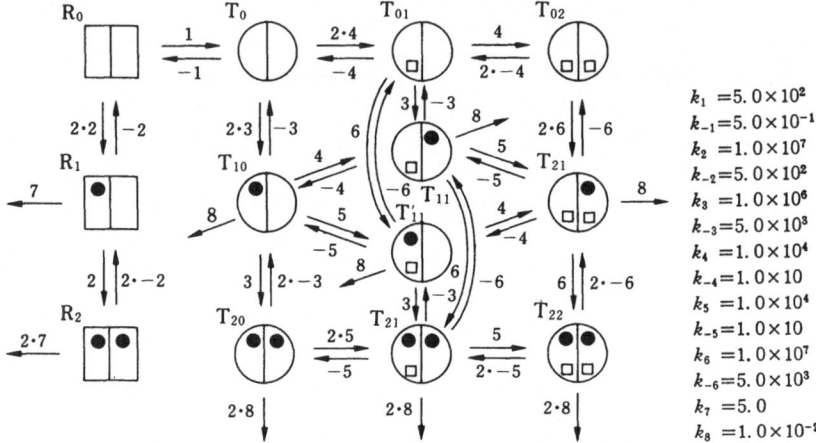

$$k_1 = 5.0 \times 10^2$$
$$k_{-1} = 5.0 \times 10^{-1}$$
$$k_2 = 1.0 \times 10^7$$
$$k_{-2} = 5.0 \times 10^2$$
$$k_3 = 1.0 \times 10^6$$
$$k_{-3} = 5.0 \times 10^3$$
$$k_4 = 1.0 \times 10^4$$
$$k_{-4} = 1.0 \times 10$$
$$k_5 = 1.0 \times 10^4$$
$$k_{-5} = 1.0 \times 10$$
$$k_6 = 1.0 \times 10^7$$
$$k_{-6} = 5.0 \times 10^3$$
$$k_7 = 5.0$$
$$k_8 = 1.0 \times 10^{-2}$$

Scheme 5.3. MWC dimeric model of allosteric enzyme. Case (4) : the R-form enzyme-species binds with substrate and the T-form species binds with substrate and inhibitor. ● : substrate ; □ : inhibitor

**Table 5.5** Steady-state concentrations of Scheme 5.3 in the open system

| Species | Concentration (mM) | | |
|---------|--------------------|--------------------|--------------------|
| $S$ | $1.0 \times 10^{-2}$ | $1.0 \times 10^{-1}$ | $1.0$ |
| $\bar{P}$ | $3.02 \times 10^{-5}$ | $7.14 \times 10^{-4}$ | $3.60 \times 10^{-2}$ |
| $\bar{R}_0$ | $6.22 \times 10^{-7}$ | $6.01 \times 10^{-7}$ | $4.37 \times 10^{-7}$ |
| $\bar{R}_1$ | $2.47 \times 10^{-7}$ | $2.38 \times 10^{-6}$ | $1.73 \times 10^{-5}$ |
| $\bar{R}_2$ | $2.44 \times 10^{-8}$ | $2.36 \times 10^{-6}$ | $1.71 \times 10^{-4}$ |
| $\bar{T}_{00}$ | $6.22 \times 10^{-4}$ | $6.01 \times 10^{-4}$ | $4.37 \times 10^{-4}$ |
| $\bar{T}_{10}$ | $2.49 \times 10^{-6}$ | $2.40 \times 10^{-5}$ | $1.75 \times 10^{-4}$ |
| $\bar{T}_{20}$ | $2.50 \times 10^{-9}$ | $2.42 \times 10^{-7}$ | $1.76 \times 10^{-5}$ |
| $\bar{T}_{01}$ | $3.73 \times 10^{-3}$ | $3.60 \times 10^{-3}$ | $2.55 \times 10^{-3}$ |
| $\bar{T}_{11}$ | $7.47 \times 10^{-6}$ | $7.21 \times 10^{-5}$ | $5.10 \times 10^{-4}$ |
| $\bar{T}'_{11}$ | $7.47 \times 10^{-6}$ | $7.21 \times 10^{-5}$ | $5.10 \times 10^{-4}$ |
| $\bar{T}_{21}$ | $1.50 \times 10^{-8}$ | $1.44 \times 10^{-6}$ | $1.02 \times 10^{-4}$ |
| $\bar{T}_{02}$ | $5.60 \times 10^{-3}$ | $5.40 \times 10^{-3}$ | $3.72 \times 10^{-3}$ |
| $\bar{T}_{12}$ | $2.24 \times 10^{-5}$ | $2.16 \times 10^{-4}$ | $1.49 \times 10^{-3}$ |
| $\bar{T}_{22}$ | $4.47 \times 10^{-8}$ | $4.31 \times 10^{-6}$ | $2.97 \times 10^{-4}$ |
| $\bar{I}$ | $3.0$ | $3.0$ | $3.0$ |

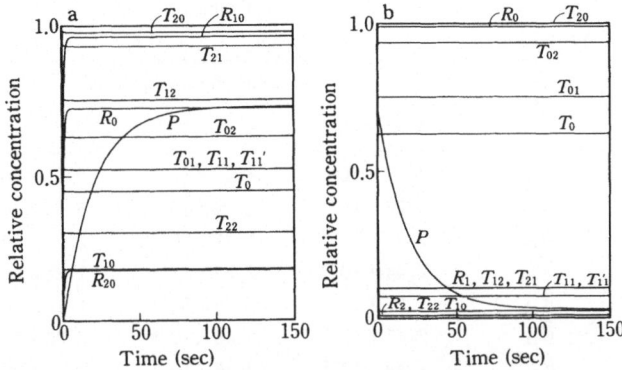

**Fig. 5.7** Transient responses of Scheme 5.3 in the open system. a) Increase in $S$ from 0.1 mM to 1.0 mM; b) Decrease in $S$ from 0.1 mM to 10$\mu$M.

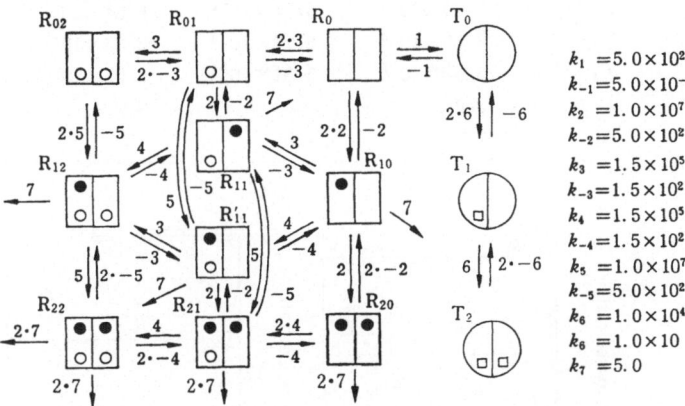

$$k_1 = 5.0 \times 10^2$$
$$k_{-1} = 5.0 \times 10^{-1}$$
$$k_2 = 1.0 \times 10^7$$
$$k_{-2} = 5.0 \times 10^2$$
$$k_3 = 1.5 \times 10^5$$
$$k_{-3} = 1.5 \times 10^2$$
$$k_4 = 1.5 \times 10^5$$
$$k_{-4} = 1.5 \times 10^2$$
$$k_5 = 1.0 \times 10^7$$
$$k_{-5} = 5.0 \times 10^2$$
$$k_6 = 1.0 \times 10^4$$
$$k_6 = 1.0 \times 10$$
$$k_7 = 5.0$$

Scheme 5.4.  MWC dimeric model of allosteric enzyme Case (5): the R-form enzyme-species binds with substrate and activator and the T-form species binds with inhibitor. ●: substrate; ○: activator; □: inhibitor

ure 5.8 illustrates the time courses of the transition from a steady state to new steady states caused by similar disturbances in substrate concentration. Compared with case (3), the R-form species reach new steady states more slowly, probably due to interaction with the positive effector.

( 6 ) *The R-form enzyme-species bind with substrate S and a positive effector A, and the T-form enzyme-species bind with substrate S and a negative effector I :*

**Table 5.6** Steady-state concentrations of Scheme 5.4 in the open system

| Species | Concentration (mM) | | |
|---|---|---|---|
| $S$ | $1.0 \times 10^{-2}$ | $1.0 \times 10^{-1}$ | $1.0$ |
| $\bar{P}$ | $1.77 \times 10^{-4}$ | $4.40 \times 10^{-3}$ | $2.79 \times 10^{-1}$ |
| $\bar{R}_{00}$ | $6.25 \times 10^{-7}$ | $6.24 \times 10^{-7}$ | $5.64 \times 10^{-7}$ |
| $\bar{R}_{10}$ | $2.47 \times 10^{-7}$ | $2.47 \times 10^{-6}$ | $2.23 \times 10^{-5}$ |
| $\bar{R}_{20}$ | $2.44 \times 10^{-8}$ | $2.44 \times 10^{-6}$ | $2.21 \times 10^{-4}$ |
| $\bar{R}_{01}$ | $1.25 \times 10^{-6}$ | $1.25 \times 10^{-6}$ | $1.14 \times 10^{-6}$ |
| $\bar{R}_{11}$ | $2.47 \times 10^{-7}$ | $2.48 \times 10^{-6}$ | $2.25 \times 10^{-5}$ |
| $\bar{R}'_{11}$ | $2.48 \times 10^{-7}$ | $2.46 \times 10^{-6}$ | $2.23 \times 10^{-5}$ |
| $\bar{R}_{21}$ | $4.86 \times 10^{-8}$ | $4.87 \times 10^{-6}$ | $4.42 \times 10^{-4}$ |
| $\bar{R}_{02}$ | $6.26 \times 10^{-7}$ | $6.30 \times 10^{-7}$ | $5.74 \times 10^{-7}$ |
| $\bar{R}_{12}$ | $2.46 \times 10^{-7}$ | $2.47 \times 10^{-6}$ | $2.25 \times 10^{-5}$ |
| $\bar{R}_{22}$ | $2.42 \times 10^{-8}$ | $2.43 \times 10^{-6}$ | $2.21 \times 10^{-4}$ |
| $\bar{T}_0$ | $6.25 \times 10^{-4}$ | $6.24 \times 10^{-4}$ | $5.64 \times 10^{-4}$ |
| $\bar{T}_1$ | $3.75 \times 10^{-3}$ | $3.74 \times 10^{-3}$ | $3.38 \times 10^{-3}$ |
| $\bar{T}_2$ | $5.62 \times 10^{-3}$ | $5.61 \times 10^{-3}$ | $5.07 \times 10^{-3}$ |
| $\bar{I}$ | $3.0$ | $3.0$ | $3.0$ |
| $\bar{A}$ | $1.0$ | $1.0$ | $1.0$ |

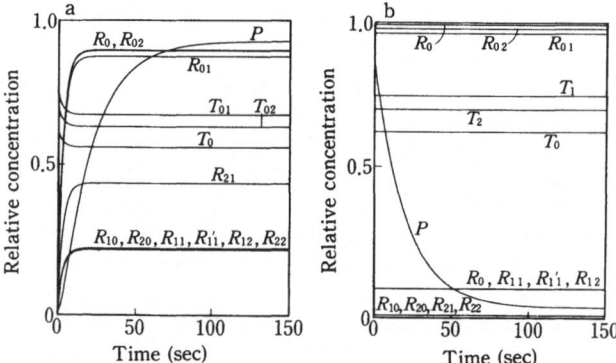

**Fig. 5.8** Transient responses of Scheme 5.4 in the open system. a) Increase in $S$ from 0.1 mM to 1.0 mM; b) Decrease in $S$ from 0.1 mM to 10μM.

The reaction scheme and rate constants are as given in Scheme 4.18. The steady-state concentrations of each chemical species are listed in Table 5.7 for the substrate concentrations of 0.01, 0.1 and 1.0 (in mM), respective-

**Table 5.7**  Steady-state concentrations of Scheme 4.18 in the open system

| Species | Concentration (mM) | | |
|---|---|---|---|
| $S$ | $1.0 \times 10^{-2}$ | $1.0 \times 10^{-1}$ | $1.0$ |
| $\bar{P}$ | $1.77 \times 10^{-4}$ | $4.24 \times 10^{-3}$ | $2.04 \times 10^{-1}$ |
| $\bar{R}_{00}$ | $6.22 \times 10^{-7}$ | $6.00 \times 10^{-7}$ | $4.13 \times 10^{-7}$ |
| $\bar{R}_{10}$ | $2.46 \times 10^{-7}$ | $2.37 \times 10^{-6}$ | $1.64 \times 10^{-5}$ |
| $\bar{R}_{20}$ | $2.43 \times 10^{-8}$ | $2.35 \times 10^{-6}$ | $1.62 \times 10^{-4}$ |
| $\bar{R}_{01}$ | $1.25 \times 10^{-6}$ | $1.21 \times 10^{-6}$ | $8.34 \times 10^{-7}$ |
| $\bar{R}_{11}$ | $2.46 \times 10^{-7}$ | $2.39 \times 10^{-6}$ | $1.65 \times 10^{-5}$ |
| $\bar{R}'_{11}$ | $2.45 \times 10^{-7}$ | $2.37 \times 10^{-6}$ | $1.64 \times 10^{-5}$ |
| $\bar{R}_{21}$ | $4.44 \times 10^{-8}$ | $4.68 \times 10^{-6}$ | $3.24 \times 10^{-4}$ |
| $\bar{R}_{02}$ | $6.24 \times 10^{-7}$ | $6.06 \times 10^{-7}$ | $4.21 \times 10^{-7}$ |
| $\bar{R}_{12}$ | $2.45 \times 10^{-7}$ | $2.38 \times 10^{-6}$ | $1.65 \times 10^{-5}$ |
| $\bar{R}_{22}$ | $2.41 \times 10^{-8}$ | $2.33 \times 10^{-6}$ | $1.62 \times 10^{-4}$ |
| $\bar{T}_{00}$ | $6.22 \times 10^{-4}$ | $6.00 \times 10^{-4}$ | $4.13 \times 10^{-4}$ |
| $\bar{T}_{10}$ | $2.49 \times 10^{-6}$ | $2.40 \times 10^{-5}$ | $1.65 \times 10^{-4}$ |
| $\bar{T}_{20}$ | $2.50 \times 10^{-9}$ | $2.41 \times 10^{-7}$ | $1.66 \times 10^{-5}$ |
| $\bar{T}_{01}$ | $3.73 \times 10^{-3}$ | $3.60 \times 10^{-3}$ | $2.41 \times 10^{-3}$ |
| $\bar{T}_{11}$ | $7.47 \times 10^{-6}$ | $7.19 \times 10^{-5}$ | $4.83 \times 10^{-4}$ |
| $\bar{T}'_{11}$ | $7.47 \times 10^{-6}$ | $7.19 \times 10^{-5}$ | $4.83 \times 10^{-4}$ |
| $\bar{T}_{21}$ | $1.50 \times 10^{-8}$ | $1.44 \times 10^{-6}$ | $9.67 \times 10^{-5}$ |
| $\bar{T}_{02}$ | $5.60 \times 10^{-3}$ | $5.39 \times 10^{-3}$ | $3.52 \times 10^{-3}$ |
| $\bar{T}_{12}$ | $2.24 \times 10^{-5}$ | $2.16 \times 10^{-4}$ | $1.41 \times 10^{-3}$ |
| $\bar{T}_{22}$ | $4.47 \times 10^{-8}$ | $4.31 \times 10^{-6}$ | $2.81 \times 10^{-4}$ |
| $\bar{I}$ | $3.0$ | $3.0$ | $3.0$ |
| $\bar{A}$ | $1.0$ | $1.0$ | $1.0$ |

ly, with $3.0$ mM of negative effector I and $1.0$ mM of positive effector A. Figure 5.9 illustrates the time courses of the transition from a steady state to new steady states caused by similar disturbances in substrate concentration. The behavior of the system is almost identical to that in case (5). The concentration of the R-form species is much higher than that in case (4), and the steady-state concentration of product P becomes about ten times higher.

## Allosteric Reaction (KNF Dimeric Model)

The treatment of one-step systems concludes with an analysis of behavior of allosteric reactions of KNF dimeric model in open system. We are con-

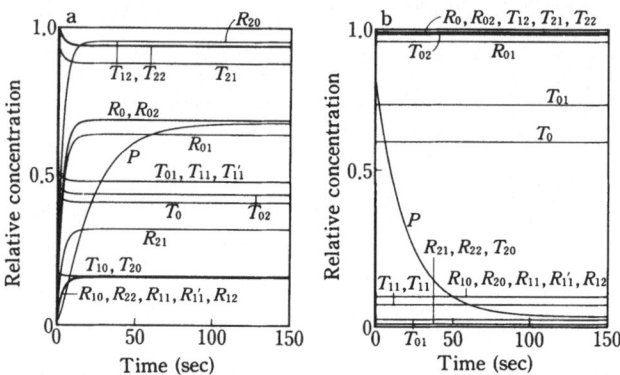

**Fig. 5.9** Transient responses of Scheme 4.18 in the open system. a) Increase in $S$ from 0.1 mM to 1.0 mM; b) Decrease in $S$ from 0.1 mM to $10\mu$M.

cerned with the behavior as compared with that in closed system and of the reactions of MWC dimeric model.

( 1 ) *Positive cooperativity with substrate $S$*:

The reaction scheme is as given in Scheme 4.20, and the rate constants specified in Table 5.8 are employed for the simulation. The steady-state concentrations of each chemical species are listed in Table 5.9 for the

**Table 5.8** Rate constants for KNF dimeric model (Scheme 4.20) in the open system

| Rate constant (unit) | Positive cooperativity | Negative cooperativity |
|---|---|---|
| $k_1$ $(M^{-1}sec^{-1})$ | $1.0 \times 10^5$ | $1.0 \times 10^5$ |
| $k_{-1}(sec^{-1})$ | $5.0$ | $5.0$ |
| $k_2$ $(M^{-1}sec^{-1})$ | $5.0 \times 10^5$ | $1.0 \times 10^4$ |
| $k_{-2}(sec^{-1})$ | $1.0$ | $1.0$ |
| $k_3$ $(sec^{-1})$ | $5.0 \times 10^{-2}$ | $5.0 \times 10^{-2}$ |

**Table 5.9** Steady-state concentrations of Scheme 4.20 with positive cooperativity in the open system

| Species | Concentration (mM) | | |
|---|---|---|---|
| $S$ | $1.0 \times 10^{-2}$ | $1.0 \times 10^{-1}$ | $1.0$ |
| $\bar{P}$ | $2.86 \times 10^{-1}$ | $1.65$ | $1.97$ |
| $\bar{E}_0$ | $7.79 \times 10^{-3}$ | $8.82 \times 10^{-4}$ | $1.17 \times 10^{-5}$ |
| $\bar{E}_s$ | $1.56 \times 10^{-3}$ | $1.76 \times 10^{-3}$ | $2.34 \times 10^{-4}$ |
| $\bar{E}_{ss}$ | $6.49 \times 10^{-4}$ | $7.35 \times 10^{-3}$ | $9.75 \times 10^{-3}$ |

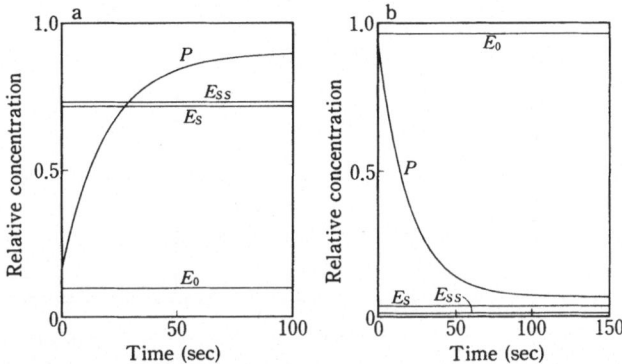

**Fig. 5. 10** Transient responses of Scheme 4.20 with positive coopera-tivity in the open system. a) Increase in $S$ from 0.1 mM to 1.0 mM; b) De-crease in $S$ from 0.1 mM to $10\mu$M.

substrate concentrations of $0.01$, $0.1$ and $1.0$ (in mM), respectively. Figure 5.10 illustrates the time courses of the transition from a steady state to new steady states caused by disturbances in substrate concentration similar to the allosteric reactions of the MWC dimeric model.

The behavior of the reaction in closed system is shown in Fig. 4.43. A noticeable difference between them is seen in that in open system each enzyme species reaches new steady states much more rapidly than product P. The steady-state concentration of product P is much higher than those in reactions of the Michaelis-Menten type and MWC dimeric model, implying that the system has a shorter dwelling-period of metabolites than other enzymatic reactions.

( 2 ) *Negative cooperativity with substrate $S$*:

The reaction scheme and rate constants are as given in Scheme 4.20 and Table 5.8, respectively. The steady-state concentrations of each chemical

**Table 5. 10** Steady-state concentrations of Scheme 4.20 with nega-tive cooperativity in the open system

| Species | Concentration (mM) | | |
|---------|--------------------|--------------------|--------------------|
| $S$ | $1.0 \times 10^{-2}$ | $1.0 \times 10^{-1}$ | $1.0$ |
| $\bar{P}$ | $1.69 \times 10^{-1}$ | $7.37 \times 10^{-1}$ | $1.42$ |
| $\bar{E}_0$ | $8.32 \times 10^{-3}$ | $3.16 \times 10^{-3}$ | $2.65 \times 10^{-4}$ |
| $\bar{E}_s$ | $1.66 \times 10^{-3}$ | $6.32 \times 10^{-3}$ | $5.31 \times 10^{-3}$ |
| $\bar{E}_{ss}$ | $1.39 \times 10^{-5}$ | $5.26 \times 10^{-4}$ | $4.42 \times 10^{-3}$ |

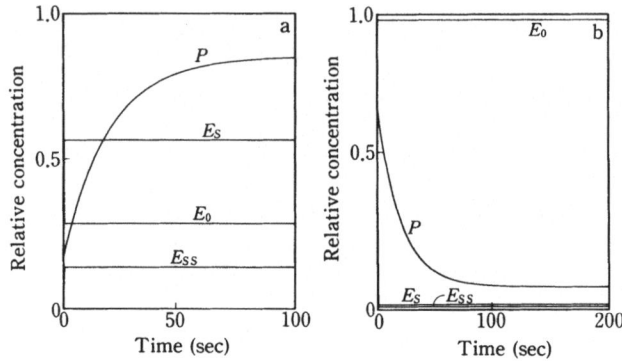

**Fig. 5.11** Transient responses of Scheme 4.20 with negative cooperativity in the open system. a) Increase in $S$ from 0.1 mM to 1.0 mM; b) Decrease in $S$ from 0.1 mM to 10$\mu$M.

species are listed in Table 5.10 for the substrate concentrations of 0.01, 0.1 and 1.0 (in mM), respectively. Figure 5.11 illustrates the time courses of the transition from a steady state to new steady states caused by similar disturbances in substrate concentration. The negative cooperativity results in higher concentration of $E_0$ and much slower transition to new steady states in product P than in case (1).

( 3 ) *Positive cooperativity with substrate S and a negative effector I :*

The reaction scheme and rate constants are as given in Scheme 4.21. The steady-state concentrations of each chemical species are listed in Table 5.11 for the substrate concentrations of 0.01, 0.1 and 1.0 (in mM), respectively, with 1.0 mM of negative effector I. Figure 5.12 illustrates the time courses of the transition from a steady state to new steady states caused by similar disturbances in substrate concentration. The negative

**Table 5.11** Steady-state concentrations of Scheme 4.21 in the open system

| Species | Concentration (mM) | | |
|---|---|---|---|
| $S$ | $1.0 \times 10^{-2}$ | $1.0 \times 10^{-1}$ | $1.0$ |
| $\bar{P}$ | $3.48 \times 10^{-3}$ | $1.62 \times 10^{-1}$ | $1.76$ |
| $\bar{E}_0$ | $9.50 \times 10^{-5}$ | $8.62 \times 10^{-5}$ | $1.04 \times 10^{-5}$ |
| $\bar{E}_S$ | $1.90 \times 10^{-5}$ | $1.73 \times 10^{-4}$ | $2.09 \times 10^{-4}$ |
| $\bar{E}_{SS}$ | $7.92 \times 10^{-6}$ | $7.23 \times 10^{-4}$ | $8.70 \times 10^{-3}$ |
| $\bar{E}_I$ | $3.80 \times 10^{-4}$ | $3.47 \times 10^{-4}$ | $4.17 \times 10^{-5}$ |
| $\bar{E}_{II}$ | $9.50 \times 10^{-3}$ | $8.67 \times 10^{-3}$ | $1.04 \times 10^{-3}$ |
| $\bar{I}$ | $1.0 \times 10^{-1}$ | $1.0 \times 10^{-1}$ | $1.0 \times 10^{-1}$ |

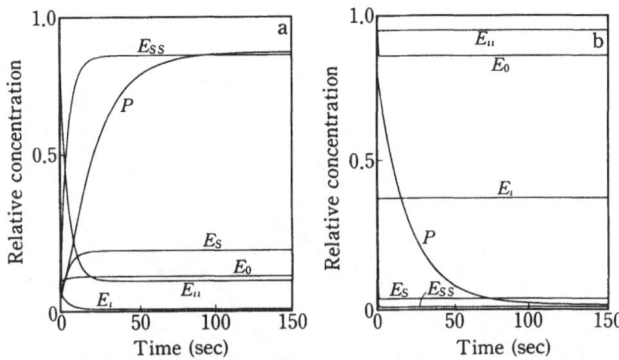

**Fig. 5. 12**  Transient responses of Scheme 4.21 in the open system. a) Increase in $S$ from 0.1 mM to 1.0 mM ; b). Decrease in $S$ from 0.1 mM to $10\mu$M.

effector induces slower transition to new steady states for all enzyme species. The behavior of product P, however, is not affected. The steady-state concentration of product P in the system with low substrate concentration is lower than in case (1), obviously due to the effect of the negative effector.

### Relationship between Substrate and Product Concentrations

The behavior of the Michaelis-Menten-type and allosteric (MWC and KNF dimeric models) reactions in one-step and open system is summarized with respect to the relationship between the substrate concentration and steady-state concentration of product P, as shown in Fig. 5.13. We can also detect in the figure the relationship between the dwelling period of metabolites in open system and the substrate concentration since product P flows out of the system at a rate of $k_0P$. Two allosteric models of a dimeric enzyme yield quite different behavior. In the MWC model $\bar{P}$ increases with higher order against an increase in the substrate concentration, while in the KNF model the increase of $\bar{P}$ tends to saturate with the substrate concentration. The KNF model with negative cooperativity [case( 2 )] behaves similarly to the Michaelis-Menten-type enzyme.

### 2.  Two-Step Systems

The study of two-step systems (*i.e.,* systems consisting of two enzymatic reactions) provides a basis for the analysis of *in vivo* catalytic and regulatory behavior of cellular metabolism, that is, the organized and controlled system of metabolic pathways operating in the cellular environments. Many metabolic pathways actually are multi-step systems consisting of

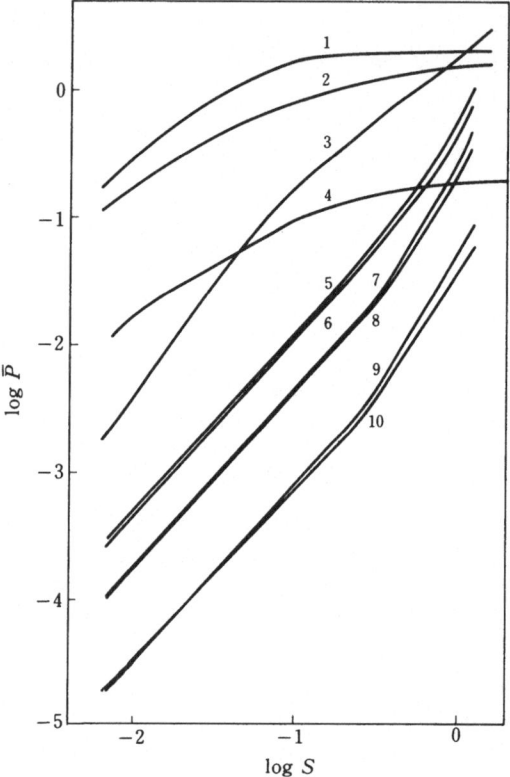

**Fig. 5.13** Relationship between the steady-state concentration in product and the substrate concentration for one-step systems in the open system. 1 : KNF model [case(1)] (Scheme 4.20); 2 : KNF model [case(2)] (Scheme 4.20); 3 : KNF model [case(3)] (Scheme 4.21); 4 : Michaelis-Menten-type reaction(Scheme 5.1); 5 : MWC model [case(1)] (Scheme 5.2); 6 : MWC model [case(2)] (Scheme 4.14); 7 : MWC model [case (5)] (Scheme 5.4); 8 : MWC model [case(6)] (Scheme 4.18); 9 : MWC model [case(3)] (Scheme 4.16); 10 : MWC model [case(4)] (Scheme 5.3).

more than two enzymatic reactions. At present, however, the methods for computer simulation employed in this book cannot be applied to multistep systems containing more than 100 chemical species, because both the numerical integration itself and the analysis of time courses become very difficult.

It is thus appropriate to determine the dynamic characteristics of metabolic systems (general multi-step systems) from the simulation of two-

and three-step systems. Moreover, actual metabolic systems include many two-step systems, and a multi-step system would possibly be represented by a two-step system in terms of the dynamic characteristics in the cellular metabolism. Hence, analysis of two-step systems would lead to much fundamental knowledge of the dynamic features of the metabolic processes.

The dynamic characteristics of metabolic systems are derived from analysis of dynamic behavior of enzymatic reaction systems in open system. It is naturally assumed that the metabolic system is not only a reaction system but also a certain control system operating in open system, in which the steady states are stable and to be maintained. In fact, allosteric enzymes are known to play significant roles in the metabolic regulation. The enzyme for the first-step reaction in metabolic pathways often is allosteric. Thus, we here examine the behavior of two-step systems in open system, with emphasis on the regulatory features of allosteric enzymes.

The most appropriate model for a two-step system would be as follows. An allosteric enzyme and a Michaelis-Menten-type enzyme respectively catalyze the first-step and second-step reactions. The reaction products of the system work as effectors for the allosteric enzyme of the first-step reaction, forming feedback loops in the system. In the following, we examine the dynamic behavior of the model with respect to the transient responses to the disturbances and the sustained oscillations. Time courses are obtained and analyzed for the cases of negative feedback and combined negative and positive feedbacks.

## Two-Step System with a Negative Feedback Loop

The most basic mechanism in metabolic regulation is the feedback inhibi-

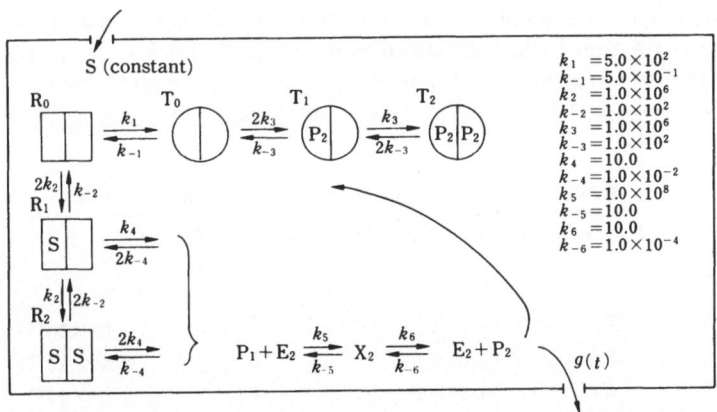

Scheme 5.5.  Two-step system with negative feedback loop in an open system

tion of the enzyme for the first-step reaction by the end product of a metabolic system. We here obtain the transient responses of the mechanism to disturbances from the simulation of a two-step system with such a negative feedback loop.

The system model is as given in Scheme 5.5. Enzyme $E_1$ for the first-step reaction is allosteric (MWC dimeric model), while enzyme $E_2$ for the second-step reaction is of the Michaelis-Menten type. Product $P_1$ of the first reaction is produced solely from the enzyme species in R-state, and substrate S does not form complexes with the enzyme species in T-state. Product $P_1$, in turn, serves as the substrate for the second reaction, product $P_2$ of which acts as an inhibitor for $E_1$ to form complexes solely with the T-form species. Substrate S is supplied to the system from the outside to hold its concentration constant at all times. End product $P_2$ of the system is taken out of the system at a constant rate of $g$ ($=10\ \mu\mathrm{Msec}^{-1}$).

One of the aims of metabolic regulation is to maintain a constant efflux rate of the end product, that is, the influx rate of the substrate to the following metabolic system, against disturbances to the system. The maintenance of a constant concentration of substrate S in the system results from regulation of the previous metabolic system producing metabolite S. The two-step system considered here performs the supply regulation of substrate $P_2$ to the next metabolic system by negative feedback.

Rate constants for the simulation are given in Scheme 5.5. The steady-state concentrations of each chemical species in the system are obtained as in Table 5.12 for $S=10\ \mathrm{mM}$ and $E_{10}=E_{20}=10\ \mu\mathrm{M}$. We examine the dynamic behavior of the system with respect to the transient responses to impulse-type disturbances in the concentration of $P_2$. The initial values for chemical species except $P_2$ are set to their respective concentrations at the steady state, and $P_2$ has an initial value of $3.32\ \mathrm{mM}$ ($=2\ \bar{P}_2$). The simulation yields the transient response of the system to the disturbance. A disturbance causing an impulse-like change in the concentration of $P_2$ is due, for

**Table 5.12**  Steady-state concentrations in Scheme 5.5

| Species | Concentration (mM) | Species | Concentration (mM) |
|---------|--------------------|---------|--------------------|
| $S$ | 10.0 | $S$ | 10.0 |
| $\bar{R}_0$ | $5.98\times10^{-8}$ | $\bar{T}_2$ | $8.05\times10^{-2}$ |
| $\bar{R}_1$ | $1.09\times10^{-5}$ | $\bar{P}_1$ | $2.22\times10^{-5}$ |
| $\bar{R}_2$ | $4.95\times10^{-4}$ | $\bar{E}_2$ | $9.00\times10^{-3}$ |
| $\bar{T}_0$ | $5.98\times10^{-5}$ | $\bar{X}_2$ | $1.00\times10^{-3}$ |
| $\bar{T}_1$ | $1.39\times10^{-3}$ | $\bar{P}_2$ | $1.16$ |

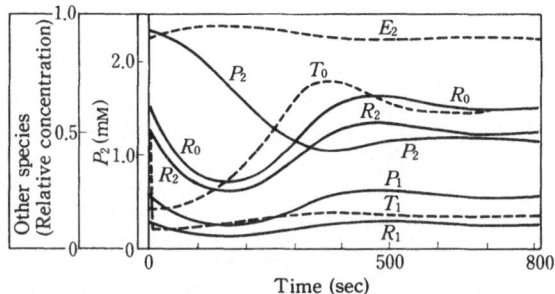

**Fig. 5.14** Transient responses of Scheme 5.5 to a disturbance (increase in $P_2$ by $\bar{P}_2$). $k_1 = 100 \text{ sec}^{-1}$, $k_{-1} = 0.1 \text{ sec}^{-1}$.

example, to excessive intake of $P_2$ in an organism, or decrease in the demand of $P_2$ as the substrate for the following metabolic system. A typical transient response is shown in Fig. 5.14.

Of the chemical species in the system, the allosteric enzyme $E_1$ undergoing negative feedback leads to the most interesting behavior. The allostericity of an MWC model results from the allosteric transition between the R- and T-form species of the enzyme. The characteristics of an allosteric enzyme have often been described by the so-called saturation functions derived from a multiple-equilibria assumption among the interactions of the enzyme with its substrate and effectors. In saturation functions, allosteric transition is represented by the allosteric constant $L$, which is an equilibrium constant for the transition [2].

On the other hand, the rate equation for the simulation employs the rate constants $k_1$ and $k_{-1}$, allowing description of the allosteric transition in a nonequilibrium state as well as at equilibrium. It is of interest to study the effects of allosteric transition rate on the behavior of the system. The effects are revealed by the simulation performed for various values of $k_1$, which represents the allosteric transition rate, with a constant $L$ ($= k_1/k_{-1}$). Corresponding to a value of $k_1$, $k_{-1}$ is chosen to have a constant $L$ ($= 1000$); $k_{-1}$ becomes 0.1, 0.5 and 10 (in $\text{sec}^{-1}$), respectively, for the values of 100, 500 and $10^4$ (in $\text{sec}^{-1}$) of $k_1$. The time courses of transient responses for $k_1$ of 500 and $10^4$ (in $\text{sec}^{-1}$) are shown in Fig. 5.15. In the case of $k_1 = 100 \text{ sec}^{-1}$ (Fig. 5.14), each chemical species returns to a steady state through damped oscillations. With $k_1 = 500 \text{ sec}^{-1}$ each chemical species changes its concentration in rapid response to the disturbance, then returns to a steady state in exponential decay. With $k_1 = 10^4 \text{ sec}^{-1}$ the system responds very quickly to the disturbance, but the return to a steady state is rather slower than in the case of $k_1 = 500 \text{ sec}^{-1}$.

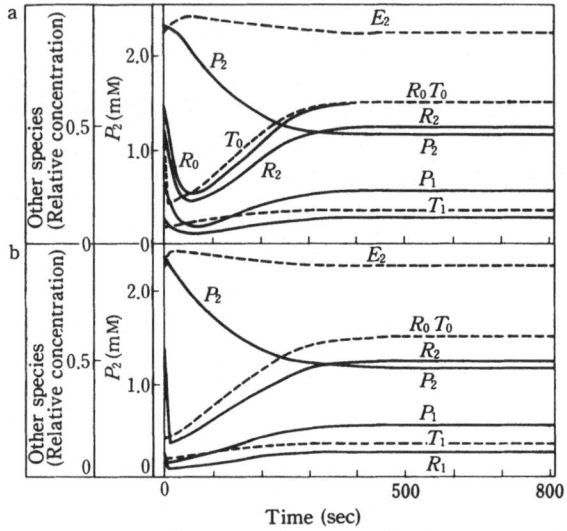

**Fig. 5.15** Transient responses of Scheme 5.5 to a disturbance (increase in $P_2$ by $\tilde{P}_2$). a) $k_1 = 500 \text{ sec}^{-1}$, $k_{-1} = 0.5 \text{ sec}^{-1}$; b) $k_1 = 10^4 \text{ sec}^{-1}$, $k_{-1} = 10 \text{ sec}^{-1}$.

A quantitative comparison of the effect of allosteric transition rate on the transient response is possible using the time taken for the system to return to a steady state after the disturbance is imposed. The transition time $T$ is defined as the shortest time in which the system attains

$$\left| 1 - \frac{P_1}{\tilde{P}_1} \right| + \left| 1 - \frac{P_2}{\tilde{P}_2} \right| \le 10^{-4} \tag{2}$$

after the concentration of $P_2$ is varied at $t = 0$. The variation in $T$ corresponding to the change in $k_1$ from 100 to $5 \times 10^6$ (in $\text{sec}^{-1}$) is shown in Fig. 5.16 for $L$ of $10^3$ and $10^4$. The transition time gets shorter with larger $L$, while the variations against $k_1$ are similar. With $L = 10^3$, $T$ becomes constant and independent of $k_1$ if $k_1 > 5 \times 10^3 \text{ sec}^{-1}$. $T$ decreases with $k_1 < 5 \times 10^3$ $\text{sec}^{-1}$, reaching a minimum value at $k_1 = 500 \text{ sec}^{-1}$. The minimum value corresponds to the boundary at which the mode of returning to a steady state changes from exponential decay to damped oscillation. With smaller $k_1$ damped oscillation arises in the response, increasing $T$ substantially.

These results infer that the regulatory characteristics of the system, especially the responsiveness to disturbance are revealed from the treatment of allostericity in terms of allosteric transition rate rather than equilibrium between the R- and T-form species of allosteric enzyme.

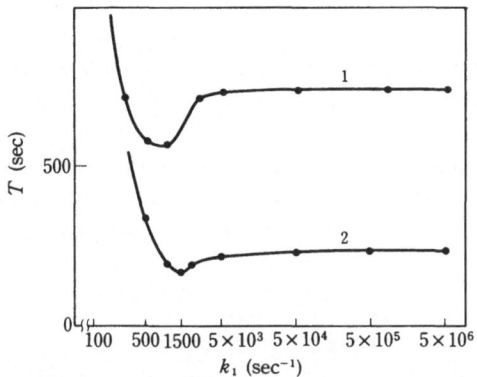

**Fig. 5.16** Relationship between transition time $T$ and rate constant $k_1$.
1: $L=10^3$; 2: $L=10^4$.

## Two-Step System with both Negative and Positive Feedback Loops

Several examples of sustained oscillation have been known in enzymatic reaction systems (and metabolic systems). We here analyze the phenomenon of sustained oscillation in such a two-step system where a positive feedback loop is added to the system with the negative feedback loop analyzed above. (Macroscopic treatment of sustained oscillation is described in Section 6.3.)

The phenomena of sustained oscillation in enzymatic reaction systems have the following characteristics [3] ;

1) The phenomenon results from periodic changes in enzymatic activity due to regulation at the molecular level.

2) From a thermodynamic point of view, sustained oscillations are temporal dissipative structures since they occur around an unstable steady-state. These oscillations are of the limit-cycle type.

3) Limit cycles arise only in thermodynamically open systems operating far from equilibrium with appropriate nonlinear kinetics. These conditions should be frequently satisfied in biochemical processes where the multiple types of feedback and the cooperativity of allosteric enzymes render kinetic equations nonlinear. Moreover, regulatory enzymes are precisely those which operate far from equilibrium.

4) The oscillatory dynamics of a complex metabolic pathway may be reduced to that of a single regulatory enzyme within the reaction sequence.

5) Periodic behavior may result from both negative and positive feedback loops. Whereas the former type of regulation is by far the most frequent in biochemical processes, the majority of known oscillatory

systems appear to depend on the latter.

Well-known examples of regulatory enzymes yielding sustained oscillations are phosphofructokinase, peroxidase and adenylate cyclase. The physiological significance of the oscillations is not yet clear despite many hypotheses proposed.

The system of Scheme 5.6 is considered as a model of a two-step system with both negative and positive feedback loops. Enzyme $E_1$ of the first-step reaction is allosteric (MWC dimeric model) so that substrate S forms

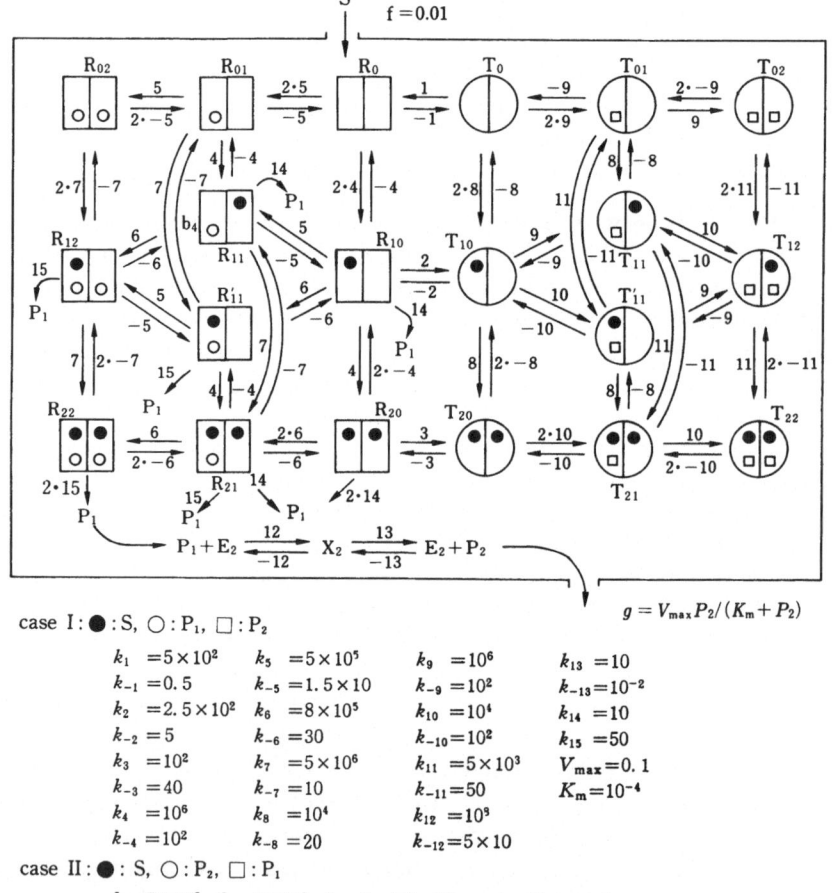

case I: ● : S, ○ : $P_1$, □ : $P_2$

| | | | |
|---|---|---|---|
| $k_1 = 5 \times 10^2$ | $k_5 = 5 \times 10^5$ | $k_9 = 10^6$ | $k_{13} = 10$ |
| $k_{-1} = 0.5$ | $k_{-5} = 1.5 \times 10$ | $k_{-9} = 10^2$ | $k_{-13} = 10^{-2}$ |
| $k_2 = 2.5 \times 10^2$ | $k_6 = 8 \times 10^5$ | $k_{10} = 10^4$ | $k_{14} = 10$ |
| $k_{-2} = 5$ | $k_{-6} = 30$ | $k_{-10} = 10^2$ | $k_{15} = 50$ |
| $k_3 = 10^2$ | $k_7 = 5 \times 10^6$ | $k_{11} = 5 \times 10^3$ | $V_{max} = 0.1$ |
| $k_{-3} = 40$ | $k_{-7} = 10$ | $k_{-11} = 50$ | $K_m = 10^{-4}$ |
| $k_4 = 10^6$ | $k_8 = 10^4$ | $k_{12} = 10^3$ | |
| $k_{-4} = 10^2$ | $k_{-8} = 20$ | $k_{-12} = 5 \times 10$ | |

case II: ● : S, ○ : $P_2$, □ : $P_1$

$k_5 = 5 \times 10^6$, $k_6 = 8 \times 10^6$, $k_9 = 3 \times 10^5$, $V_{max} = 1$, $K_m = 10^{-3}$.

The other rate constants are the same as in case I.

Scheme 5.6. Two-step system with both positive and negative feedback loop in an open system complexes with both R- and T-form species of the enzyme and transforms to product $P_1$ only through the enzyme-substrate complexes in R-state. A positive effector interacts solely with the R-form species, while a negative effector affects only the T-form species. Enzyme $E_2$ of the second-step reaction is of the Michaelis-Menten type, producing end product $P_2$ from substrate $P_1$. Substrate S is supplied to the system from the outside at a constant rate of $f$ $(=10\,\mu M\ sec^{-1})$, while end product $P_2$ is consumed as the substrate for the following metabolic process, taken out of the system at an efflux rate of $g(t) = V_{max}P_2/(K_m+P_2)$.

Two cases of feedback interactions are possible for the system model:
case I: $P_1$ and $P_2$ act as positive and negative effectors, respectively.
case II: $P_1$ and $P_2$ act as negative and positive effectors, respectively.
The system with the rate constants and parameters given in Scheme 5.6 yields sustained oscillations of concentrations in every chemical species for both cases I and II. The time courses for cases I and II are shown in Figs. 5.17 and 5.18, respectively. Substrate S and the R-form species of $E_1$

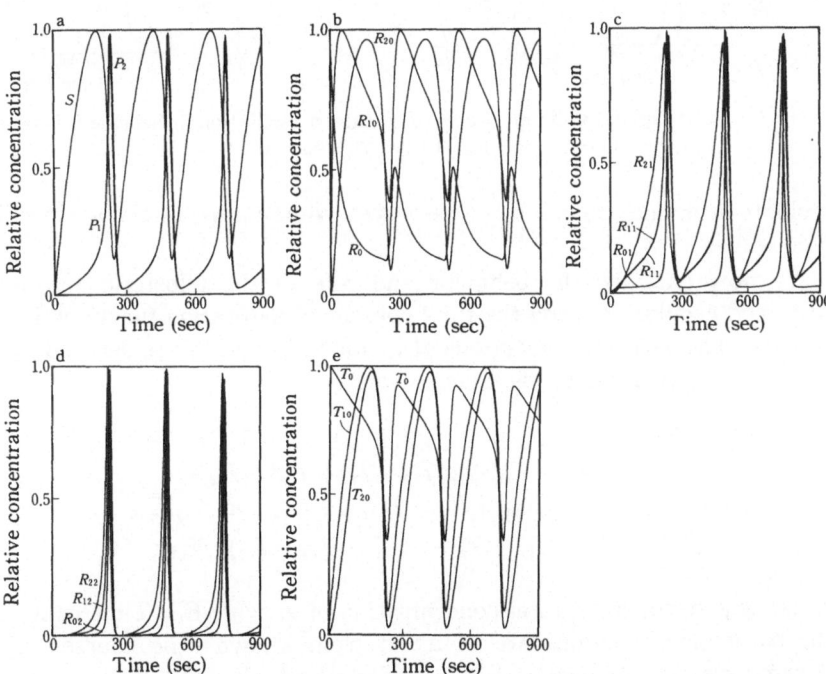

**Fig. 5.17** Time course of sustained oscillation in Scheme 5.6 (case I). Initial condition: $R_0=10$ nM, $T_0=9.99\mu$M.

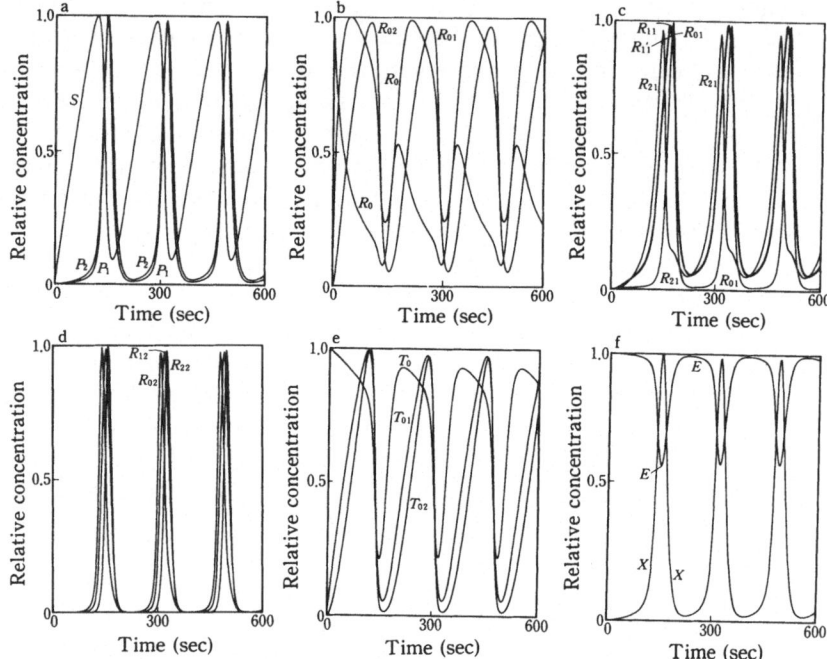

**Fig. 5.18**  Time course of sustained oscillation in Scheme 5.6 (case II). Initial condition : $R_0 = 10$ nM, $T_0 = 9.99 \mu$M.

oscillate in smooth, saw-tooth-type waves, while $P_1$ and $P_2$ change in spike-type waves.

We now examine the behavior and role of the allosteric enzyme in sustained oscillation using the time courses of saturation functions for the ligands. The saturation functions of $E_1$ for S, $P_1$ and $P_2$ are denoted by $Y_S$, $Y_{P_1}$ and $Y_{P_2}$, respectively and defined as

$$
\begin{aligned}
Y_S &= [R_{10} + R_{11} + R_{11}' + R_{12} + 2(R_{20} + R_{21} + R_{22}) \\
&\quad + T_{10} + T_{11} + T_{11}' + 2(T_{20} + T_{21} + T_{22})]/E_{10} \\
Y_{P_1} &= [R_{01} + R_{11} + R_{11}' + R_{21} + 2(R_{02} + R_{12} + R_{22})]/E_{10} \\
Y_{P_2} &= [T_{01} + T_{11} + T_{11}' + T_{21} + 2(T_{02} + T_{12} + T_{22})]/E_{10},
\end{aligned}
\right\} \quad (3)
$$

where $E_{10}$ denotes the total concentration of enzyme $E_1$. Time courses of the saturation functions are obtained from known time courses of the enzyme species. As seen in Fig. 5.19 for case I, $Y_{P_1}$ and $Y_{P_2}$ vary in symmetrical spike-type waves, while $Y_S$ displays a complicated pattern with an inflection point at the maximum concentration of free substrate. During

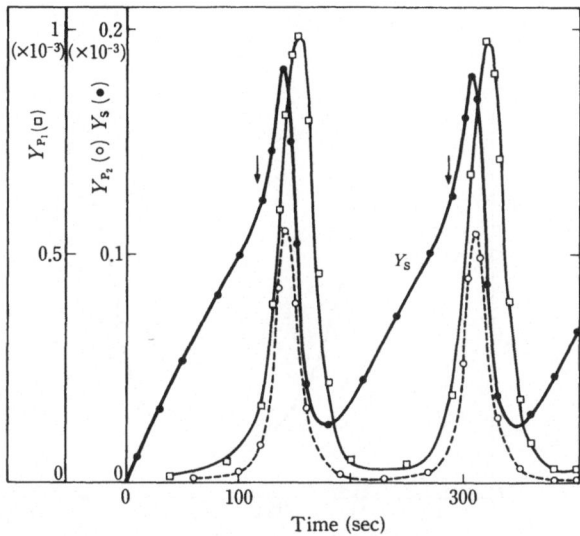

**Fig. 5. 19** Time course of saturation functions in case I. The substrate concentration becomes maximum at the time indicated by the arrow.

one period of oscillation in $Y_s$, binding of S with $R_0$ and $T_0$ leads to a gradual increase in $Y_s$ without noticeable effect of effectors in the first 90 sec. In the next 28 sec the positive effector affects the system increasing the R-form species considerably. In the last 44 sec the allosteric transition competes with the effect of the positive effector, rapidly decreasing the R - form species.

As described above, the allosteric transition rate affects the dynamic behavior in open system more characteristically than the allosteric transition equilibrium. The effect of allosteric transition rate on the period of sustained oscillation is summarized in Fig. 5.20 for case I and in Fig. 5.21 for case II. In both cases the period becomes shortest at $k_1 = 50 \ \text{sec}^{-1}$, increasing monotonously with larger or smaller value of $k_1$. In case I the system yields damped oscillations for $k_1$ less than $0.1 \ \text{sec}^{-1}$ or larger than $2 \times 10^3 \ \text{sec}^{-1}$. In case II damped oscillations also appear for $k_1$ less than $0.5 \ \text{sec}^{-1}$, but the computer simulation cannot evaluate the behavior for $k_1$ larger than $10^4 \ \text{sec}^{-1}$ because the period becomes extremely long.

The simulation also reveals that the oscillatory phenomena in the system are affected by the affinity of the effector. Increase in the affinity of a positive effector causes a monotonous decrease in the period, and finally damped oscillation for both cases I and II. Increase in the affinity of a

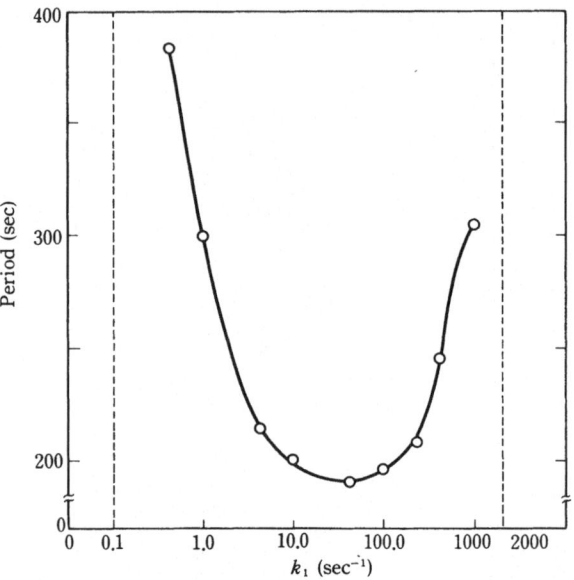

**Fig. 5. 20**  Relationship between period of sustained oscillation and rate constant $k_1$ in Scheme 5.6 (case I). $E_1 = E_2 = 10$ mM ; $k_1/k_3 = 2$, $k_1/k_5 = 5$ for the variation in $k_1$.

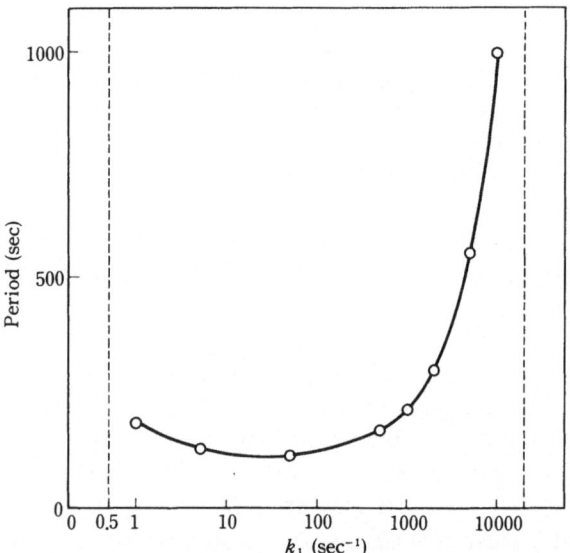

**Fig. 5. 21**  Relationship between period of sustained oscillation and rate constant $k_1$ in Scheme 5.6 (case II). Conditions are the same as in Fig. 5.20.

negative effector results in a monotonous increase in the period in both cases.

## 5.2 Branched Biosynthetic Pathways

In metabolic processes a rather small number of compounds are utilized as starting substances for the biosynthesis of many kinds of metabolites. This is accomplished by a great number of branched biosynthetic pathways, as seen in Fig. 5.22 for biosynthesis of amino acids. The dynamic behavior and regulatory mechanism in branched biosynthetic pathways are very important and interesting factors in the analysis of metabolic systems. In this section, employing computer simulation, we analyze the basic properties of branched reaction systems, which Umbarger derived from classifying and modeling the biosynthetic pathways of amino acids.

### 1. Regulatory Mechanisms in Branched Reaction Systems

Umbarger derived 6 types of regulatory mechanisms in branched reaction systems from the analysis of branched biosynthetic pathways of amino acids [4]. Each type of mechanism is described in Table 5.13.

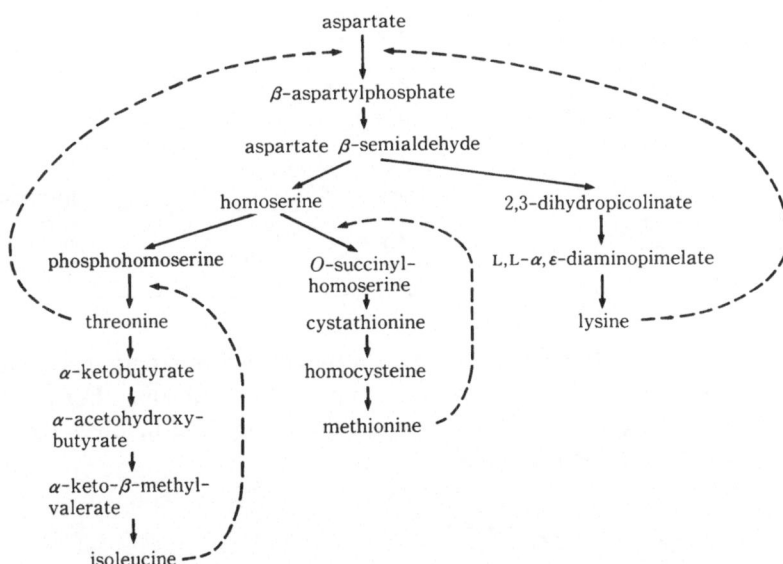

**Fig. 5.22** A branched biosynthetic pathway. Broken lines indicate the feedback inhibition by the end products.

**Table 5.13** Regulatory mechanisms for branched biosynthetic pathways

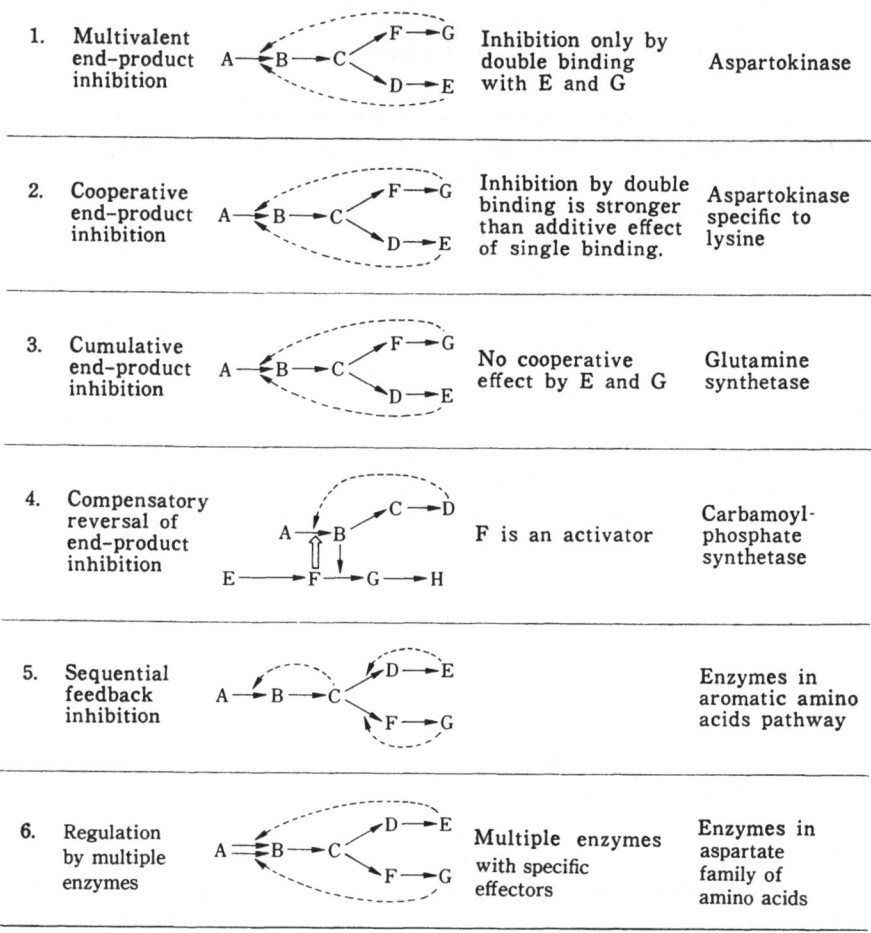

| | | | |
|---|---|---|---|
| 1. | Multivalent end-product inhibition | Inhibition only by double binding with E and G | Aspartokinase |
| 2. | Cooperative end-product inhibition | Inhibition by double binding is stronger than additive effect of single binding. | Aspartokinase specific to lysine |
| 3. | Cumulative end-product inhibition | No cooperative effect by E and G | Glutamine synthetase |
| 4. | Compensatory reversal of end-product inhibition | F is an activator | Carbamoyl-phosphate synthetase |
| 5. | Sequential feedback inhibition | | Enzymes in aromatic amino acids pathway |
| 6. | Regulation by multiple enzymes | Multiple enzymes with specific effectors | Enzymes in aspartate family of amino acids |

The regulatory mechanism in an actual branched reaction system is formed of a combination of these basic types of mechanisms. For example, the biosynthetic pathway from pyruvic acid in Fig. 5.23 involves the following sequential feedback inhibition:

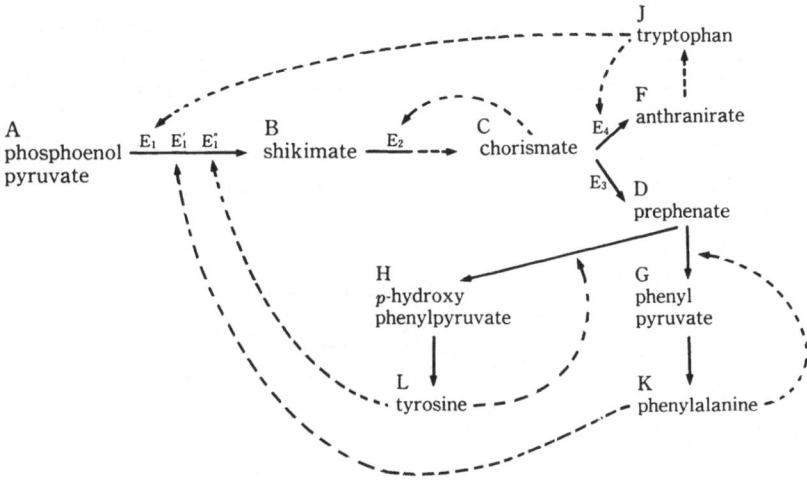

**Fig. 5. 23**  Regulatory mechanism in a branched biosynthetic pathway.

It is noted that the system is also provided with the control of multiple enzymes $E_1$ by K and L, which compensatorily enables the regulation of concentration in pool A deficient in the sequential feedback inhibition. The system can control the distribution ratio as well, since at least one of the enzymatic activities is regulated immediately after branching.

The reaction schemes of branched reaction systems are reduced to the model represented in Scheme 5.7. In this basic system substrate S in a pool

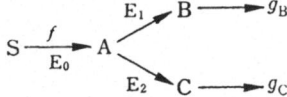

Scheme 5. 7.   Basic scheme of branched reaction system

is converted to a metabolite A at a rate $f$ through the enzymatic reaction of $E_0$. The enzymatic reactions catalyzed by enzymes $E_1$ and $E_2$ have a common substrate A, constituting the branched paths to produce metabolites B and C, respectively. Products B and C are consumed at the respective rates of $g_B$ and $g_C$.

## 2.   Dynamic Behavior of Basic Regulatory Mechanisms

Of the six types of basic regulatory mechanisms classified in Table 5.13, three are considered basic and are analyzed here ; multivalent end-product inhibition, cumulative end-product inhibition, and sequential feedback inhibition. Cooperative end-product inhibition and regulation by multiple enzymes are reduced to basic mechanisms with respect to their dynamic behavior. The regulatory function of compensatory reversal of end-product inhibition must be treated in association with a larger system of physiological significance.

In the analysis of dynamic behavior the branched reaction system is assumed to have a large pool of the starting metabolite (for example, aspar-

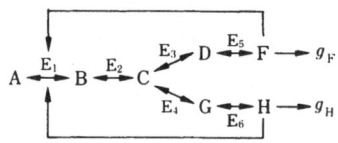

$E_1$-system:

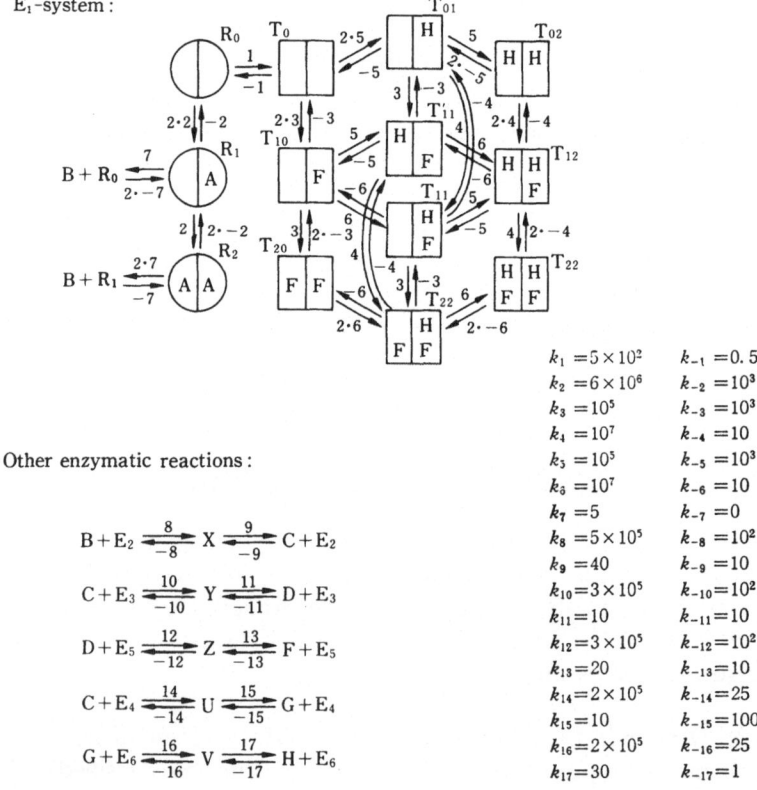

Other enzymatic reactions:

$$B+E_2 \underset{-8}{\overset{8}{\rightleftharpoons}} X \underset{-9}{\overset{9}{\rightleftharpoons}} C+E_2$$

$$C+E_3 \underset{-10}{\overset{10}{\rightleftharpoons}} Y \underset{-11}{\overset{11}{\rightleftharpoons}} D+E_3$$

$$D+E_5 \underset{-12}{\overset{12}{\rightleftharpoons}} Z \underset{-13}{\overset{13}{\rightleftharpoons}} F+E_5$$

$$C+E_4 \underset{-14}{\overset{14}{\rightleftharpoons}} U \underset{-15}{\overset{15}{\rightleftharpoons}} G+E_4$$

$$G+E_6 \underset{-16}{\overset{16}{\rightleftharpoons}} V \underset{-17}{\overset{17}{\rightleftharpoons}} H+E_6$$

| | |
|---|---|
| $k_1 = 5 \times 10^2$ | $k_{-1} = 0.5$ |
| $k_2 = 6 \times 10^6$ | $k_{-2} = 10^3$ |
| $k_3 = 10^5$ | $k_{-3} = 10^3$ |
| $k_4 = 10^7$ | $k_{-4} = 10$ |
| $k_5 = 10^5$ | $k_{-5} = 10^3$ |
| $k_6 = 10^7$ | $k_{-6} = 10$ |
| $k_7 = 5$ | $k_{-7} = 0$ |
| $k_8 = 5 \times 10^5$ | $k_{-8} = 10^2$ |
| $k_9 = 40$ | $k_{-9} = 10$ |
| $k_{10} = 3 \times 10^5$ | $k_{-10} = 10^2$ |
| $k_{11} = 10$ | $k_{-11} = 10$ |
| $k_{12} = 3 \times 10^5$ | $k_{-12} = 10^2$ |
| $k_{13} = 20$ | $k_{-13} = 10$ |
| $k_{14} = 2 \times 10^5$ | $k_{-14} = 25$ |
| $k_{15} = 10$ | $k_{-15} = 100$ |
| $k_{16} = 2 \times 10^5$ | $k_{-16} = 25$ |
| $k_{17} = 30$ | $k_{-17} = 1$ |

Scheme 5.8. Multivalent and cumulative end-product inhibitions

tate in Fig. 5.22) so that quick replenishment keeps its concentration from varying during the operation. We also assume that the enzymes relevant to regulation in the basic regulatory mechanisms are allosteric and represented by the MWC dimeric models.

**Multivalent End-Product Inhibition**

A system model is given in Scheme 5.8. The concentration of metabolite A is fixed at 10 mM for the simulation. The rates of $g_F$ and $g_H$ for consumption or demand of the respective end products F and H are expressed by

$$\left.\begin{array}{l} g_F = V_{\text{max,F}} F/(K_{\text{m,F}} + F) \\ g_H = V_{\text{max,H}} H/(K_{\text{m,H}} + H) \end{array}\right\}, \tag{4}$$

where $V_{\text{max,F}} = V_{\text{max,H}} = 10 \ \mu\text{M sec}^{-1}$ and $K_{\text{m,F}} = K_{\text{m,H}} = 0.1 \ \text{mM}$ are used as the standard values. The following two types of disturbance are imposed on the system to examine its transient responses:

1) Change in the rates of demand for end products F and H: Increase in the rates corresponds to assignment of larger values to $V_{\text{max,F}}$ and $V_{\text{max,H}}$.
2) Change in the concentration of pool A.

The time courses of the transient responses to disturbances (1) and (2) are shown in Figs. 5.24 and 5.25, respectively. The system reaches new steady states through damped oscillations in both cases. No noticeable

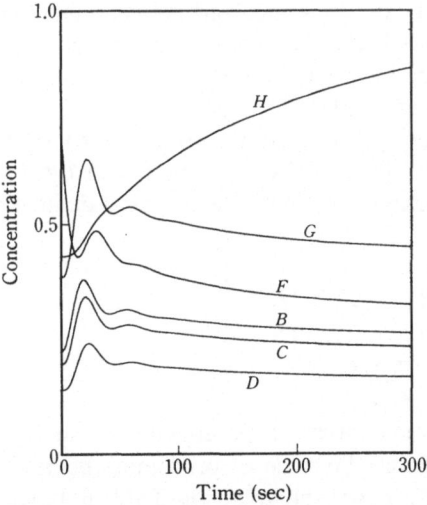

**Fig. 5.24** Transient responses of multivalent end-product inhibition (Scheme 5.8) to a disturbance (twofold increase in $V_{\text{max,F}}$). Full scale (1.0) on the ordinate: $B = 40\mu\text{M}$, $C = 80\mu\text{M}$, $D = 60\mu\text{M}$, $F = 0.1\text{mM}$, $G = 20\mu\text{M}$, $H = 1.2 \ \text{mM}$.

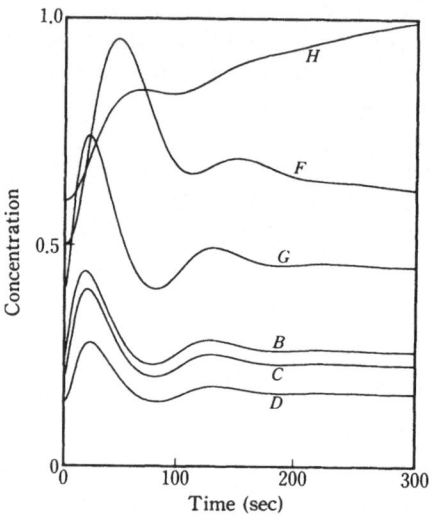

**Fig. 5. 25** Transient response of multivalent end-product inhibition (Scheme 5.8) to a disturbance (twofold increase in $A$). Full scale (1.0) on the ordinate : $F = 0.15$ mM, $H = 0.9$ mM. Others are the same as in Fig. 5.24.

phase-difference (time delay) is recognized between intermediates B, C, D and G, while end products F and H display some difference with the intermediates. The difference between F and the intermediates is particularly remarkable as seen in Fig. 5.24.

The steady-state values for the system to take after various disturbances are imposed are listed in Table 5.14. $Y_A$ to $Y_{FH}$ in the table denote the saturation functions of enzyme $E_1$ with each ligand defined as

$$\left.\begin{array}{l} Y_A = (R_1 + 2R_2)/2E_{10} \\ Y_F = (T_{10} + 2T_{20} + T_{11}' + T_{21})/2E_{10} \\ Y_H = (T_{01} + 2T_{02} + T_{11}' + T_{12})/2E_{10} \\ Y_{FH} = (T_{21} + T_{12} + 2T_{22})/2E_{10} , \end{array}\right\} \tag{5}$$

where $E_{10}$ is the total concentration of enzyme $E_1$ in the system. The saturation function indicates the ratio of protomers bound with a specified ligand to total protomers. From the values in Table 5.14 we have $Y_{FH} \gg Y_F$ or $Y_H$, which means that the inhibition is mostly due to the simultaneous binding with both F and H. With a two-fold increase in $V_{max,F}$, i.e., an increase in demand of one of the end products, $Y_A$ increases and $Y_{FH}$ decreases. Inversely, with a decrease in $V_{max,F}$ by half, the response to the

**Table 5.14**  Steady-state values of multivalent end-product inhibition (Scheme 5.8)

| Disturbances \\ Species | Standard | $V_{max,F} \times \frac{1}{2}$ | $V_{max,F} \times 2$ | $V_{max,H} \times \frac{1}{2}$ | $V_{max,H} \times 2$ | $A \times \frac{1}{2}$ | $A \times 2$ |
|---|---|---|---|---|---|---|---|
| $A$ | 10 | 10 | 10 | 10 | 10 | 5 | 20 |
| $B$ | 0.00916 | 0.00735 | 0.0101 | 0.00538 | 0.0138 | 0.0080 | 0.00994 |
| $C$ | 0.0161 | 0.0128 | 0.0178 | 0.00931 | 0.0248 | 0.0140 | 0.0175 |
| $D$ | 0.00869 | 0.00697 | 0.0096 | 0.00507 | 0.0133 | 0.00757 | 0.00945 |
| $F$ | 0.0738 | 0.214 | 0.0306 | 0.0334 | 0.178 | 0.059 | 0.0855 |
| $G$ | 0.00794 | 0.0064 | 0.00875 | 0.00471 | 0.0119 | 0.00695 | 0.00861 |
| $H$ | 0.533 | 0.214 | 1.18 | 1.64 | 0.164 | 0.284 | 0.983 |
| $Y_A$ | 0.127 | 0.102 | 0.139 | 0.075 | 0.188 | 0.111 | 0.137 |
| $Y_F$ | 0.013 | 0.003 | 0.001 | 0.000 | 0.004 | 0.002 | 0.001 |
| $Y_H$ | 0.025 | 0.003 | 0.021 | 0.023 | 0.003 | 0.009 | 0.009 |
| $Y_{FH}$ | 0.686 | 0.730 | 0.516 | 0.762 | 0.597 | 0.547 | 0.762 |

$V_{max,F} = V_{max,H} = 10 \ \mu M \ sec^{-1}$ (standard)

**Table 5.15**  Responses of production rate to disturbances in multivalent end-product inhibition

| Disturbances \\ Rate | $g_F$ $(\times 10^{-2})$ | $g_H$ $(\times 10^{-2})$ | $f = g_F + g_H$ |
|---|---|---|---|
| Standard | 0.425 (100) | 0.842 (100) | 1.267 (100) |
| $V_{max,F} \times \frac{1}{10}$ | 0.093 ( 22) | 0.350 ( 42) | 0.443 ( 35) |
| $''\quad \times \frac{1}{2}$ | 0.341 ( 80) | 0.682 ( 81) | 1.023 ( 81) |
| $''\quad \times 2$ | 0.469 (110) | 0.9222(110) | 1.391 (110) |
| $''\quad \times 10$ | 0.568 (134) | 0.979 (116) | 1.547 (122) |
| $V_{max,H} \times \frac{1}{10}$ | 0.157 ( 37) | 0.098 ( 12) | 0.255 ( 20) |
| $''\quad \times \frac{1}{2}$ | 0.251 ( 59) | 0.471 ( 56) | 0.722 ( 57) |
| $''\quad \times 2$ | 0.640 (151) | 1.241 (147) | 1.881 (148) |
| $''\quad \times 10$ | 0.909 (214) | 1.718 (204) | 2.627 (207) |
| $A \quad \times \frac{1}{2}$ | 0.371 ( 87) | 0.740 ( 88) | 1.111 ( 88) |
| $''\quad \times 2$ | 0.461 (109) | 0.908 (108) | 1.369 (108) |

( ) : Value relative to the standard

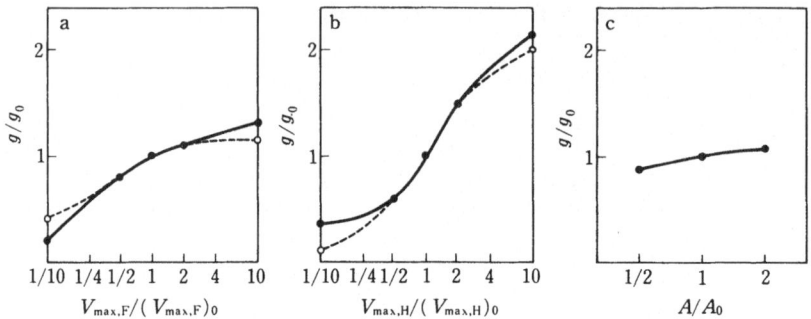

**Fig. 5. 26** Responses of multivalent end-product inhibition (Scheme 5.8) to disturbances. Solid line : $g_F/g_{F0}$ ; Broken line : $g_H/g_{H0}$. The subscript 0 indicates a value at $A = 10\,\text{mM}$ and $V_{max,F} = V_{max,H} = 10\,\mu\text{M}\,\text{sec}^{-1}$. a) Change in demand of $F$ ($V_{max,F}$) ; b) Change in demand of $H$ ($V_{max,H}$) ; c) Change in $A$.

disturbance is represented by a decrease in $Y_A$ and increase in $Y_{FH}$.

Changes of production rates in the branches are shown in Table 5.15 and Fig. 5.26 for the disturbances. The patterns of change in $g_F$ and $g_H$ are similar for small disturbances, implying that the change in demand of one of the end products affects the demand of the other. For larger disturbances, however, the regulatory feature is revealed that the branch at which the demand is varied has a larger change than the other branch, which has a suppressed effect. This is due to the fact that the control of $f$ is indirectly related to the distributive ratio in the branches.

**Cumulative End-Product Inhibition**

A system model is also given in Scheme 5.8, in which the following rate constants are used :

$$k_3 = k_4 = k_5 = k_6 = 10^6 \text{ M}^{-1} \text{ sec}^{-1}$$
$$k_{-3} = k_{-4} = k_{-5} = k_{-6} = 10^2 \text{ sec}^{-1}$$

so that the individual affinity of enzyme $E_1$ to a single ligand F or H is comparable to the combined affinity to F and H, and both ligands can inhibit the enzyme independently.

Other conditions and disturbances are the same as for the preceding case of multivalent end-product inhibition. The responses of the system to the disturbances are analyzed in a similar manner. The time course of the response to a two-fold increase in $V_{max,F}$ is shown in Fig. 5.27. Similar to the preceding case, intermediates B, C, D and G follow almost the same damped oscillations with short time-delay to reach new steady states. The transition of F and H to new steady states is monotonous, responding to the

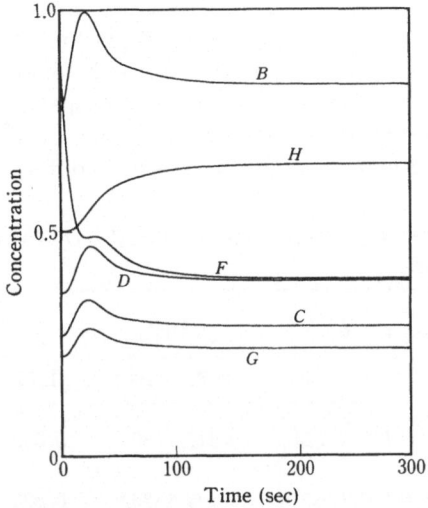

**Fig. 5. 27** Transient response of cumulative end-product inhibition (Scheme 5.8) to a disturbance (twofold increase in $V_{max,F}$). Full scale (1.0) on the ordinate: $B=10\mu M$, $C=50\mu M$, $D=20\mu M$, $F=60\mu M$, $G=30\mu M$, $H=0.5$ mM.

**Table 5. 16** Steady-state values of cumulative end-product inhibition (Scheme 5.8)

| Disturb-ances Species | Stand-ard | $V_{max,F}\times\frac{1}{2}$ | $V_{max,F}\times2$ | $V_{max,H}\times\frac{1}{2}$ | $V_{max,H}\times2$ | $A\times\frac{1}{2}$ | $A\times2$ |
|---|---|---|---|---|---|---|---|
| $A$ | 10 | 10 | 10 | 10 | 10 | 5 | 20 |
| $B$ | 0. 00773 | 0. 0064 | 0. 0083 | 0. 00441 | 0. 0114 | 0. 00601 | 0. 00901 |
| $C$ | 0. 0135 | 0. 0111 | 0. 0145 | 0. 00759 | 0. 0203 | 0. 0104 | 0. 0158 |
| $D$ | 0. 00731 | 0. 00604 | 0. 0078 | 0. 00414 | 0. 0109 | 0. 00567 | 0. 00855 |
| $F$ | 0. 0559 | 0. 146 | 0. 0238 | 0. 0257 | 0. 112 | 0. 0387 | 0. 0718 |
| $G$ | 0. 00672 | 0. 00557 | 0. 00720 | 0. 00386 | 0. 00985 | 0. 00525 | 0. 00782 |
| $H$ | 0. 252 | 0. 147 | 0. 326 | 0. 489 | 0. 108 | 0. 128 | 0. 485 |
| $Y_A$ | 0. 107 | 0. 089 | 0. 115 | 0. 062 | 0. 157 | 0. 084 | 0. 125 |
| $Y_F$ | 0. 091 | 0. 218 | 0. 040 | 0. 033 | 0. 214 | 0. 112 | 0. 063 |
| $Y_H$ | 0. 410 | 0. 220 | 0. 546 | 0. 619 | 0. 205 | 0. 370 | 0. 422 |
| $Y_{FH}$ | 0. 229 | 0. 321 | 0. 130 | 0. 159 | 0. 231 | 0. 143 | 0. 303 |

change in demand of F. Although the concentration of F decreases in Fig. 5. 27, the production rate $g_F$ increases to 107% of the initial rate owing to a two-fold increase in $V_{max,F}$. The production rate $g_H$ increases to 107% as well. The steady-state values after the disturbances are listed in Table

5.16. The saturation functions $Y_F$, $Y_H$ and $Y_{FH}$ are all comparable, demonstrating that the individual ligands F and H also inhibit the enzyme. It is also revealed that the system prevails against the disturbances by the changes in $Y_A$ and inhibitory roles among the enzyme species. The changes of production rate in each branch are shown in Table 5.17 and Fig. 5.28. Comparison with Fig. 5.26 indicates that there is little difference in dynamic

**Table 5.17** Responses of production rate to disturbances in cumulative end-product inhibition

| Rate<br><br>Disturbances | $g_F$<br>$(\times 10^{-2})$ | $g_H$<br>$(\times 10^{-2})$ | $f = g_F + g_H$ |
|---|---|---|---|
| Standard | 0.359 (100) | 0.716 (100) | 1.075 (100) |
| $V_{max,F} \times \dfrac{1}{10}$ | 0.088 ( 25) | 0.217 ( 30) | 0.305 ( 28) |
| $''\quad \times \dfrac{1}{2}$ | 0.297 ( 83) | 0.593 ( 83) | 0.890 ( 83) |
| $''\quad \times 2$ | 0.385 (107) | 0.765 (107) | 1.150 (107) |
| $''\quad \times 10$ | 0.401 (112) | 0.798 (112) | 1.199 (112) |
| $V_{max,H} \times \dfrac{1}{10}$ | 0.057 ( 16) | 0.093 ( 13) | 0.150 ( 14) |
| $''\quad \times \dfrac{1}{2}$ | 0.205 ( 57) | 0.415 ( 58) | 0.620 ( 58) |
| $''\quad \times 2$ | 0.529 (147) | 1.037 (145) | 1.566 (146) |
| $''\quad \times 10$ | 0.639 (193) | 1.337 (187) | 2.030 (189) |
| $A\quad \times \dfrac{1}{2}$ | 0.279 ( 78) | 0.562 ( 79) | 0.841 ( 78) |
| $''\quad \times 2$ | 0.418 (116) | 0.829 (116) | 1.247 (116) |

( ) : Value relative to the standard

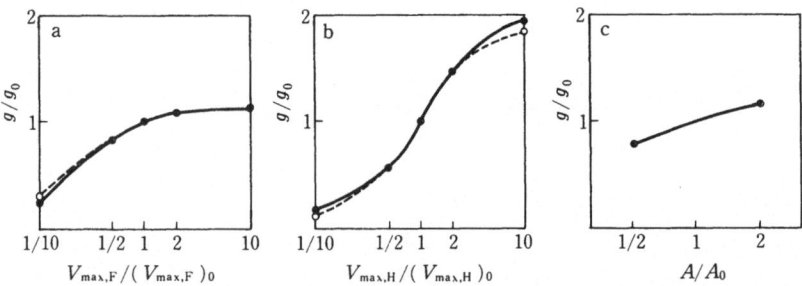

**Fig. 5.28** Responses of cumulative end-product inhibition (Scheme 5.8) to disturbances. Notations are the same as in Fig. 5.26.

behavior between the multivalent and cumulative end-product inhibitions.

## Sequential Feedback Inhibition

A system model is given in Scheme 5.9 in which enzymes $E_3$ and $E_4$ in the branched reactions as well as enzyme $E_1$ in the first-step reaction are regulated by feedback inhibition. In this case $V_{max,F} = V_{max,H} = 20\ \mu M\ sec^{-1}$

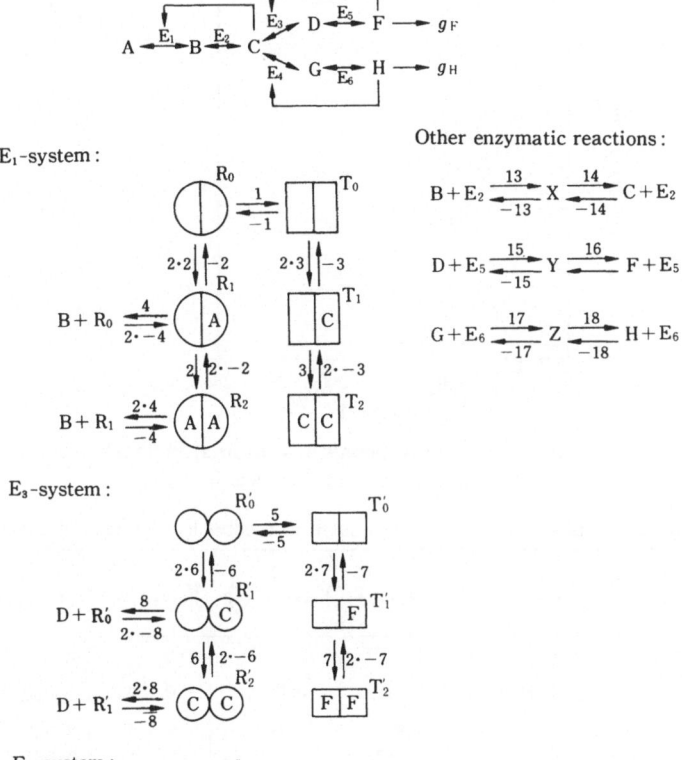

Scheme 5.9. Sequential feedback inhibition

and $K_{m,F} = K_{m,H} = 1.0$ mM are chosen as the standard values for equation (4). The time course of the response to a two-fold increase in $V_{max,F}$ is shown in Fig. 5.29. Intermediates B, C, D and G all reach new steady states monotonously without oscillation. This also happens against the disturbance

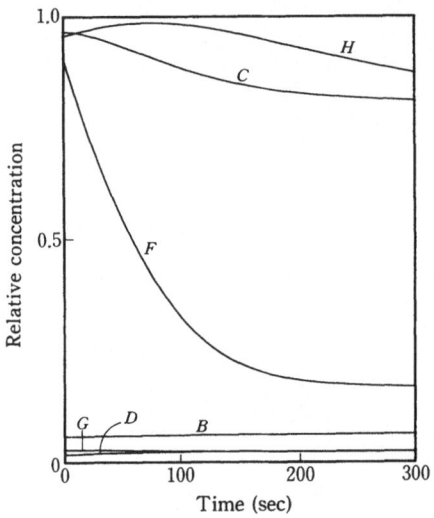

**Fig. 5. 29** Transient responses of sequential feedback inhibition (Scheme 5.9) to a disturbance (twofold increase in $V_{max,F}$).

**Table 5. 18** Steady-state values of sequential feedback inhibition (Scheme 5.9)

| Disturbances / Species | Standard | $V_{max,F} \times \frac{1}{2}$ | $V_{max,F} \times 2$ | $V_{max,H} \times \frac{1}{2}$ | $V_{max,H} \times 2$ | $A \times \frac{1}{2}$ | $A \times 2$ |
|---|---|---|---|---|---|---|---|
| $A$ | 10 | 10 | 10 | 10 | 10 | 5 | 20 |
| $B$ | 0. 0116 | 0. 0106 | 0. 0123 | 0. 0106 | 0. 0124 | 0. 00679 | 0. 0162 |
| $C$ | 0. 965 | 1. 15 | 0. 851 | 1. 16 | 0. 846 | 0. 474 | 1. 86 |
| $D$ | 0. 0108 | 0. 00775 | 0. 0131 | 0. 120 | 0. 010 | 0. 00635 | 0. 0149 |
| $F$ | 0. 906 | 2. 00 | 0. 395 | 1. 078 | 0. 790 | 0. 408 | 1. 67 |
| $G$ | 0. 0113 | 0. 0124 | 0. 0105 | 0. 0080 | 0. 0137 | 0. 00669 | 0. 0154 |
| $H$ | 0. 955 | 1. 13 | 0. 840 | 2. 120 | 0. 412 | 0. 432 | 1. 75 |
| $Y_{1A}$ | 0. 482 | 0. 438 | 0. 511 | 0. 437 | 0. 513 | 0. 296 | 0. 636 |
| $Y_{1C}$ | 0. 250 | 0. 296 | 0. 221 | 0. 298 | 0. 219 | 0. 223 | 0. 233 |
| $Y_{3C}$ | 0. 475 | 0. 342 | 0. 566 | 0. 521 | 0. 441 | 0. 290 | 0. 629 |
| $Y_{3F}$ | 0. 245 | 0. 435 | 0. 120 | 0. 244 | 0. 243 | 0. 203 | 0. 228 |
| $Y_{4C}$ | 0. 492 | 0. 533 | 0. 456 | 0. 350 | 0. 584 | 0. 302 | 0. 641 |
| $Y_{4H}$ | 0. 249 | 0. 247 | 0. 247 | 0. 441 | 0. 120 | 0. 210 | 0. 227 |

in pool A. Thus, this regulatory mechanism is considered more stable than the preceding two.

The steady-state values after the disturbances are listed in Table 5.18. Saturation functions for the three allosteric enzymes $E_1$, $E_3$ and $E_4$ are defined as follows:

$$
\left.
\begin{aligned}
Y_{1A} &= (R_1 + 2R_2)/2E_{10} \\
Y_{1C} &= (T_1 + 2T_2)/2E_{10} \\
Y_{3C} &= (R_1' + 2R_2')/2E_{30} \\
Y_{3F} &= (T_1' + 2T_2')/2E_{30} \\
Y_{4C} &= (R_1'' + 2R_2'')/2E_{40} \\
Y_{4H} &= (T_1'' + 2T_2'')/2E_{40}
\end{aligned}
\right\}
\tag{6}
$$

At the steady state before the disturbances each of the regulatory enzymes saturates its specific binding sites with the substrate and inhibitor by about 50% and 25%, respectively. Increase (or decrease) in the demand of one of the end products causes $Y_{1A}$ to increase (or decrease) and $Y_{1C}$ to decrease (or increase), respectively, resulting in the increase (or decrease) of $f$.

**Table 5.19** Responses of production rate to disturbances in sequential feedback inhibition

| Rate<br>Disturbances | $g_F$<br>$(\times 10^{-2})$ | $g_H$<br>$(\times 10^{-2})$ | $f = g_F + g_H$ |
|---|---|---|---|
| Standard | 0.950 (100) | 0.977 (100) | 1.927 (100) |
| $V_{max,F} \times \dfrac{1}{4}$ | 0.385 ( 41) | 1.115 (114) | 1.500 ( 78) |
| $''\quad \times \dfrac{1}{2}$ | 0.667 ( 70) | 1.060 (109) | 1.727 ( 90) |
| $''\quad \times 2$ | 1.132 (119) | 0.913 ( 93) | 2.045 (106) |
| $''\quad \times 10$ | 1.280 (135) | 0.857 ( 88) | 2.137 (111) |
| $V_{max,H} \times \dfrac{1}{4}$ | 1.093 (115) | 0.389 ( 40) | 1.482 ( 77) |
| $''\quad \times \dfrac{1}{2}$ | 1.038 (109) | 0.680 ( 70) | 1.718 ( 89) |
| $''\quad \times 2$ | 0.883 ( 93) | 1.167 (119) | 2.050 (106) |
| $''\quad \times 10$ | 0.825 ( 87) | 1.321 (135) | 2.146 (111) |
| $A \quad \times \dfrac{1}{2}$ | 0.580 ( 61) | 0.604 ( 62) | 1.184 ( 61) |
| $''\quad \times 2$ | 1.250 (132) | 1.272 (130) | 2.522 (131) |

( ) : Value relative to the standard

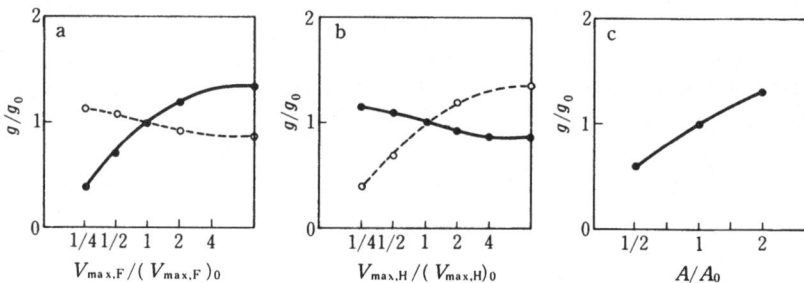

**Fig. 5.30** Responses of sequential feedback inhibition (Scheme 5.9) to disturbances. Notations are the same as in Fig. 5.26.

Similar changes are detected in the enzyme in the branch imposed with the disturbance. Smaller changes are found in the other branch.

Responding to the change in concentration of A, the saturation functions with the substrates vary considerably for each enzyme, indicating that the change in $A$ directly contributes to the variation in production rates of F and H. These are readily revealed in Table 5.19 and Fig. 5.30. The change in demand of an end product is compensated by the change in its production rate without affecting the production rate of the other end product. Changes in $g_F$ and $g_H$ due to the disturbances in $A$ are larger than those in the preceding two regulatory mechanisms, for the levels of end products do not directly regulate $f$. They directly control the activities of enzymes $E_3$ and $E_4$ at the first-step reactions in each branch, regulating the distributive ratio for the branches.

**Comparison of Basic Regulatory Mechanisms**

Multivalent and cumulative end-product inhibitions regulate only the flow rate $f$ from the pool, leaving a strong effect of the disturbances in one branch on the other branch. Sequential feedback inhibition regulates the enzymatic activities immediately after branching to control each branch separately. Effective control of $f$ is not always obtained through intermediate C.

The regulation of $f$ against the disturbances is compared in Fig. 5.31 for the basic regulatory mechanisms. Analysis in this section leads to the following conclusions:

1) Multivalent end-product inhibition yields the least change in $f$ against the concentration change of A.
2) Sequential feedback inhibition responds most effectively to the change in demand of either end product.
3) The objectives of a regulatory mechanism in a branched reaction sys-

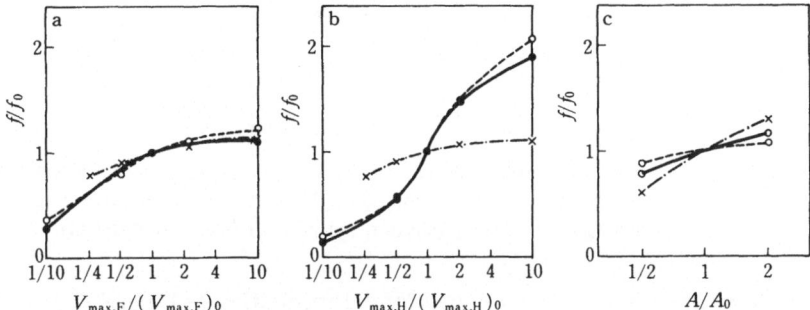

**Fig. 5.31** Comparison of the three regulatory mechanisms with respect to responses of $f$ to disturbances. $\bigcirc$ : Multivalent end-product inhibition (Scheme 5.8) ; $\bullet$ : Cumulative end-product inhibition (Scheme 5.8) ; $\times$ : Sequential feedback inhibition (Scheme 5.9). a) Change in $V_{max,F}$ ; b) Change in $V_{max,H}$ ; c) Change in $A$. Other notations are the same as in Fig. 5.26.

tem are to maintain a stable production of end products against disturbances and to respond flexibly to exogenous demand.

## 5.3 Coupled Reaction Systems

It commonly occurs in metabolic pathways that some enzymatic reactions are integrated to function properly in the cellular environments. A representative example is seen in the coupled reactions with high-energy compounds as intermediates in biosynthetic pathways. Most biosynthetic processes apparently proceed in the direction of an increase in free energy which violates thermodynamic laws. These reactions, however, are always coupled with reactions releasing free energy through high-energy compounds like ATP.

Moreover, in *in vitro* or laboratory (test-tube) systems coupled reaction

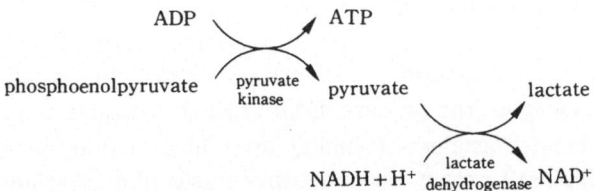

**Fig. 5.32** Coupled enzyme assay for pyruvate kinase. Time course of production of pyruvate is measured by absorbance of NADH (at 340 nm).

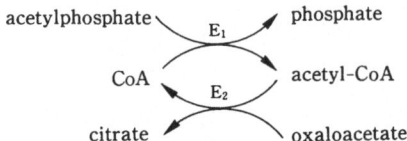

**Fig. 5. 33** Measurement of CoA by enzymatic cycling method. The amount of CoA is determined by the measurement of citrate accumulated through the cycling reaction between CoA and acetyl-CoA.

systems are frequently employed for determination of kinetic parameters of enzymes and ultra-microquantitative analysis of some biological substances in methods like coupled enzyme assay (Fig. 5.32) and enzymatic cycling method (Fig. 5.33). In this section we analyze the time courses of two substrates-two products reactions essential to coupled reactions, and then the dynamic characteristics of the coupled enzyme assay and enzymatic cycling method.

## 1. Two Substrates-Two Products Reactions

The Michaelis-Menten mechanism has so far in this book adopted the simplest two-step mechanism of single substrate-single product reaction. We now consider a reversible reaction converting two substrates, A and B, to two products, P and Q, as an elementary step of coupled reaction:

$$A+B \overset{(E)}{\rightleftharpoons} P+Q.$$

This type is common in oxidation-reduction and transfer reactions. Four principal mechanisms of the reactions have been proposed with the names of random Bi Bi, ordered Bi Bi, ping-pong Bi Bi and Theorell-Chance. These have been investigated in detail by Cleland [5] and Segel [6] with respect to the steady state kinetics and their identification using the steady state relationships. We here examine the time courses of these mechanisms to characterize the dynamic aspects of two substrate-two products reactions in closed system.

### Random Bi Bi Mechanism

The reaction mechanism is as given in Scheme 5.10. Two substrates bind with enzyme E in a random order, forming a three-body complex EAB. The complex undergoes the process, EAB ⇌ EPQ, to convert to complex EPQ, which dissociates products P and Q again in a random order. This mechanism is applied to many phosphotransferases like creatine kinase.

The time course of a reaction with the rate constants given in Scheme 5.10 is shown in Fig. 5.34. The rate-limiting step in the forward reaction is at the step EAB → EPQ. For 10 sec after the start of reaction, in which

$$E + A \underset{k_{-1}}{\overset{k_1}{\rightleftarrows}} EA \qquad E + B \underset{k_{-2}}{\overset{k_2}{\rightleftarrows}} EB \qquad EA + B \underset{k_{-3}}{\overset{k_3}{\rightleftarrows}} EAB$$

$$EB + A \underset{k_{-4}}{\overset{k_4}{\rightleftarrows}} EAB \qquad EAB \underset{k_{-5}}{\overset{k_5}{\rightleftarrows}} EPQ \qquad EPQ \underset{k_{-6}}{\overset{k_6}{\rightleftarrows}} EQ + P$$

$$EPQ \underset{k_{-7}}{\overset{k_7}{\rightleftarrows}} EP + Q \qquad EQ \underset{k_{-8}}{\overset{k_8}{\rightleftarrows}} E + Q \qquad EP \underset{k_{-9}}{\overset{k_9}{\rightleftarrows}} E + P$$

| | | | | |
|---|---|---|---|---|
| $k_1 = 5 \times 10^5$ | $k_3 = 10^6$ | $k_5 = 5 \times 10^3$ | $k_7 = 30$ | $k_9 = 20$ |
| $k_{-1} = 2.5$ | $k_{-3} = 100$ | $k_{-5} = 3$ | $k_{-7} = 10^3$ | $k_{-9} = 2 \times 10^3$ |
| $k_2 = 3.5 \times 10^5$ | $k_4 = 2.5 \times 10^6$ | $k_6 = 10^4$ | $k_8 = 10$ | |
| $k_{-2} = 30$ | $k_{-4} = 1.5 \times 10^2$ | $k_{-6} = 0.1$ | $k_{-8} = 5 \times 10^2$ | |

Scheme 5.10.  Random Bi Bi mechanism

the condition of excess substrate is valid, each enzyme species is in a steady state with almost constant concentration, resulting in a linear change in the concentrations of substrates and products. In addition, A and B follow an identical time course despite their difference in affinity to the enzyme. This is also the case for P and Q.

When the substrate concentrations are comparable to the enzyme concentration, the system follows a time course as shown in Fig. 5.35. From the early stage of reaction (0.1 sec) each enzyme species varies its concentration greatly without undergoing steady states. The concentration of A is

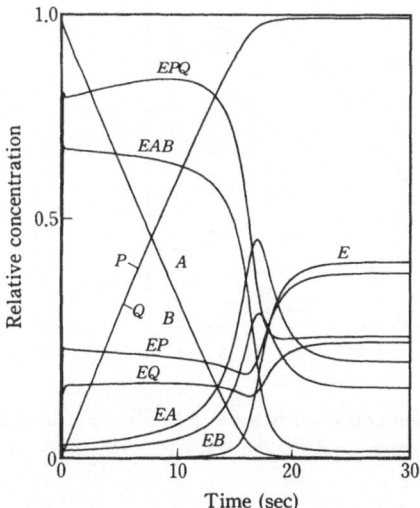

**Fig. 5. 34**  Time course of Scheme 5.10(I).  Initial condition : $A = B = 5$ mM, $E = 0.1$ mM.

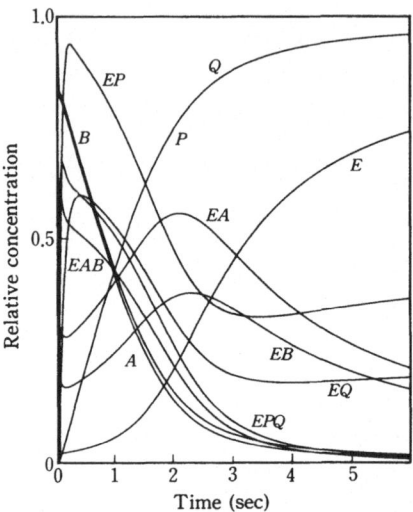

**Fig. 5. 35** Time course of Scheme 5.10(II). Initial condition: $A = B = 0.5$ mM, $E = 0.1$ mM.

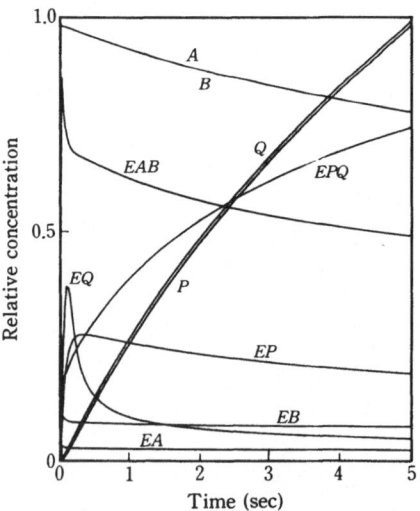

**Fig. 5. 36** Time course of Scheme 5.10(III). Initial condition: $A = B = 5$ mM, $E = 0.1$ mM. $k_{-6} = 8.0 \times 10^5$; $k_{-7} = 1.0 \times 10^5$; $k_{-8} = 5.0 \times 10^5$; $k_{-9} = 6.0 \times 10^5$ (in $\text{M}^{-1} \text{ sec}^{-1}$).

lower than that of B during the progress of reaction because A has a higher affinity to the enzyme than B.

The rate constants $k_{-6}$, $k_{-7}$, $k_{-8}$, and $k_{-9}$ in Scheme 5.10 have small values in the two cases above, in which the substrate concentrations decrease rapidly to zero. In contrast, Fig. 5.36 illustrates the time course of a system with these rate constants at large values. In the early stage of reaction $(0 \sim 2 \, \text{sec})$ complex EPQ increases its concentration, producing the products until equilibration of the system. A steady state as seen in Fig. 5.34, however, is not observable in this case.

## Ordered Bi Bi Mechanism

The reaction mechanism is as given in Scheme 5.11, in which binding of substrates A and B to enzyme E is ordered in formation of a three-body complex EAB so that formation of a complex EA always precedes the binding of B to the enzyme. The order of dissociation of P and Q is also definite. This mechanism is associated with many dehydrogenases requiring

$$E+A \underset{k_{-1}}{\overset{k_1}{\rightleftarrows}} EA \qquad EA+B \underset{k_{-2}}{\overset{k_2}{\rightleftarrows}} EAB \underset{k_{-3}}{\overset{k_3}{\rightleftarrows}} EPQ$$

$$EPQ \underset{k_{-4}}{\overset{k_4}{\rightleftarrows}} EQ+P \qquad EQ \underset{k_{-5}}{\overset{k_5}{\rightleftarrows}} E+Q$$

$$k_1 = 3 \times 10^5 \qquad k_2 = 2 \times 10^5 \qquad k_3 = 2 \qquad k_4 = 10 \qquad k_5 = 20$$
$$k_{-1} = 25 \qquad k_{-2} = 10 \qquad k_{-3} = 10^3 \qquad k_{-4} = 10^2 \qquad k_{-5} = 5 \times 10^2$$

Scheme 5.11. Ordered Bi Bi mechanism

coenzyme NAD. Incidentally, both random and ordered Bi Bi mechanisms are called single-substitution reaction due to the simultaneous binding of substrates A and B with the active site of an enzyme to form a three-body complex.

The time course of a reaction with the rate constants given in Scheme 5.11 is shown in Fig. 5.37. Immediately after the start of reaction, the system attains a steady state, in which EAB and EPQ have constant concentrations. Substrates A and B follow an identical time course. At the completion of reaction (about 36 sec) the concentrations of three-body complexes change stepwise. The rate-limiting step in the forward reaction is at the step EAB → EPQ, implying that most enzyme species take the form of EAB in a steady state.

A reaction with initial respective concentrations of $0.5 \, \text{mM}$ and $0.1 \, \text{mM}$ for the substrates and enzyme yields a time course in Fig. 5.38. The complex of the enzyme with A and B is very rapidly formed immediately after the start of reaction, resulting in apparently discontinuous changes in A and B. At $0.1 \, \text{sec}$ the concentrations of A and B reach $0.40 \, \text{mM}$ and $0.41$ mM, respectively, gradually reducing thereafter without attaining steady states. Products P and Q are produced with an initial time-delay.

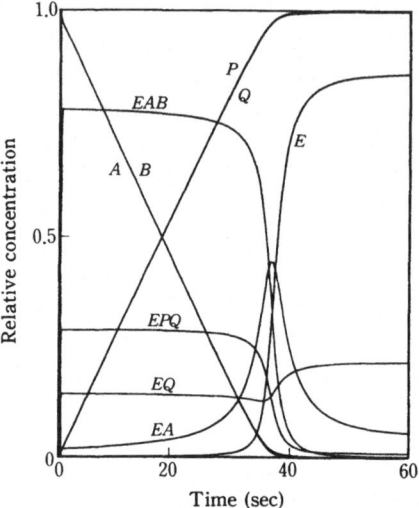

**Fig. 5.37**  Time course of Scheme 5.11(I).  Initial condition: $A = B = 5$ mM, $E = 0.1$ mM.

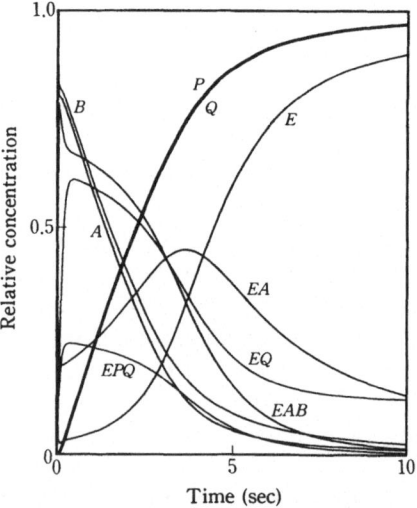

**Fig. 5.38**  Time course of Scheme 5.11(II).  Initial condition: $A = B = 0.5$ mM, $E = 0.1$ mM.

A reaction with large values for $k_{-4}$ and $k_{-5}$ in the reverse reaction proceeds on a time course as in Fig. 5.39. Similar to the case of random Bi Bi mechanism (Fig. 5.36), the effect of reverse reaction arises at a very

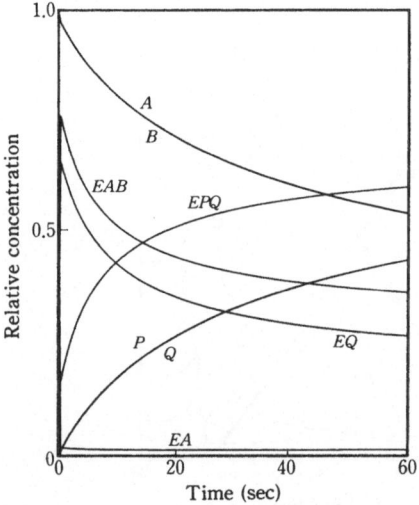

**Fig. 5. 39**  Time course of Scheme 5.11 (III). Initial condition : $A = B = 5$ mM, $E = 0.1$ mM. $k_{-4} = 1.0 \times 10^5$ ; $k_{-5} = 5.0 \times 10^5$ (in $M^{-1}$ sec$^{-1}$).

early stage (within 2 sec), causing the production rates to deviate from linearity.

**Ping-Pong Bi Bi Mechanism**

In this mechanism of Scheme 5.12 the first substrate A first binds with enzyme E to form a complex EA, which then is converted to a complex FP to dissociate into product P and modified enzyme F. The second substrate

$$E + A \xrightleftharpoons[k_{-1}]{k_1} EA \xrightleftharpoons[k_{-2}]{k_2} FP \xrightleftharpoons[k_{-3}]{k_3} F + P$$

$$F + B \xrightleftharpoons[k_{-4}]{k_4} FB \xrightleftharpoons[k_{-5}]{k_5} EQ \xrightleftharpoons[k_{-6}]{k_6} E + Q$$

$k_1 = 3 \times 10^6$  $k_2 = 5$  $k_3 = 30$  $k_4 = 3.5 \times 10^5$  $k_5 = 1$  $k_6 = 35$

$k_{-1} = 25$  $k_{-2} = 1$  $k_{-3} = 2.5 \times 10^3$  $k_{-4} = 30$  $k_{-5} = 0.5$  $k_{-6} = 3 \times 10^3$

Scheme 5. 12.  Ping-pong Bi Bi mechanism

B binds with F to form a complex FB, which is converted to complex EQ to dissociate into E and product Q. This mechanism is different from the preceding two in that a three-body complex is not formed and that the reaction from A to P proceeds without participation of the second substrate B. The mechanism thus is called a double substitution reaction ; a well-known example of this mechanism is aspartate transaminase.

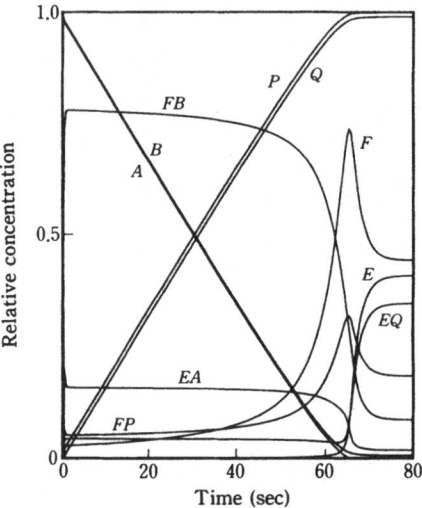

**Fig. 5.40**   Time course of Scheme 5.12(I)   Initial condition: $A = B =$ 5 mM, $E = 0.1$ mM.

The time course of a reaction with the rate constants given in Scheme 5.12 is shown in Fig. 5.40.  The concentrations of P and Q become 0.14 mM and 0.06 mM, respectively, at 1.0 sec after the start of reaction, that is, about twice as much P is produced as Q.   As the reaction advances, however, the concentrations of P and Q become almost equal, for example, $P = 3.87$ mM and $Q = 3.79$ mM at 50.5 sec.  During about 60 sec after the start of reaction, no enzyme species vary their concentrations with time, staying in a steady state.  The rate-limiting step in the forward reaction is at the step FB → EQ, keeping most enzyme species in the FB-form.

A reaction with comparable concentrations of substrates and enzyme yields a time course in Fig. 5.41.  In contrast to the case of excess substrates (Fig. 5.40), the reaction distinctly displays a burst in production of P and a time delay in production of Q.  The production rate of P is 0.27 mMsec$^{-1}$ between the start of reaction and 0.2 sec, reducing by three quarters to 0.067 mMsec$^{-1}$ at 0.6 sec,   which is maintained until almost the end of reaction.   The production rate of Q has a time delay of 0.22 sec, then reaches 0.065 mMsec$^{-1}$.  The system apparently is in a steady state between 0.6 sec and about 7 sec (the end of reaction), because the enzyme species do not vary their concentrations during that period.

A reaction with large rate constants of $k_{-3} = 250$ and $k_{-6} = 300$ (in mM$^{-1}$ sec$^{-1}$) in the reverse reaction follows a time course in Fig. 5.42.  The

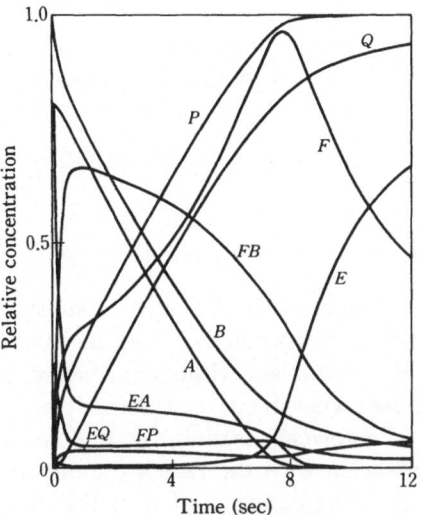

**Fig. 5. 41** Time course of Scheme 5.12(II). Initial condition: $A=B=0.5$ mM, $E=0.1$ mM

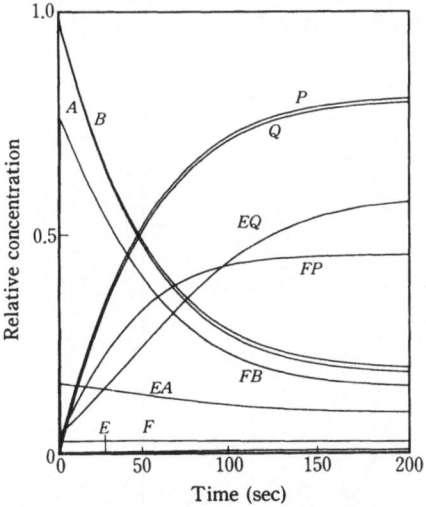

**Fig. 5. 42** Time course of Scheme 5.12(III). Initial condition: $A=B=5$ mM, $E=0.1$ mM. $k_{-3}=2.5\times10^5$; $k_{-6}=3.0\times10^5$ (in M$^{-1}$ sec$^{-1}$).

concentration of complex EA remains almost constant until the end of reaction. Complex FB, however, changes its concentration sharply. The concentrations of EQ and FP also vary considerably to let the system stay in a nonsteady state.

## Theorell-Chance Mechanism

This mechanism of Scheme 5.13 was proposed by Theorell and Chance [7] for alcohol dehydrogenase. Substrate A binds with the enzyme to form a

$$E+A \underset{k_{-1}}{\overset{k_1}{\rightleftharpoons}} EA \qquad EA+B \underset{k_{-2}}{\overset{k_2}{\rightleftharpoons}} EQ+P \qquad EQ \underset{k_{-3}}{\overset{k_3}{\rightleftharpoons}} E+Q$$

$$k_1=3\times10^5 \qquad k_{-1}=25 \qquad k_2=5\times10^3 \qquad k_{-2}=0.1 \qquad k_3=2 \qquad k_{-3}=50$$

Scheme 5.13. Theorell-Chance mechanism

complex EA, which then reacts with the second substrate B in a bi-molecular process to produce the products. The mechanism thus corresponds to the ordered Bi Bi mechanism without formation of a three-body complex. In general, precise experiments can detect the three-body complexes, leaving now only a historical meaning for this mechanism.

The time course of a reaction with the rate constants given in Scheme 5.13 is shown in Fig. 5.43. The concentration of EA reaches a peak at about 30 sec, when the steady state in the early stage of reaction disappears. The rate-limiting step in the forward reaction is at the step $EQ \rightarrow E+Q$. A similar time course is obtained with a reaction in which the rate-limiting step is at the step $EA+B \rightarrow EQ+P$.

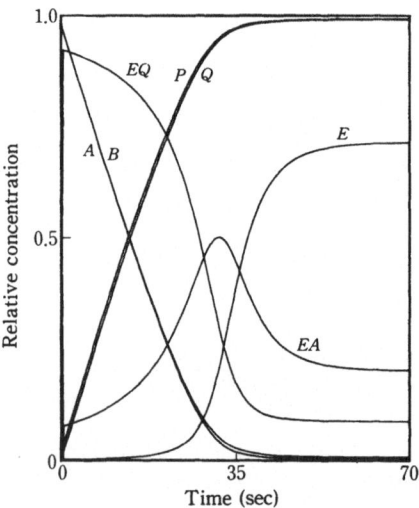

**Fig. 5.43** Time course of Scheme 5.13. Initial condition : $A=B=5$ mM, $E=0.5$ mM.

## 2. Coupled Enzyme Assay

Kinetic parameters such as Michaelis constant and maximum velocity in enzymatic reaction can be determined readily and quickly if the substrate and product have specific colors or absorb lights of specific wavelengths. In fact, many reactions of oxidation-reduction enzymes require association with coenzyme NAD so that the progress curve of reaction is easily obtained by spectrophotometric measurement of the concentration change of reduced coenzyme NADH which has absorbance at 340 nm. Furthermore, the progress curve of reaction in which neither substrates nor products absorb specific lights can also be readily obtained by coupling the reaction with one having light‑absorbing substance as a substrate or product. This method is called coupled enzyme assay, an example of which is given in Fig. 5.32.

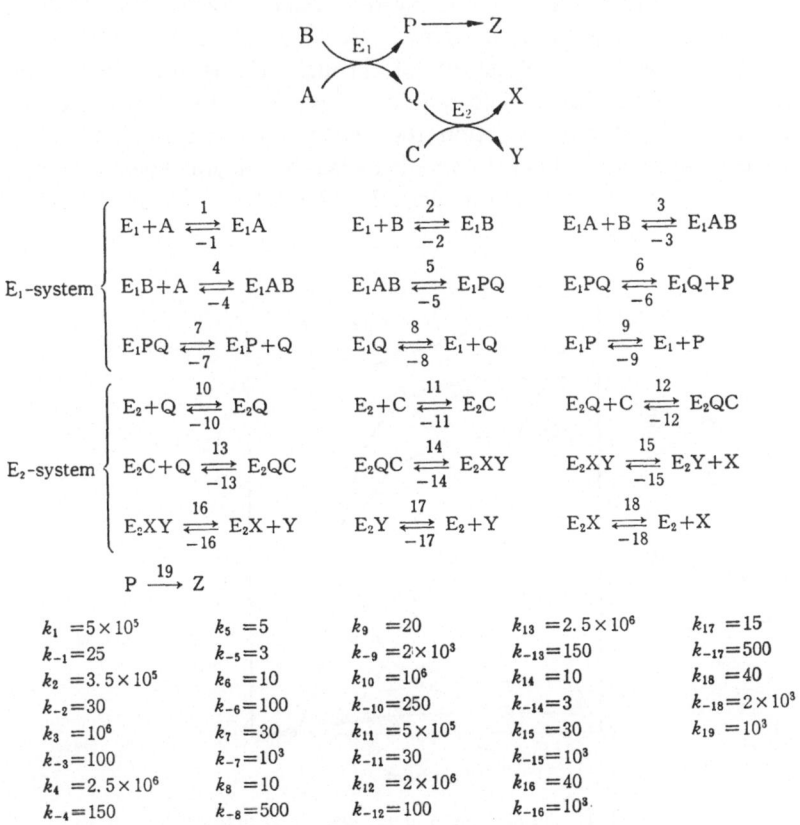

$E_1$-system $\begin{cases}
E_1+A \underset{-1}{\overset{1}{\rightleftarrows}} E_1A & E_1+B \underset{-2}{\overset{2}{\rightleftarrows}} E_1B & E_1A+B \underset{-3}{\overset{3}{\rightleftarrows}} E_1AB \\
E_1B+A \underset{-4}{\overset{4}{\rightleftarrows}} E_1AB & E_1AB \underset{-5}{\overset{5}{\rightleftarrows}} E_1PQ & E_1PQ \underset{-6}{\overset{6}{\rightleftarrows}} E_1Q+P \\
E_1PQ \underset{-7}{\overset{7}{\rightleftarrows}} E_1P+Q & E_1Q \underset{-8}{\overset{8}{\rightleftarrows}} E_1+Q & E_1P \underset{-9}{\overset{9}{\rightleftarrows}} E_1+P
\end{cases}$

$E_2$-system $\begin{cases}
E_2+Q \underset{-10}{\overset{10}{\rightleftarrows}} E_2Q & E_2+C \underset{-11}{\overset{11}{\rightleftarrows}} E_2C & E_2Q+C \underset{-12}{\overset{12}{\rightleftarrows}} E_2QC \\
E_2C+Q \underset{-13}{\overset{13}{\rightleftarrows}} E_2QC & E_2QC \underset{-14}{\overset{14}{\rightleftarrows}} E_2XY & E_2XY \underset{-15}{\overset{15}{\rightleftarrows}} E_2Y+X \\
E_2XY \underset{-16}{\overset{16}{\rightleftarrows}} E_2X+Y & E_2Y \underset{-17}{\overset{17}{\rightleftarrows}} E_2+Y & E_2X \underset{-18}{\overset{18}{\rightleftarrows}} E_2+X
\end{cases}$

$$P \xrightarrow{19} Z$$

| | | | | |
|---|---|---|---|---|
| $k_1 =5\times10^5$ | $k_5 =5$ | $k_9 =20$ | $k_{13} =2.5\times10^6$ | $k_{17} =15$ |
| $k_{-1}=25$ | $k_{-5}=3$ | $k_{-9}=2\times10^3$ | $k_{-13}=150$ | $k_{-17}=500$ |
| $k_2 =3.5\times10^5$ | $k_6 =10$ | $k_{10} =10^6$ | $k_{14} =10$ | $k_{18} =40$ |
| $k_{-2}=30$ | $k_{-6}=100$ | $k_{-10}=250$ | $k_{-14}=3$ | $k_{-18}=2\times10^3$ |
| $k_3 =10^6$ | $k_7 =30$ | $k_{11} =5\times10^5$ | $k_{15} =30$ | $k_{19} =10^3$ |
| $k_{-3}=100$ | $k_{-7}=10^3$ | $k_{-11}=30$ | $k_{-15}=10^3$ | |
| $k_4 =2.5\times10^6$ | $k_8 =10$ | $k_{12} =2\times10^6$ | $k_{16} =40$ | |
| $k_{-4}=150$ | $k_{-8}=500$ | $k_{-12}=100$ | $k_{-16}=10^3$ | |

Scheme 5.14. Model of coupled enzyme assay

We here examine the dynamic behavior and accuracy of coupled enzyme assay in the determination of Michaelis constant $K_m$ and maximum velocity $V_{max}$ for enzyme $E_1$ catalyzing a two substrates-two products reaction in the random Bi Bi mechanism. The model and rate constants for coupled enzyme assay are as given in Scheme 5.14. The assay is performed to determine $K_{m,B}$ and $V_{max,B}$ of $E_1$ for substrate B under the excess condition of substrate A. Product Q works as a common intermediate, coupling the $E_2$-system with the $E_1$-system. Substrate C is supplied in excess in the $E_2$-system. Product P is converted to Z at a high rate in an irreversible unimolecular reaction, because the accumulation of P would cause undesirable effects on the system through reversible reaction. The rate-limiting steps are at the respective steps of isomerization of three-body complexes, i.e., $E_1AB \to E_1PQ$ in the $E_1$-system and $E_2QC \to E_2XY$ in the $E_2$-system.

In order to have the standard values of $K_{m,B}$ and $V_{max,B}$ for later comparison, the time course of the $E_1$-system alone in Scheme 5.14 is obtained under the excess condition of substrate A (Fig. 5.44). The total concentration of enzyme $E_1$ is set as $10\ \mu M$ throughout the following discussion on calculation of $V_{max,B}$. The reaction stays in a steady state until about 55 sec after the start of reaction, that is, the concentration of substrate B decreases linearly to about 0.4 mM because the enzyme species do not vary their concentrations during this period. The data of $\{B(t),\ dP(t)/dt\}$ at

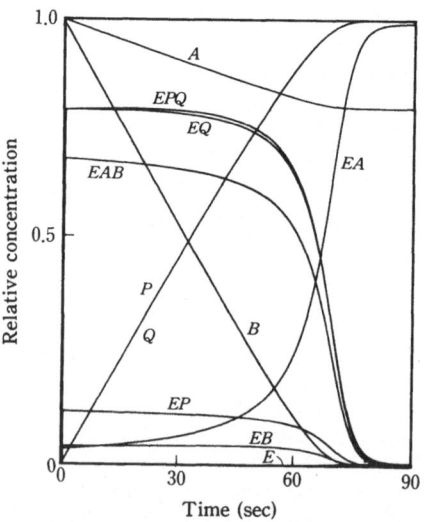

**Fig. 5.44** Time course of $E_1$-system of Scheme 5.14. Initial condition: $A = 10$ mM, $B = 2$ mM, $E_1 = 10\mu M$.

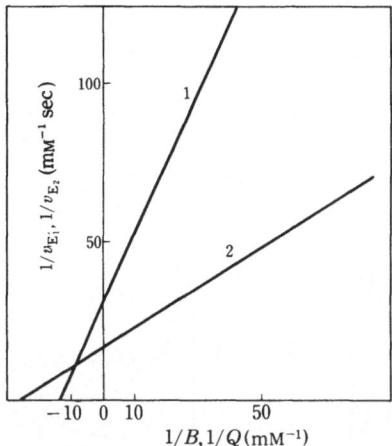

**Fig. 5. 45**  Lineweaver - Burk plot for Scheme 5.14. Initial condition: $A = 10$ mM, $B = 2$ mM, $E_1 = 10\mu$M for $E_1$-system; $C = 10$ mM, $Q = 2$ mM, $E_2 = 10\mu$M for $E_2$-system. 1: $E_1$-system, inverse of $v_{E_1}$ ($= dP/dt$) vs. $1/B$; 2: $E_2$-sysem, inverse of $v_{E_2}$ ($= dX/dt$) vs. $1/Q$.

each time-step obtained from the time course yield a Lineweaver–Burk plot in Fig. 5.45. The linear regression line gives

$$K_{m,B} = 72.6\ \mu\text{M} \quad \text{and} \quad V_{max,B} = 32.4\ \mu\text{Msec}^{-1} \tag{7}$$

with correlation coefficient 0.99999.

For the $E_2$-system alone, $K_{m,Q}$ and $V_{max,Q}$ for substrate Q are similarly determined with the total concentration in enzyme $E_2$ of $10\ \mu$M. A Lineweaver–Burk plot is drawn again in Fig. 5.45, leading to

$$K_{m,Q} = 38.7\ \mu\text{M} \quad \text{and} \quad V_{max,Q} = 60.6\ \mu\text{Msec}^{-1} \tag{8}$$

from the linear regression line with correlation coefficient 0.99997.

In a coupled enzyme assay, the enzymatic reaction for which kinetic parameters are to be determined (*i.e.*, the $E_1$-system in this discussion) must be rate-limiting in the coupled reaction system. The comparison of $V_{max,B}$ and $V_{max,Q}$ given in equations (7) and (8) thus implies that the concentration of enzyme $E_2$ must be at least $5.4\ \mu$M for $10\ \mu$M of enzyme $E_1$. The coupled enzyme assay for determination of $K_{m,B}$ and $V_{max,B}$ is simulated with variation in the concentration of enzyme $E_2$. Figure 5.46 illustrates the time course of the system with $5\ \mu$M of $E_2$, in which the $E_2$-system is rate-limiting because the maximum velocities are 32.4 and 30.3 (in $\mu$Msec$^{-1}$) for the $E_1$- and $E_2$-systems, respectively. End products X and Y of the $E_2$-system are produced with a time delay of 4 sec and at a slower rate than

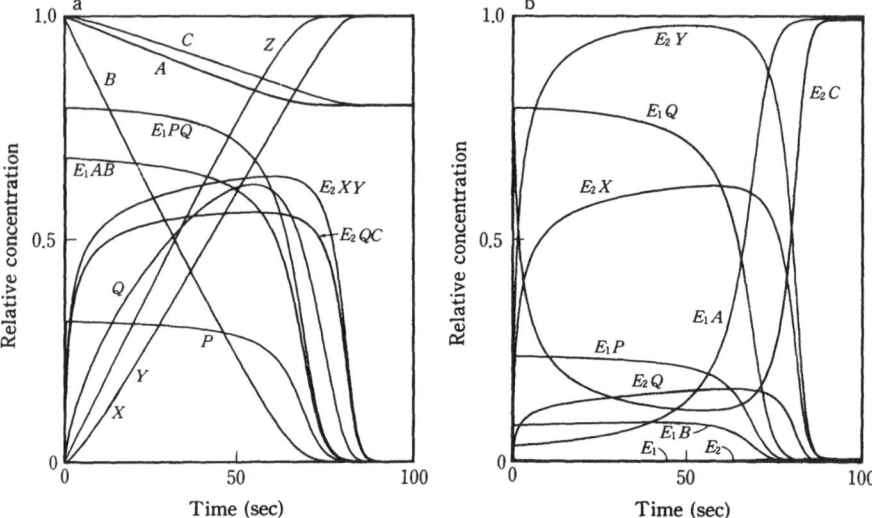

**Fig. 5.46** Time course of Scheme 5.14(I). Initial condition: $A = 10$ mM, $B = 2$ mM, $E_1 = 10\mu$M, $C = 10$ mM, $E_2 = 5\mu$M.

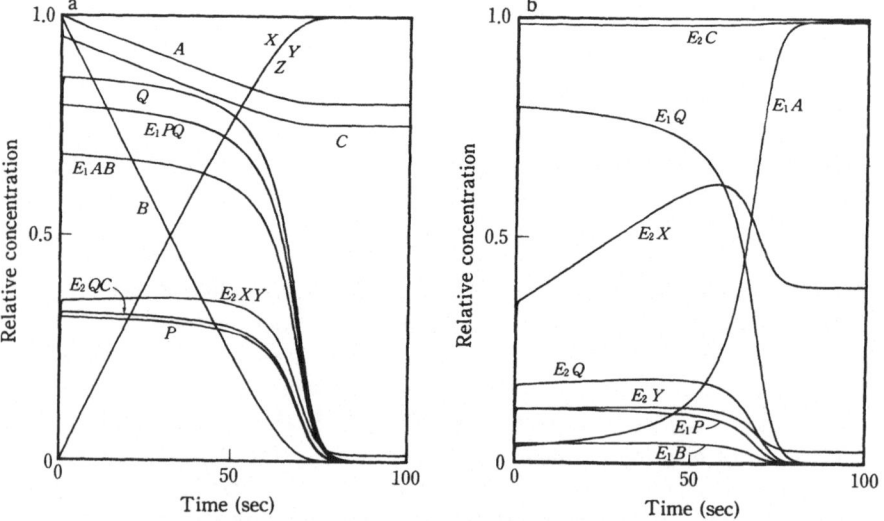

**Fig. 5.47** Time course of Scheme 5.14(II). Initial condition: $A = 10$ mM, $B = 2$ mM, $E_1 = 10\mu$M, $C = 10$ mM, $E_2 = 0.5$ mM.

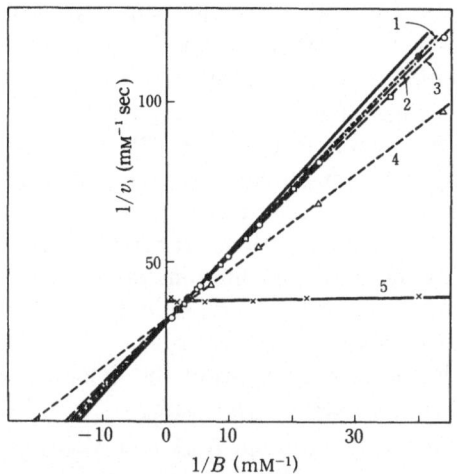

**Fig. 5.48** Lineweaver-Burk plot for Scheme 5.14. Solid line yields the true $K_{m,B}$ and $V_{max,B}$. $v_Y$ is obtained from simulation. Total concentration of $E_2$: 1: 0.5 mM; 2: 0.1 mM; 3: 50 $\mu$M; 4: 10 $\mu$M; 5: 5 $\mu$M.

end product Z of the $E_1$-system. Common intermediate Q is accumulated gradually in the system, reaching a maximum concentration of 0.311 mM around the end of reaction.

The time course of the system with 0.5 mM of $E_2$ is shown in Fig. 5.47. Maximum velocity of the $E_2$-system becomes 3.03 mM sec$^{-1}$ ($V_{max,Q} = 3.03$ mMsec$^{-1}$), about 100 times higher than that of the $E_2$-system. End products Z, X and Y of both systems follow an identical progress-curve. Common intermediate Q remains almost constant (about 0.4 $\mu$M) until about 50 sec after the start of reaction.

In Fig. 5.48 Lineweaver-Burk plots are drawn using the data from simulation of the coupled enzyme assay with variation in the concentrations of $E_2$ between 0.005 and 0.5 (in mM). In case (5), in which the $E_1$-system is

**Table 5.20** Determination of $K_{m,B}$ and $V_{max,B}$ by coupled enzyme assay

| Case | $E_2$(mM) | $v_Y^{-1} = aB^{-1} + b$ | | $K_{m,B}$ (mM) | $V_{max,B}$ (mM sec$^{-1}$) |
|---|---|---|---|---|---|
| | | $a$ | $b$ | | |
| 1 | 0.5 | 2.12 | 30.8 | 0.0689 | 0.0325 |
| 2 | 0.1 | 2.06 | 31.1 | 0.0663 | 0.0322 |
| 3 | 0.05 | 2.02 | 31.1 | 0.0650 | 0.0322 |
| 4 | 0.01 | 1.52 | 31.7 | 0.0478 | 0.0315 |
| 5 | 0.005 | — | — | — | — |

not rate-limiting in the assay system, the plot is nearly parallel to the $1/B$-axis, offering no regression line for $K_{m,B}$ and $V_{max,B}$. In other cases the regression lines are obtained with correlation coefficients better than 0.999.

Table 5.20 lists the values of $K_{m,B}$ and $V_{max,B}$ of the $E_1$-system obtained from the Lineweaver-Burk plots in Fig. 5.48. The accuracy in determination of $V_{max,B}$ is equally good for all cases $(1) \sim (4)$ in which $V_{max,Q}$ of the $E_2$-system ranges from about 2 to 100 times higher than $V_{max,B}$. This is also suggested from almost equal intersections on the $1/v$-axis for the plots in Fig. 5.48. On the other hand, $K_{m,B}$ has different values in each case. In case (4) the error amounts to about 34%. Increase in $V_{max,Q}$ of the $E_2$-system is still ineffective in determining the true value.

The following conclusion is derived from the dynamic analysis of the model in Scheme 5.14 for coupled enzyme assay:

1) If the $E_1$-system is not rate-limiting, $K_{m,B}$ and $V_{max,B}$ cannot be determined.
2) As long as the $E_1$-system is rate-limiting, $V_{max,B}$ is determined with comparative accuracy.
3) The true value of $K_{m,B}$, however, is rather difficult to be determined even with the higher maximum velocity of the $E_2$-system.

## 3. Enzymatic Cycling Method

A system for the enzymatic cycling method consists of a combination of two enzymatic reactions with amplifying capacity. Cycling reaction is called linear or exponential, depending on the mode of amplification. A linear cycling reaction, as shown in Fig. 5.33, is employed in a method of ultra-microanalysis, which is usually called the enzymatic cycling method and is currently applied mainly to neurochemistry. The behavior of similar reaction systems is treated macroscopically in Section 6.1.

Scheme 5.15 provides a model for the enzymatic cycling method, in which the reactions proceed with excess substrates A and X to measure the accumulated amount of product D or Y for quantitative determination of the very small amount of substances B and C in the system. The time course of an assay is shown in Fig. 5.49. The total concentration of each of the enzymes in the $E_1$- and $E_2$-systems is set equal to $10\ \mu M$ (i.e., $E_{10} = E_{20} = 10\ \mu M$). A similar procedure (as in Fig. 5.45) is employed to determine Michaelis constants $K_{m,B}$ and $K_{m,C}$ and maximum velocities $V_{max,B}$ and $V_{max,C}$; we have

$$\left. \begin{array}{l} K_{m,B} = 5.09\,\mu M, \quad V_{max,B} = 50.4\,\mu M\,sec^{-1} \\ K_{m,C} = 5.49\,\mu M, \quad V_{max,C} = 27.0\,\mu M\,sec^{-1} \end{array} \right\} \tag{9}$$

for the $E_1$- and $E_2$-systems, respectively.

$E_1$-system
$$
\begin{cases}
E_1+A \underset{-1}{\overset{1}{\rightleftharpoons}} E_1A & E_1+B \underset{-2}{\overset{2}{\rightleftharpoons}} E_1B & E_1A+B \underset{-3}{\overset{3}{\rightleftharpoons}} E_1AB \\
E_1B+A \underset{-4}{\overset{4}{\rightleftharpoons}} E_1AB & E_1AB \underset{-5}{\overset{5}{\rightleftharpoons}} E_1CD & E_1CD \underset{-6}{\overset{6}{\rightleftharpoons}} E_1D+C \\
E_1CD \underset{-7}{\overset{7}{\rightleftharpoons}} E_1C+D & E_1D \underset{-8}{\overset{8}{\rightleftharpoons}} E_1+D & E_1C \underset{-9}{\overset{9}{\rightleftharpoons}} E_1+C
\end{cases}
$$

$E_2$-system
$$
\begin{cases}
E_2+C \underset{-10}{\overset{10}{\rightleftharpoons}} E_2C & E_2+X \underset{-11}{\overset{11}{\rightleftharpoons}} E_2X & E_2C+X \underset{-2}{\overset{12}{\rightleftharpoons}} E_2CX \\
E_2X+C \underset{-13}{\overset{13}{\rightleftharpoons}} E_2CX & E_2CX \underset{-14}{\overset{14}{\rightleftharpoons}} E_2BY & E_2BY \underset{-15}{\overset{15}{\rightleftharpoons}} E_2Y+B \\
E_2BY \underset{-16}{\overset{16}{\rightleftharpoons}} E_2B+Y & E_2Y \underset{-17}{\overset{17}{\rightleftharpoons}} E_2+Y & E_2B \underset{-18}{\overset{18}{\rightleftharpoons}} E_2+B
\end{cases}
$$

| | | | | |
|---|---|---|---|---|
| $k_1 = 10^6$ | $k_5 = 10$ | $k_9 = 20$ | $k_{13} = 2.5 \times 10^6$ | $k_{17} = 10$ |
| $k_{-1} = 250$ | $k_{-5} = 3$ | $k_{-9} = 200$ | $k_{-13} = 15$ | $k_{-17} = 5$ |
| $k_2 = 5 \times 10^5$ | $k_6 = 50$ | $k_{10} = 5 \times 10^5$ | $k_{14} = 5$ | $k_{18} = 20$ |
| $k_{-2} = 30$ | $k_{-6} = 100$ | $k_{-10} = 25$ | $k_{-14} = 30$ | $k_{-18} = 2$ |
| $k_3 = 2 \times 10^6$ | $k_7 = 30$ | $k_{11} = 3.5 \times 10^5$ | $k_{15} = 50$ | |
| $k_{-3} = 10$ | $k_{-7} = 100$ | $k_{-11} = 30$ | $k_{-15} = 100$ | |
| $k_4 = 2.5 \times 10^6$ | $k_8 = 10$ | $k_{12} = 10^6$ | $k_{16} = 30$ | |
| $k_{-4} = 15$ | $k_{-8} = 500$ | $k_{-12} = 10$ | $k_{-16} = 10$ | |

Scheme 5.15. Enzymatic cycling method

The system attains a steady state immediately after the start of reaction, remaining in that state until almost the end of reaction. During this period, common intermediates B and C as well as all enzyme species keep their concentrations virtually constant, letting the reaction velocity in the $E_1$-system be almost equal to that in the $E_2$-system to maintain a steady-state rate in cycling between B and C. The concentrations of products D and Y follow an identical time course, increasing linearly with time.

In steady states the temporal change in concentration of products D and Y is represented by

$$k_c Z t, \tag{10}$$

where $k_c$ denotes the rate (in sec$^{-1}$) of cycling between B and C, and $Z = B + C$. $k_c$ is called the cycling rate or amplifying rate, indicating how

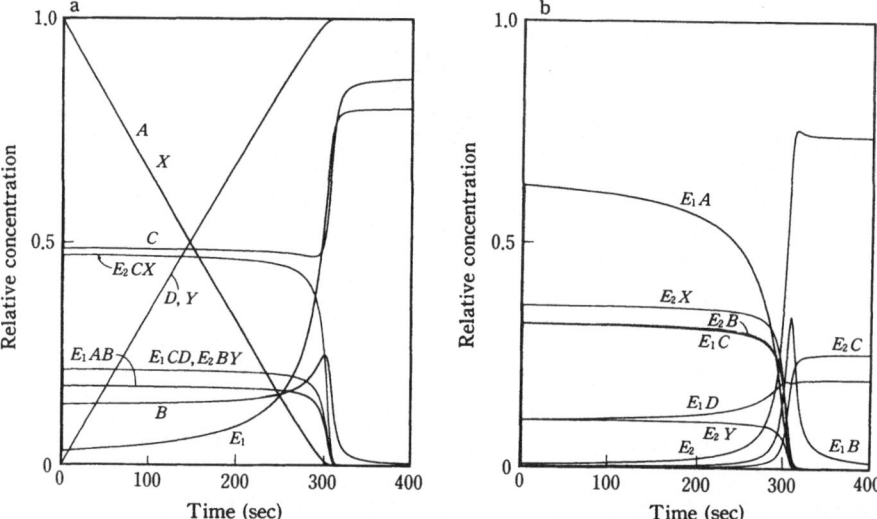

**Fig. 5. 49** Time course of Scheme 5.15. Initial condition: $A = 5$ mM, $C = 20\mu$M, $X = 5$ mM, $E_1 = E_2 = 10\mu$M.

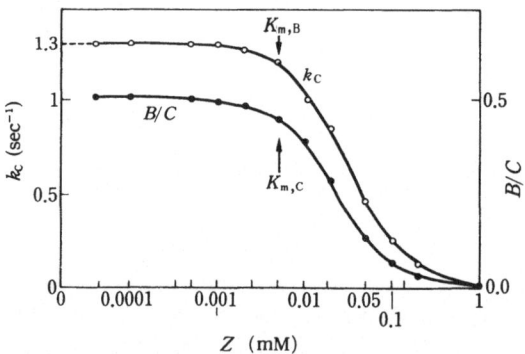

**Fig. 5. 50** Cycling rate of Scheme 5. 15(I). Initial condition: $A = X = 5$ mM, $E_1 = E_2 = 10\mu$M. $K_{m,B} = 5.09\mu$M; $K_{m,C} = 5.49\mu$M.

many fold amount of $Z$ the production of D or Y in a unit time corresponds to. From a simulation such as in Fig. 5.49, $k_C$ in expression (10) and concentration ratio $B/C$ are derived for steady states with various values of $Z$. The results are summarized in Fig. 5.50.

When the amount of $Z$ in the system is larger than either of the Michaelis constants $K_{m,B}$ and $K_{m,C}$, both $B/C$ and $k_C$ are dependent on $Z$. On the other hand, with $Z$ less than either $K_{m,B}$ and $K_{m,C}$, $k_C$ and $B/C$ are virtually constant and independent of $Z$. This implies that, if the system contains an amount of $Z$ less than $K_{m,B}$ and $K_{m,C}$, measurement of the accumulated amount of D or Y leads to an accurate determination of $Z$ in the system using the relationship (10) and $k_C$, which is evaluated separately as described below.

The most essential condition for the enzymatic cycling method is maintenance of a steady state for a long period with excess substrates A and X in the system. Under this condition measurement is carried out more easily due to the increased concentration in the products. The relationship between cycling rate and kinetic parameters is derived using the $S\text{-}v$ relationship in quasi-steady-states as follows. The $S\text{-}v$ relationship for enzyme $E_1$ and substrate B in Scheme 5.15 employs not the Michaelis-Menten equation in common use but the Cha-Cha equation,

$$v=\frac{k_1 E_{10} B}{K_{m,B}+E_{10}+B},$$                               (11)

which is valid in the case of high enzyme concentration [8]. $k_1 E_{10}$ corresponds to the maximum velocity $V_{max,B}$. When the cycling reactions are in a steady state, both the reaction rates of the $E_1$- and $E_2$-systems are equal to $k_C Z$, that is,

$$k_C Z=\frac{k_1 E_{10} B}{K_{m,B}+E_{10}+B}=\frac{k_2 E_{20} C}{K_{m,C}+E_{20}+C}.$$                               (12)

Accurate measurement in the enzymatic cycling method is obtained in the

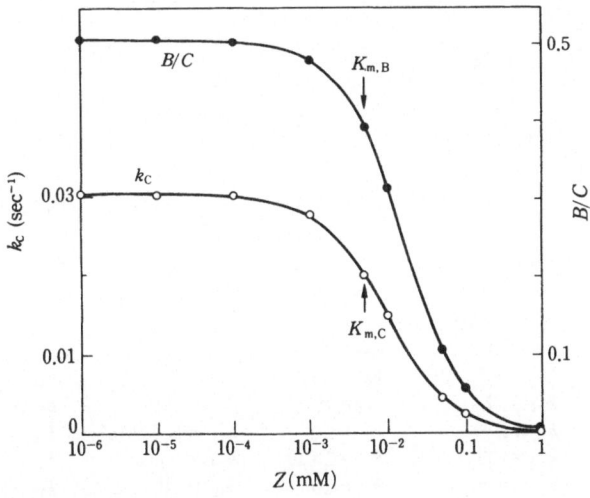

**Fig. 5.51** Cycling rate of Scheme 5.15(II). Initial condition: $A=X=5$ mM, $E_1=E_2=0.1\mu$M.

range where $k_C$ is independent of $Z$, that is, under the condition that the amount of $Z$ is negligible compared to $K_{m,B}$ and $K_{m,C}$, *i.e.*,

$$Z \ll K_{m,B} \quad \text{and} \quad Z \ll K_{m,C}. \tag{13}$$

Under this condition we have the relationship,

$$\frac{1}{k_C} = \frac{1}{k_1} + \frac{1}{k_2} + \frac{K_{m,B}}{k_1 E_{10}} + \frac{K_{m,C}}{k_2 E_{20}}, \tag{14}$$

from equation (12) and $B + C = Z$. Substitution of the kinetic parameters of equation (9) into equation (14) yields $k_C = 1.15 \, \text{sec}^{-1}$, which is in good agreement with the value in Fig. 5.50.

On the other hand, when the enzyme concentrations are sufficiently lower than the respective Michaelis constants, the first and second terms on the right-hand side of equation (14) are negligible, leading to the equation,

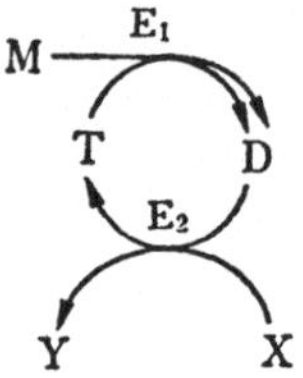

$E_1$-system

$$E_1 + M \underset{-1}{\overset{1}{\rightleftharpoons}} E_1M \qquad E_1 + T \underset{-2}{\overset{2}{\rightleftharpoons}} E_1T \qquad E_1M + T \underset{-3}{\overset{3}{\rightleftharpoons}} E_1MT$$

$$E_1T + M \underset{-4}{\overset{4}{\rightleftharpoons}} E_1MT \qquad E_1MT \underset{-5}{\overset{5}{\rightleftharpoons}} E_1DD \qquad E_1DD \underset{-6}{\overset{6}{\rightleftharpoons}} E_1D + D$$

$$E_1D \underset{-7}{\overset{7}{\rightleftharpoons}} E_1 + D$$

$E_2$-system

$$E_2 + D \underset{-8}{\overset{8}{\rightleftharpoons}} E_2D \qquad E_2 + X \underset{-9}{\overset{9}{\rightleftharpoons}} E_2X \qquad E_2D + X \underset{-10}{\overset{10}{\rightleftharpoons}} E_2DX$$

$$E_2X + D \underset{-11}{\overset{11}{\rightleftharpoons}} E_2DX \qquad E_2DX \underset{-12}{\overset{12}{\rightleftharpoons}} E_2TY \qquad E_2TY \underset{-13}{\overset{13}{\rightleftharpoons}} E_2Y + T$$

$$E_2TY \underset{-14}{\overset{14}{\rightleftharpoons}} E_2T + Y \qquad E_2Y \underset{-15}{\overset{15}{\rightleftharpoons}} E_2 + Y \qquad E_2T \underset{-16}{\overset{16}{\rightleftharpoons}} E_2 + T$$

| | | | |
|---|---|---|---|
| $k_1 = 5 \times 10^5$ | $k_5 = 5$ | $k_9 = 5 \times 10^5$ | $k_{13} = 50$ |
| $k_{-1} = 25$ | $k_{-5} = 3$ | $k_{-9} = 30$ | $k_{-13} = 10^3$ |
| $k_2 = 3.5 \times 10^5$ | $k_6 = 10$ | $k_{10} = 2 \times 10^6$ | $k_{14} = 40$ |
| $k_{-2} = 30$ | $k_{-6} = 10^3$ | $k_{-10} = 100$ | $k_{-14} = 10^3$ |
| $k_3 = 10^6$ | $k_7 = 20$ | $k_{11} = 2.5 \times 10^6$ | $k_{15} = 15$ |
| $k_{-3} = 100$ | $k_{-7} = 2 \times 10^3$ | $k_{-11} = 150$ | $k_{-15} = 500$ |
| $k_4 = 2.5 \times 10^5$ | $k_8 = 10^6$ | $k_{12} = 10$ | $k_{16} = 40$ |
| $k_{-4} = 150$ | $k_{-8} = 250$ | $k_{-12} = 3$ | $k_{-16} = 2 \times 10^3$ |

Scheme 5.16. Exponential cycling reaction

$$\frac{1}{k_C} = \frac{K_{m,B}}{k_1 E_{10}} + \frac{K_{m,C}}{k_2 E_{20}} = \frac{K_{m,B}}{V_{max,B}} + \frac{K_{m,C}}{V_{max,C}}. \tag{15}$$

The term $K_{m,B}/V_{max,B}$ is equal to the inverse of the pseudolinear rate constant,

$$\lim_{B \to 0} \frac{\partial v(B)}{\partial B} = \frac{V_{max,B}}{K_{m,B}},$$

which is derived from the Michaelis-Menten equation,

$$v = \frac{V_{max,B} B}{K_{m,B} + B}, \tag{16}$$

at a very low concentration of B.

A reaction with the total concentration of $0.1\,\mu$M for each enzyme in the $E_1$- and $E_2$-systems in Scheme 5.15 follows the time course similar to that in Fig. 5.49. $k_C$ and $B/C$ thus obtained for various values of $Z$ are shown in Fig. 5.51. For a concentration of $Z$ less than $0.1\,\mu$M (*i.e.*, total enzyme concentration), $k_C$ has a constant value of $0.031\,\text{sec}^{-1}$, which is in good agreement with the value of $0.0329\,\text{sec}^{-1}$ obtained from a substitution of values of equation (9) into equation (15).

We conclude this section with an analysis of the exponential cycling reaction. A model is given in Scheme 5.16 in which each reaction in cycling reactions through T and D increases the amounts of T and D, resulting in

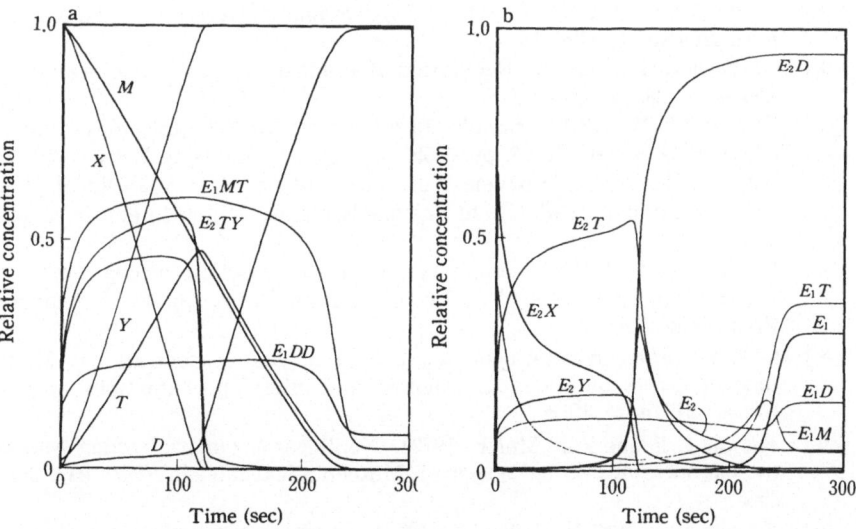

**Fig. 5.52** Time course of Scheme 5.16. Initial condition : $M = 5$ mM, $T = 0.1$ mM, $X = 5$ mM, $E_1 = E_2 = 10\,\mu$M.

a corresponding promotion of the progress throughout the system. The time course of the reaction in Scheme 5.16 is shown in Fig. 5.52. Substrate X reduces its concentration at about a twofold higher rate than substrate M. Product Y is pseudolinearly accumulated up to 120 sec, when X is all consumed. During that period a common intermediate T behaves similarly to Y. After 120 sec, however, T starts to decrease, while another common intermediate D rapidly increases its concentration.

A reaction system exponentially amplifying its production is considered to play an important role in *in vivo* regulatory mechanisms. Knoop *et al.* [9,10] have experimentally analyzed a cycling reaction system consisting of myokinase and pyruvate kinase under the condition of pseudolinear rate law, and applied it to quantitative microanalysis in a way similar to linear cycling reactions. They suggest that this mechanism operates in the Krebs-Kornberg cycle and the Calvin-Bassham cycle in plants and microorganisms. More extensive analysis on the cycling systems, however, remains to be done both in experimental studies and by simulation.

## References

[ 1 ]  Walter, C. (1976). "Enzyme Reactions and Enzyme Systems," Marcel Dekker, New York.

[ 2 ]  Levitzki, A. (1978). "Quantitative Aspects of Allosteric Mechanisms," Springer-Verlag, Berlin.

[ 3 ]  Goldbeter, A. and S.R. Caplan (1976). Oscillatory enzymes. *Annu. Rev. Biophys. Bioeng.*, **5**, 449-476.

[ 4 ]  Umbarger, H.E. (1969). Regulation of amino acid metabolism. *Annu. Rev. Biochem.*, **38**, 323-370.

[ 5 ]  Cleland, W.W. (1970). Steady state kinetics. *In* "The Enzymes," 3 rd ed., ed. by P.D. Boyer, Vol.2, pp.1-65, Academic Press, New York.

[ 6 ]  Segel, I.H. (1975). "Enzyme Kinetics. Behavior and Analysis of Rapid Equilibrium and Steady-State Enzyme Systems," Wiley-Interscience, New York.

[ 7 ]  Fromm, H.J. (1979). Summary of kinetic reaction mechanisms. *In* "Methods in Enzymology," ed. by D.L. Purich, Vol.63, pp.42-53, Academic Press, New York.

[ 8 ]  Cha, S. (1970). Kinetic behavior at high enzyme concentrations. Magnitude of errors of Michaelis - Menten and other approxmations. *J. Biol. Chem.*, **245**, 4814-4818.

[ 9 ]  Knoop, L.E. and R.P. Meich (1972). Nonlinear enzymatic cycling systems : the exponential cycling system. 1. Mathematical models. *ibid.*, **247**, 3558-3563.

[10]  Knoop, L.E. and R.P. Meich (1972). Nonlinear enzymatic cycling systems : the exponential cycling system. II. Experimental cycling system. *ibid*, **247**, 3564-3570.

# CHAPTER 6

# Macroscopic Analysis of Enzyme Systems

The microscopic analysis of enzyme systems is demonstrated in the preceding chapters to be a very effective procedure for precise and detailed observation of the dynamic behavior of every chemical species in the system.
However, such analysis of complex and large-scale systems demands an
extremely long computer time (CPU time) for simulation because of the
great increase in the number of variables and parameters of a system model. Their analysis thus remains practically impossible at present. Nevertheless, the dynamic analysis of these enzyme systems is still feasible with
their macroscopic treatment.

In a macroscopic analysis the intrinsically complex system is characterized through macroscopic modeling, which formulates the essential functions of the system by a simplified mathematical expression. Although the
behavior of every process in the system cannot be discussed in detail with
respect to their molecular mechanisms, the macroscopic model can ade-
quately represent the global characteristics of the system as a whole, which
are obtained from the computer simulation. That is, the model is not structurally valid but it is very effective for a comparison of systems according
to a certain specific behavior. The required CPU time for simulation is
consumed less than for microscopic analysis, and above all the model makes
it easier to determine which parameter is essentially correlated to the
behavior of a system. Thus, in the dynamic analysis of enzyme systems,
a generally efficient procedure is that a macroscopic analysis is performed to reveal the characteristics of the system as a whole, followed by a
microscopic analysis of the processes most affecting the behavior of the
system.

---

This chapter was written by Masahiro Okamoto and Katsuya Hayashi.

In this chapter several enzyme systems with peculiar and important characteristics are analyzed macroscopically. The time courses of cyclic reaction systems are compared to examine the relationship between rate constants and behavior. The system scheme to generate the desired behavior is determined by comparison of the time courses from the computer simulation of probable models for the reaction system with delay, oscillatory reaction system, feedback system with constant output, two-factor system, and system with threshold, respectively.

## 6.1 Cyclic Reaction System

We consider the cyclic reaction system of Scheme 6.1, in which A and C represent the substrates or products, and B and D are the different enzyme-substrate complexes. In this scheme, it is assumed that the identical enzyme

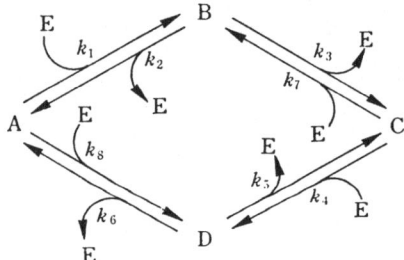

Scheme 6.1. Cyclic reaction system

E can catalyze both forward and reverse reactions between A and C. We examine the concentration ratio of $A$ and $C$ at a steady state to reveal a feature of cyclic reaction [1]. In the rate equation for the scheme we can put $dA/dt = dB/dt = dC/dt = dE/dt = 0$ at a steady state. From these algebraic equations, we can obtain the ratio $\bar{C}/\bar{A}$ as

$$\frac{\bar{C}}{\bar{A}} = \frac{k_1 + k_8}{k_4 + k_7} \frac{k_3 + k_5 F}{k_2 + k_6 F},\tag{1}$$

where

$$F = \frac{k_2 k_4 + k_2 k_8}{k_1 k_5 + k_1 k_6 + k_5 k_8 - k_4 k_6}.\tag{2}$$

Now we choose the standard values of rate constants as

$$k_1 = k_4 = 5 \times 10^5 \text{ M}^{-1} \text{ sec}^{-1}, \quad k_2 = k_5 = 5 \text{ sec}^{-1}, \quad k_3 = 0.5 \text{ sec}^{-1},$$
$$k_6 = 0.25 \text{ sec}^{-1}, \quad k_7 = k_8 = 10^{-5} \text{ sec}^{-1}.\tag{3}$$

Using the standard values, the value of $\bar{C}/\bar{A}$ is calculated to be 1.05 from equations (1) and (2), which indicates that the concentrations of A and C are almost equally partitioned at steady state, though the values of $k_3$ and $k_6$ are different twofold.

Of the standard values, the values of $k_7$ and $k_8$ are very small. If we ignore the reaction steps of C → B and A → D, we can postulate Scheme

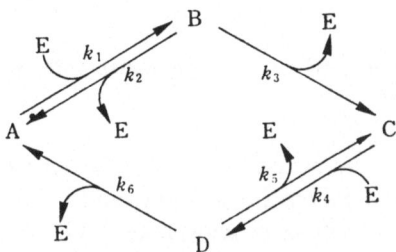

Scheme 6.2. Cyclic reaction system with irreversible steps

6.2, for which the ratio is represented by

$$\frac{\bar{C}}{\bar{A}} = \frac{k_5 + k_6}{k_2 + k_3} \frac{k_3}{k_6} \frac{k_1}{k_4} \tag{4}$$

Substituting the standard values of equation (3) into equation (4), we obtain $\bar{C}/\bar{A} = 1.91$, which is almost twice as much as that for Scheme 6.1. This result implies that we should simplify the reaction model very carefully: Elimination of a reaction step from the model because the specified step has very small rate-constant may lead to the wrong conclusion. Equation (4) can be transformed into

$$\frac{\bar{C}}{\bar{A}} = \frac{k_1}{k_2 + k_3} \frac{k_5 + k_6}{k_4} \frac{k_3 E_0}{k_6 E_0} = \frac{K_{m,b}}{K_{m,a}} \frac{V_{max,a}}{V_{max,b}}, \tag{5}$$

where $K_{m,a}$ and $K_{m,b}$ are the Michaelis constant and $V_{max,a}$ and $V_{max,b}$ the maximum velocity of steps A → C and C → A, respectively:

$$K_{m,a} = \frac{k_2 + k_3}{k_1}, \quad K_{m,b} = \frac{k_5 + k_6}{k_4} \tag{6}$$

$$V_{max,a} = k_3 E_0, \quad V_{max,b} = k_6 E_0, \tag{7}$$

where $E_0$ is the total concentration of enzyme E. From equation (5), the ratio $\bar{C}/\bar{A}$ may be directly estimated using the measured ratios of $K_{m,b}/K_{m,a}$ and $V_{max,a}/V_{max,b}$.

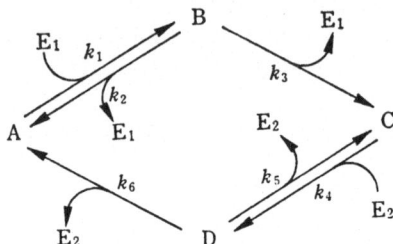

Scheme 6.3. Cyclic reaction system with two different enzymes

We now assume in the cyclic reaction of Scheme 6.3 that two different enzymes, $E_1$ and $E_2$, catalyze steps $A \to C$ and $C \to A$, respectively. The reaction system of D-fructose-6-phosphate (F6P) and D-fructose-1,6-bisphosphate (FDP) is a typical example for Scheme 6.3; $E_1$ and $E_2$ correspond to phosphofructokinase and fructose-1,6-bisphosphatase, respectively. The rate equation for Scheme 6.3 may be written as

$$
\left.
\begin{aligned}
\frac{dA}{dt} &= -k_1 E_1 A + k_2 B + k_6 D \\[4pt]
\frac{dB}{dt} &= \phantom{-}k_1 E_1 A - (k_2 + k_3) B \\[4pt]
\frac{dC}{dt} &= -k_4 E_2 C + k_3 B + k_5 D \\[4pt]
\frac{dD}{dt} &= \phantom{-}k_4 E_2 C - (k_5 + k_6) D \\[4pt]
\frac{dE_1}{dt} &= -k_1 E_1 A + (k_2 + k_3) B \\[4pt]
\frac{dE_2}{dt} &= -k_4 E_2 C + (k_5 + k_6) D \\[4pt]
E_{10} &= E_1 + B \\
E_{20} &= E_2 + D \\
S_0 &= A + B + C + D .
\end{aligned}
\right\} \tag{8}
$$

From equation (8) the ratio $\bar{C}/\bar{A}$ can be derived as follows:

$$
\frac{\bar{C}}{\bar{A}} = \frac{k_5 + k_6}{k_2 + k_3} \frac{k_3}{k_6} \frac{k_1}{k_4} \frac{\{E_{10} - B\}}{\left\{ E_{20} - \dfrac{k_3}{k_6} B \right\}} \tag{9}
$$

As for $B$ in equation (9), the following cubic equation for $B$ can be obtained:

$$\left(k_1k_3k_4+\frac{k_1k_3{}^2k_4}{k_6}\right)B^3+\left[-k_3k_4(k_2+k_3)-k_1k_4k_6E_{20}-k_1k_3k_4E_{10}\right.$$

$$\left.-k_1k_3(k_5+k_6)-k_1k_3k_4E_{20}-k_1k_3{}^2\frac{k_4}{k_6}E_{10}-k_1k_3k_4S_0\right]B^2 \tag{10}$$

$$+[k_4k_6(k_2+k_3)E_{20}+k_1k_4k_6E_{10}E_{20}+k_1k_3(k_5+k_6)E_{10}$$

$$+k_1k_3k_4E_{10}E_{20}+k_1k_4k_6E_{20}S_0+k_1k_3k_4E_{10}S_0]B$$

$$-k_1k_4k_6E_{10}E_{20}S_0=0.$$

Furthermore, equation (9) is rewritten as

$$\frac{\bar{C}}{\bar{A}}=\frac{K_{m,b}}{K_{m,a}}\frac{V_{max,a}-k_3B}{V_{max,b}-k_3B} \tag{11}$$

Solving equation (10) to substitute the obtained real root into equations (9) or (11), we can estimate the value of $\bar{C}/\bar{A}$. We choose here the standard values of parameters as

$$k_1=k_4=5\times10^5 \text{ M}^{-1}\text{ sec}^{-1}, \quad k_2=k_5=5 \text{ sec}^{-1}, \quad k_3=0.5 \text{ sec}^{-1}, \tag{12}$$
$$k_6=0.25 \text{ sec}^{-1}, \quad E_{10}=E_{20}=E_0=10^{-6} \text{ M}, \quad S_0=10^{-4} \text{ M}.$$

By using these values, $\bar{C}/\bar{A}$ and kinetic parameters $K_m$ and $V_{max}$ are calculated as

$$K_{m,a}=\frac{k_2+k_3}{k_1}=\frac{5.5}{5\times10^5}=1.1\times10^{-5}$$

$$K_{m,b}=\frac{k_5+k_6}{k_4}=\frac{5.25}{5\times10^5}=1.05\times10^{-5}$$

$$V_{max,a}=5\times10^{-7} \tag{13}$$

$$V_{max,b}=2.5\times10^{-7}=\frac{1}{2}V_{max,a}$$

$$\bar{C}/\bar{A}=10.1.$$

Figure 6.1 exhibits the computed time course of every chemical species in Scheme 6.3. As seen from equations (9) and (10), $\bar{C}/\bar{A}$ may vary with the change in values of $k_i$ $(i=1,2,\cdots,6)$, $E_{10}$, $E_{20}$ and $S_0$. Therefore, we are concerned with the effects of each parameter in Scheme 6.3 on $\bar{C}/\bar{A}$.

**Effect of $k_1/k_2$ $(=k_4/k_5)$**

The rate constants $k_1$ and $k_2$ (or, $k_4$ and $k_5$) predominantly govern the affinity of enzyme to substrate. The effects of change in $k_1/k_2$ $(=k_4/k_5)$ with fixed values of other kinetic parameters are summarized in Tables 6.1 and 6.2. It obviously follows that $\bar{C}/\bar{A}$ varies its value remarkably with the change in $k_1/k_2$, even if the ratios of $K_{m,b}/K_{m,a}$ and $V_{max,a}/V_{max,b}$ are fixed at certain values.

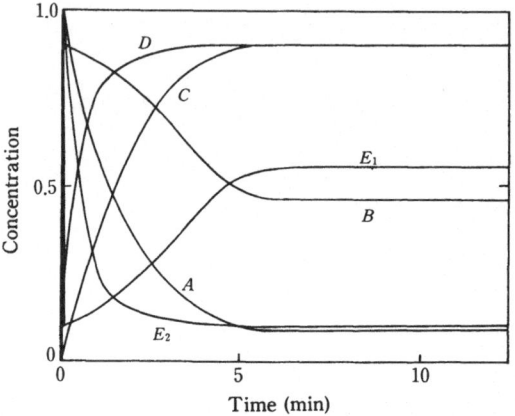

**Fig. 6.1** Simulated time course of Scheme 6.3. Full scale (1.0) on the ordinate: $100\mu M$ ($A$ and $C$); $1.0\mu M$ ($E_1$, $E_2$, $B$ and $D$).

**Table 6.1** Effect of $k_1(=k_4)$ on $\bar{C}/\bar{A}$

| $k_1(k_4)$ | $k_2(k_5)$ | $k_1/k_2(k_4/k_5)$ | $K_{m,a}$ | $K_{m,b}$ | $V_{max,a}$ | $V_{max,b}$ | $\bar{C}/\bar{A}$ |
|---|---|---|---|---|---|---|---|
| $5\times10^3$ | 5 | $10^3$ | $1.1\times10^{-3}$ | $1.05\times10^{-3}$ | $5\times10^{-7}$ | $2.5\times10^{-7}$ | 1.97 |
| $5\times10^4$ | 5 | $10^4$ | $1.1\times10^{-4}$ | $1.05\times10^{-4}$ | $5\times10^{-7}$ | $2.5\times10^{-7}$ | 2.56 |
| $5\times10^5$ | 5 | $10^5$ | $1.1\times10^{-5}$ | $1.05\times10^{-5}$ | $5\times10^{-7}$ | $2.5\times10^{-7}$ | 10.1 |
| $5\times10^6$ | 5 | $10^6$ | $1.1\times10^{-6}$ | $1.05\times10^{-6}$ | $5\times10^{-7}$ | $2.5\times10^{-7}$ | 90.5 |

**Table 6.2** Effect of $k_2(=k_5)$ on $\bar{C}/\bar{A}$

| $k_1(k_4)$ | $k_2(k_5)$ | $k_1/k_2(k_4/k_5)$ | $K_{m,a}$ | $K_{m,b}$ | $V_{max,a}$ | $V_{max,b}$ | $\bar{C}/\bar{A}$ |
|---|---|---|---|---|---|---|---|
| $5\times10^5$ | 0.5 | $10^6$ | $2.0\times10^{-6}$ | $1.5\times10^{-6}$ | $5\times10^{-7}$ | $2.5\times10^{-7}$ | 49.8 |
| $5\times10^5$ | 5 | $10^5$ | $1.1\times10^{-5}$ | $1.05\times10^{-5}$ | $5\times10^{-7}$ | $2.5\times10^{-7}$ | 10.1 |
| $5\times10^5$ | 50 | $10^4$ | $1.01\times10^{-4}$ | $1.01\times10^{-4}$ | $5\times10^{-7}$ | $2.5\times10^{-7}$ | 2.71 |
| $5\times10^5$ | 500 | $10^3$ | $1.00\times10^{-3}$ | $1.00\times10^{-3}$ | $5\times10^{-7}$ | $2.5\times10^{-7}$ | 2.07 |

## Effect of $k_3$ and $k_6$

The rate constants $k_3$ and $k_6$ may govern the overall reaction rate. The estimated values of $\bar{C}/\bar{A}$ are listed in Tables 6.3 and 6.4. As seen in Table 6.3, $\bar{C}/\bar{A}$ increases its value in proportion to $k_3/k_6$. Table 6.4 shows the contribution of changes in $k_3$ and $k_6$ to $\bar{C}/\bar{A}$ with the ratio $k_3/k_6$ fixed at 2.0. Thus, it is clear that small values of $k_3$ and $k_6$ result in a large value of $\bar{C}/\bar{A}$, even though the ratio $k_3/k_6$ is fixed at a certain value.

**Table 6.3**  Effect of $k_3/k_6$ on $\bar{C}/\bar{A}$

| $k_3$ | $k_6$ | $k_3/k_6$ | $K_{m,a}$ | $K_{m,b}$ | $V_{max,a}$ | $V_{max,b}$ | $\bar{C}/\bar{A}$ |
|---|---|---|---|---|---|---|---|
| 0.5 | 0.05 | 10 | $1.1 \times 10^{-5}$ | $1.01\times10^{-5}$ | $5\times10^{-7}$ | $5 \times10^{-8}$ | 89.3 |
| 0.5 | 0.1 | 5 | $1.1 \times10^{-5}$ | $1.02\times10^{-5}$ | $5\times10^{-7}$ | $1 \times10^{-7}$ | 39.7 |
| 0.5 | 0.25 | 2 | $1.1 \times10^{-5}$ | $1.05\times10^{-5}$ | $5\times10^{-7}$ | $2.5\times10^{-7}$ | 10.1 |
| 0.5 | 0.4 | 1.25 | $1.1 \times10^{-5}$ | $1.08\times10^{-5}$ | $5\times10^{-7}$ | $4 \times10^{-7}$ | 2.89 |
| 0.5 | 0.45 | 1.11 | $1.1 \times10^{-5}$ | $1.09\times10^{-5}$ | $5\times10^{-7}$ | $4.5\times10^{-7}$ | 1.73 |

**Table 6.4**  Effect of $k_3$ and $k_6$ on $\bar{C}/\bar{A}$

| $k_3$ | $k_6$ | $k_3/k_6$ | $K_{m,a}$ | $K_{m,b}$ | $V_{max,a}$ | $V_{max,b}$ | $\bar{C}/\bar{A}$ |
|---|---|---|---|---|---|---|---|
| 0.005 | 0.0025 | 2 | $1.00\times10^{-5}$ | $1.00\times10^{-5}$ | $5\times10^{-9}$ | $2.5\times10^{-9}$ | 11.0 |
| 0.05 | 0.025 | 2 | $1.01\times10^{-5}$ | $1.01\times10^{-5}$ | $5\times10^{-8}$ | $2.5\times10^{-8}$ | 10.9 |
| 0.5 | 0.25 | 2 | $1.1 \times10^{-5}$ | $1.05\times10^{-5}$ | $5\times10^{-7}$ | $2.5\times10^{-7}$ | 10.1 |
| 5.0 | 2.5 | 2 | $2 \times10^{-5}$ | $1.5 \times10^{-5}$ | $5\times10^{-6}$ | $2.5\times10^{-6}$ | 5.70 |
| 50.0 | 25.0 | 2 | $1.1 \times10^{-4}$ | $6 \times10^{-5}$ | $5\times10^{-5}$ | $2.5\times10^{-5}$ | 1.65 |

## Effect of $E_{10}$ $(=E_{20}=E_0)$ and $S_0$

It is well known that the most predominant factor affecting the steady state behavior is the ratio of the initial concentration of substrate to that of enzyme (*i.e.*, $S_0/E_0$), and that if the ratio is below a certain limit, the system cannot be in quasi-steady state during the progress of enzymatic reaction, regardless of the values of kinetic parameters of the reaction. Furthermore, the ratio $S_0/E_0$ is thought to affect the magnitude of $\bar{C}/\bar{A}$ as seen in equations (9) and (10). Table 6.5 demonstrates the effects of $S_0/E_0$ on $\bar{C}/\bar{A}$ for Scheme 6.3. The values in Table 6.5 (I) are obtained with a fixed value of $E_0$ and those in Table 6.5 (II) are with a fixed value of $S_0$. The table reveals that $\bar{C}/\bar{A}$ increases its value remarkably with the increase in $S_0/E_0$, under which the enzyme is saturated with the substrate. However, even if the ratio $S_0/E_0$ is fixed at a certain value, the value of $\bar{C}/\bar{A}$ can vary with the change in individual values of $E_0$ or $S_0$, as shown in Table 6.5 (III).

We can find several cyclic reaction systems in metabolic pathways. Newsholme and Start [2] first pointed out the physiological significance of the cyclic reaction system (they called it the "substrate cycle system" or "futile cycles"). The ratio $\bar{C}/\bar{A}$ is regarded as an index of amplification of metabolite in the cyclic reaction system. Therefore, the results in this section would be valuable for understanding the mechanism of amplification in substrate cycles. Finally, for a reaction system forming a cycle with two different enzymes, it is impossible to estimate uniquely the apparent equi-

**Table 6.5**  Effect of $E_0$ and $S_0$ on $\bar{C}/\bar{A}$

|        | $E_0(\times 10^{-6})$ | $S_0(\times 10^{-6})$ | $S_0/E_0$ | $\bar{C}/\bar{A}$ |
|--------|-----------------------|-----------------------|-----------|-------------------|
| (I)    | 1.0                   | 1000                  | 1000      | 91.7              |
|        | 1.0                   | 100                   | 100       | 10.1              |
|        | 1.0                   | 10                    | 10        | 2.52              |
|        | 1.0                   | 1.0                   | 1.0       | 1.96              |
|        | 1.0                   | 0.1                   | 0.1       | 1.91              |
| (II)   | 0.1                   | 100                   | 1000      | 10.2              |
|        | 1.0                   | 100                   | 100       | 10.1              |
|        | 10                    | 100                   | 10        | 9.01              |
|        | 100                   | 100                   | 1.0       | 3.09              |
|        | 1000                  | 100                   | 0.1       | 1.98              |
| (III)  | 1.0                   | 1000                  | 1000      | 91.7              |
|        | 0.1                   | 100                   | 1000      | 10.2              |
|        | 10                    | 100                   | 10        | 2.52              |
|        | 1.0                   | 10                    | 10        | 9.01              |
|        | 1.0                   | 1.0                   | 1         | 1.96              |
|        | 100                   | 100                   | 1         | 3.09              |
|        | 1.0                   | 0.1                   | 0.1       | 1.91              |
|        | 1000                  | 100                   | 0.1       | 1.98              |

librium constant $K'$ of the process,

$$A \underset{}{\overset{K'}{\rightleftharpoons}} C, \tag{14}$$

from the values of $K_m$ and $V_{max}$, even if the reaction occurs in closed system (see equation (11)). Accurate estimation of the apparent equilibrium constant $K'$ and the ratio $\bar{C}/\bar{A}$ requires the values of rate constants in addition to the values of $K_m$ and $V_{max}$.

## 6.2  Reaction System with Induction Period

The time courses of reaction products reveal some features characteristic to the reaction mechanisms. In principle, therefore, the reaction mechanism can be estimated through analysis of the experimental time course. A simple and one-step reaction exhibits an exponential time course, while a complicated reaction follows correspondingly complex time course. In this section, we consider the reaction which yields a time course with an induction period (or time delay, or dead time). Reaction schemes are sought which can realize the induction period in the time course.

## 1. Induction Period

We define the reaction with induction period as a reaction which gives rise
to a time course with an inflection point (see Fig. 6.2). A typical example
of induction period is given in Fig. 6.3 [3]. The lysozyme-catalyzed reaction of oligosaccharide $(GlcNAc-MurNAc)_2$ of cell wall exhibits about 10
min of the induction period without a clear inflection point, because of a rapid increase in the product $(GlcNAc-MurNAc)_1$ after the induction period.

First, we examine the possibility of induction period in a first-order reac-

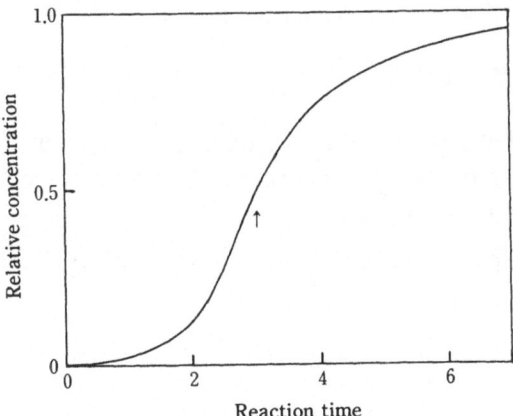

**Fig. 6.2** Time course of reaction with induction period. Arrow indicates the inflection point.

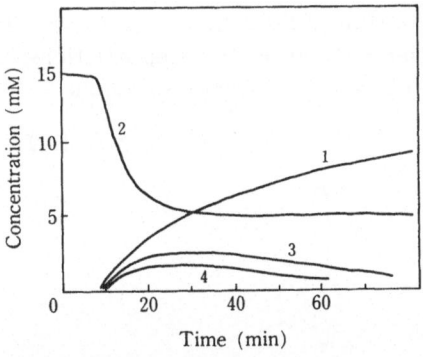

**Fig. 6.3** Time course of lysozyme-catalyzed reaction of $(GlcNAc-MurNAc)_2$. Numbers within the figure indicate $n$-value (size) of $(GlcNAc-MurNAc)_n$. GlcNAc : $N$-Acetylglucosamine residue ; MurNAc : $N$-Ace-
tylmuramic acid residue.

tion such as

$$X_1 \xrightarrow{k} X_2, \quad X_1(t_0) = X_0, \quad X_2(t_0) = 0, \quad t \in [t_0, \infty], \tag{15}$$

where $X_1$ is the reactant, $X_2$ the product and $k$ the rate constant, respectively. The solution of the rate equation of the reaction is readily obtained as

$$X_1(t) = X_0 e^{-kt}, \quad X_2(t) = X_0 (1 - e^{-kt}). \tag{16}$$

If the time course has an inflection point, the time $t$ at which $d^2 X_2/dt^2 = 0$ holds must exist. The second derivative becomes

$$\frac{d^2 X_2}{dt^2} = -X_0 k_1^2 e^{-kt}, \tag{17}$$

indicating that a finite time $t$ never leads to $d^2 X_2/dt^2 = 0$. It is thus concluded that the first-order reaction essentially has no induction period.

For a bimolecular reaction,

$$X_1 + X_2 \xrightarrow{k} X_3, \quad X_1(t_0) = X_{10}, \quad X_2(t_0) = X_{20}, \quad X_3(t) = 0, \tag{18}$$

$X_3(t)$ is represented by

$$X_3(t) = \frac{X_{10} X_{20} (e^{(X_{10} - X_{20})kt} - 1)}{X_{10} e^{(X_{10} - X_{20})kt} - X_{20}} \tag{19}$$

and the second derivative becomes

$$\frac{d^2 X_3}{dt^2} = -\frac{X_{10} X_{20} (X_{10} - X_{20})^3 k^2 e^{(X_{10} - X_{20})kt} (X_{10} e^{(X_{10} - X_{20})kt} + X_{20})}{(X_{10} e^{(X_{10} - X_{20})kt} - X_{20})^3}. \tag{20}$$

There exists again no finite time $t$ allowing $d^2 X_3/dt^2 = 0$. Therefore, the induction period may not be realized by one-step reactions.

Then, we consider a two-step linear reaction such as

$$X_1 \xrightarrow{k_1} X_2 \xrightarrow{k_2} X_3, \quad X_1(t_0) = X_0, \quad X_2(t_0) = X_3(t_0) = 0. \tag{21}$$

The rate equation is expressed by

$$\left. \begin{aligned} \frac{dX_1}{dt} &= -k_1 X_1 \\[2mm] \frac{dX_2}{dt} &= k_1 X_1 - k_2 X_2 \\[2mm] \frac{dX_3}{dt} &= k_2 X_2. \end{aligned} \right\} \tag{22}$$

Substituting the solution of the first equation into the second equation, we obtain

$$\frac{dX_2}{dt} = k_1 X_0 e^{-k_1 t} - k_2 X_2. \tag{23}$$

The solution of equation (23) is given by

$$X_2(t) = \frac{k_1 X_0}{k_2 - k_1} \{e^{(k_2 - k_1)t} - 1\} e^{-k_2 t}. \tag{24}$$

From the conservation law we have

$$X_3(t) = X_0 - \{X_1 + X_2\}$$
$$= X_0 \left\{ 1 - \frac{k_2}{k_2 - k_1} e^{-k_1 t} + \frac{k_1}{k_2 - k_1} e^{-k_2 t} \right\} \tag{25}$$

The second derivative of $X_3$ with respect to $t$ becomes

$$\frac{d^2 X_3}{dt^2} = \frac{k_1 k}{k_2 - k_1} \left\{ -k_2 e^{-k_1 t} + k_1 e^{-k_2 t} \right\} \tag{26}$$

From $d^2 X_3/dt^2 = 0$, we get

$$t = \begin{cases} \dfrac{1}{k_1 - k_2} \ln \dfrac{k_1}{k_2} & k_1 \neq k_2 \\[2mm] \dfrac{1}{k_1} & k_1 = k_2. \end{cases} \tag{27}$$

Hence, the two-step linear reaction obviously exhibits the induction period.

It is known that the autocatalytic reaction generally proceeds with the induction period. We here analyze the simplest autocatalytic reaction represented by Scheme 6.4:

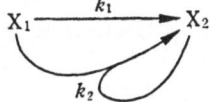

Scheme 6.4. Autocatalytic reaction

The rate equation of the system may be written as

$$\left. \begin{aligned} \frac{dX_1}{dt} &= -k_1 X_1 - k_2 X_1 X_2, \quad X_1(t_0) = X_0, \\ \frac{dX_2}{dt} &= k_1 X_1 + k_2 X_1 X_2, \quad X_2(t_0) = 0. \end{aligned} \right\} \tag{28}$$

From the conservation law, $X_2 = X_0 - X_1$, we obtain

$$\frac{dX_1}{dt} = k_2 X_1^2 - (k_1 + k_2 X_0) X_1. \tag{29}$$

The solution of equation (29) is given by

$$X_1(t) = \frac{X_0(k_1 + k_2 X_0)}{k_2 X_0 + k_1 e^{(k_1 + k_2 X_0)t}}$$                                    (30)

and $X_2$ may be derived from the conservation law:

$$X_2(t) = \frac{X_0(e^{\alpha t} - 1)}{k_2 X_0 + k_1 e^{\alpha t}}, \quad \alpha = k_1 + k_2 X_0$$                      (31)

The second derivative of $X_2$ with respect to $t$ becomes

$$\frac{d^2 X_2}{dt^2} = \frac{k_1 X_0 \alpha e^{\alpha t}(k_2 X_0 - k_1 e^{\alpha t})}{(k_2 X_0 + k_1 e^{\alpha t})^3},$$              (32)

leading to the necessary condition for the existence of inflection point such that

$$k_2 X_0 = k_1 e^{\alpha t}, \quad \text{or} \quad t = \frac{1}{\alpha} \ln \frac{k_2 X_0}{k_1}.$$                  (33)

Since $t$ is positive and then $k_2 X_0 > k_1$, the induction period may appear if the rate of autocatalytic reaction is larger than that of the spontaneous reaction.

## 2. Enzyme System with Induction Period

We consider the induction period in an enzymatic reaction represented by Scheme 6.5. From the results described above, it is naturally expected that

$$\text{E} + \text{S} \underset{k_{-1}}{\overset{k_{+1}}{\rightleftharpoons}} \text{ES}_1 \underset{k_{-2}}{\overset{k_{+2}}{\rightleftharpoons}} \text{ES}_2 \overset{k_{+3}}{\longrightarrow} \text{E} + \text{P}$$

$$k_{+1} = 5 \times 10^5, \quad k_{-1} = 5, \quad k_{+2} = 10^3, \quad k_{-2} = 10^2, \quad k_{+3} = 0.16$$

Scheme 6.5. Michaelis-Menten-type reaction with two complexes

the time course of this enzymatic reaction will exhibit a distinct induction-period due to its complex reaction scheme. The result of computer simulation, however, reveals in Fig. 6.4 that the induction period is not observable with respect to the product formation. This may result from large values of $k_1$ and $k_2$ for the complex formation; the complex is formed instantaneously and hence the product is accumulated rapidly in the early stage of reaction. For enzymatic reactions in general, the induction period appears in too early a stage to be observed by usual experimental techniques. Nevertheless, some enzymatic reactions demonstrate the distinct induction-period. In lysozyme-catalyzed reaction, the reaction scheme is extremely complicated as described in Section 4.1. This complexity is the

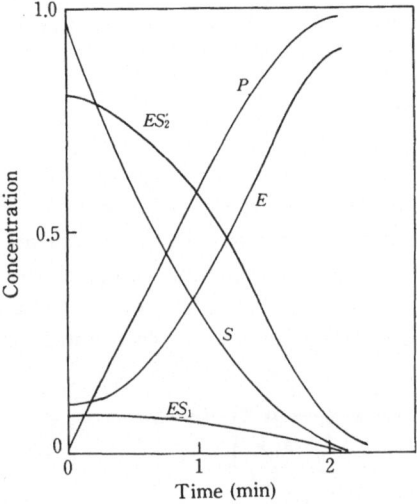

**Fig. 6. 4**  Time course of Scheme 6.5. Initial condition : $S_0 = 10\mu M$, $E_0 =$ 1.0$\mu M$. Full scale (1.0) on the ordinate : $S = P = 10\mu M$, $E = ES_1 = ES_2 =$ 1.0$\mu M$.

origin of a distinct induction–period as shown in Fig. 6.3. A hysteretic enzyme which changes its structure during the reaction also yields an induction period (see Section 4.2).

Another example is the activation process of pepsinogen to active pepsin. As described in detail in Section 8.1, pepsinogen autocatalytically releases

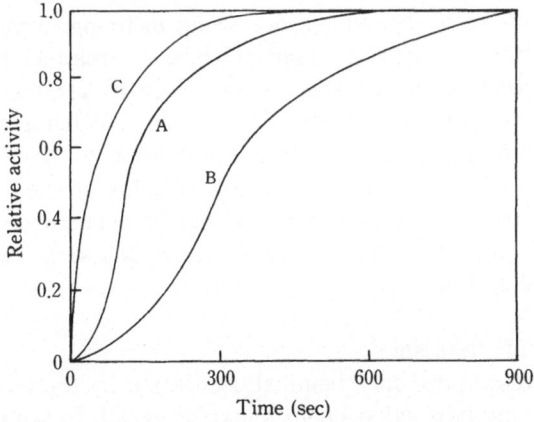

**Fig. 6. 5**  Time course of activation of pepsinogen. A : Activation at 15°C ; B : Activation at 0°C ; C : Exponential curve.

**Table 6.6**   Rate constants for activation reaction of pepsinogen

| $k_1(\text{sec}^{-1})$ | $k_2(\text{sec}^{-1})$ | $k_3(\text{sec}^{-1})$ | $k_4(\text{M}^{-1}\text{sec}^{-1})$ |
|---|---|---|---|
| $1.4 \times 10^{-3}$ | $2.3 \times 10^{-3}$ | $1.6 \times 10^{-3}$ | $2.54 \times 10^{-3}$ |

a peptide fragment composed of about 40 amino-acid residues to convert to pepsin. The formation of pepsin proceeds with the induction period as seen in Fig. 6.5. The estimated reaction scheme and values of rate constants for the activation process are shown in Scheme 6.6 and Table 6.6, respectively [4]. The most responsible step for appearance of the induction period may

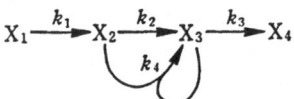

Scheme 6.6.   Activation reaction of pepsinogen

be the autocatalytic process with rate constant $k_4$. The fact that $k_4 > k_2$ may be consistent with the necessary condition that the rate of the autocatalytic process must be larger than that of the spontaneous process.

In conclusion, it should be stated that the induction period in a time course is not a special phenomenon but an outcome common to complicated reactions.

## 6.3   Oscillatory Reaction System

The various oscillatory phenomena occurring in living organisms such as secretion of hormone, cardiovascular pulse and circadian rhythm would fundamentally stem from the specific behavior of chemical events [5]. Therefore, the quantitative explanation of the mechanism for *in vivo* chemical oscillation may provide some information for understanding the mechanism by which the highly organized oscillation are generated. In this section, we consider the conditions for oscillation occurrence in a chemical reaction system and the relationship between chemical oscillation and feedback control system.

### 1.   Oscillatory Mechanism
Higgins [6] has explained the chemical oscillation by the two-body mechanism in which only two independent dynamic variables are taken into account. In general, the mathematical equation of chemical reactions may be written as

$$\left.\begin{array}{l} \dfrac{dx}{dt} = f(x,y,z,\cdots\cdots\} = v_x \\[2mm] \dfrac{dy}{dt} = g(x,y,z,\cdots\cdots\} = v_y \\[2mm] \vdots \end{array}\right\} \tag{34}$$

If $v_x$, the net rate of production of $x$, depends on the concentration of other chemical species (effector) such as $z$, then we can define the following three modes of action of $z$ for $v$:

$$\frac{\partial v_x}{\partial z} > 0: \quad v_x \text{ is activated by } z \tag{35}$$

$$\frac{\partial v_x}{\partial z} = 0: \quad v_x \text{ is independent of } z \tag{36}$$

$$\frac{\partial v_x}{\partial z} < 0: \quad v_x \text{ is inhibited by } z. \tag{37}$$

In the case of $z = x$, the reaction with relation (35) is said to be self-activating and the one with relation (37) to be self-inhibiting. Thus, the kinetic behavior is strongly dependent on the partial derivative of the net rate; the array of all possible partial derivatives is referred to as the characteristic of reaction scheme.

In the two-body mechanism, the characteristic may be expressed in the matrix form as

$$\begin{bmatrix} \dfrac{\partial v_x}{\partial x} & \dfrac{\partial v_x}{\partial y} \\[3mm] \dfrac{\partial v_y}{\partial x} & \dfrac{\partial v_y}{\partial y} \end{bmatrix}, \tag{38}$$

where $\partial v_x/\partial x$ and $\partial v_y/\partial y$ are called self-coupling, and $\partial v_x/\partial y$ and $\partial v_y/\partial x$ cross-coupling. By considering the sign and magnitude in the characteristic matrix (equation (38)), Higgins [6] has deduced the conditions for oscillation in the two-body mechanism. The results are summarized as follows:
a) The self-coupling terms should be of opposite signs.
b) The cross-coupling terms should also be of opposite signs.
c) The magnitude of product of cross-coupling terms should be greater than that of self-coupling terms:

$$\left|\frac{\partial v_x}{\partial y}\right|\left|\frac{\partial v_y}{\partial x}\right| > \left|\frac{\partial v_x}{\partial x}\right|\left|\frac{\partial v_y}{\partial y}\right|. \tag{39}$$

Figures 6.6 and 6.7 present several oscillatory schemes which satisfy the above conditions. From these conditions, we can further predict that a

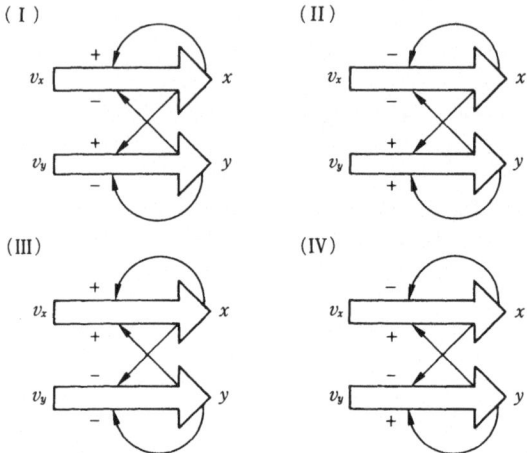

**Fig. 6.6** Oscillatory schemes with two-body mechanism. The signs $+$ and $-$ indicate the activation and inhibition, respectively.

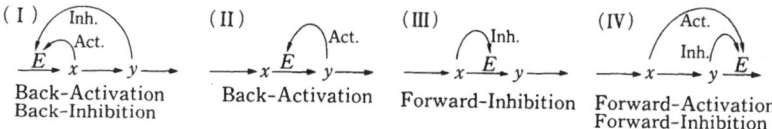

**Fig. 6.7** Oscillatory schemes with two reactants. (I), (II), (III) and (IV) correspond to those in Fig. 6.6.

scheme involving only back inhibition (negative feedback control with the smallest loop size) cannot yield sustained oscillations. On the other hand, there would be a possibility that a system with large feedback loop could generate oscillations in the concentrations of intermediates or product of the system.

Many types of models for enzymatic oscillation have been proposed on the basis of negative feedback, elucidating the oscillatory conditions for negative feedback systems. Most of these models, however, are of the Yates-Pardee type as shown in Scheme 6.7, in which the end product inhibits cooperatively the first enzyme in a reaction sequence. In this scheme, $S_0$

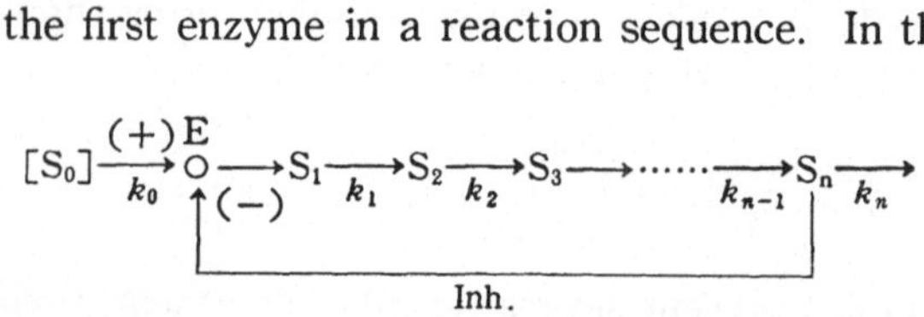

$$[S_0] \xrightarrow{\ \ (+)E\ \ }_{k_0} O \xrightarrow{\ \ }_{\substack{\uparrow \\ (-)}} S_1 \xrightarrow{\ \ }_{k_1} S_2 \xrightarrow{\ \ }_{k_2} S_3 \longrightarrow \cdots\cdots \xrightarrow{\ \ }_{k_{n-1}} S_n \xrightarrow{\ \ }_{k_n}$$

Inh.

Scheme 6.7. System with negative feedback loop

represents the pool of substrate which is kept at constant concentration, and $S_n$ denotes the end product. The activity of enzyme E is regulated by negative feedback in the mode of parametric regulation. The step indicated by a small circle with arrows is called the summing point, at which $(+)$ corresponds to the flux of substrate and $(-)$ to inhibition of the enzymatic activity. According to Morales and McKay [7], the control mode at the summing point in Scheme 6.7 is given by

$$v = \frac{v_0}{K_m + \alpha (S_n)^p},$$                                      (40)

where $v$ and $v_0$ represent the reaction rates of allosteric enzyme with and without the feedback inhibition, respectively, and $K_m$ is a Michaelis constant. The constants $p$ and $\alpha$ specify the number of allosteric effectors per enzyme molecule (or Hill coefficent) and the feedback constant (gain) controlling the weight of contribution of $S_n$ to the summing point, respectively. The rate equation for Scheme 6.7 may be written as

$$\left. \begin{aligned}
\frac{dS_1}{dt} &= \frac{k_0 S_0}{1 + \alpha (S_n)^p} - k_1 S_1 \\
\frac{dS_2}{dt} &= k_1 S_1 - k_2 S_2 \\
&\vdots \\
\frac{dS_n}{dt} &= k_{n-1} S_{n-1} - k_n S_n,
\end{aligned} \right\}$$                                                       (41)

where $K_m = 1.0$ is assumed. Morales and McKay [7] and Walter [8] have deduced the necessary condition for sustained oscillation in equation (41). The computer simulation indicates that the relationship,

$$p > \frac{1}{\left( \cos \frac{\pi}{n} \right)^n},$$                           (42)

should be satisfied for the oscillation of reactants in Scheme 6.7. It follows from inequality (42) that the values of $p$ yielding the sustained oscillation should be $p \geq 9$ for $n = 3$ and $p \geq 2$ for $n = 10$. In the case of $n = 4$, inequality (42) leads to $p \geq 5$, indicating that the presence of five or more binding sites to allosteric effector $(S_n)$ per enzyme molecule is required for the system to oscillate. Thus, inequality (42) provides the general relationship between the feature of an allosteric enzyme and the sustained oscillation of a negative feedback system.

The oscillatory behavior of non-Yates-Pardee type, in which the end product inhibits an enzyme other than the first-step enzyme in the system, has not been studied in detail. We here consider the behavior of Scheme 6.8. It is expected that the time-delay element would be essential for the

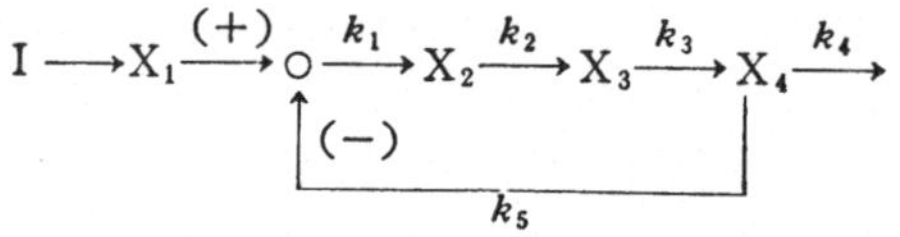

$$I \longrightarrow X_1 \xrightarrow{(+)} O \xrightarrow{k_1} X_2 \xrightarrow{k_2} X_3 \xrightarrow{k_3} X_4 \xrightarrow{k_4}$$

Scheme 6.8. System with a time-delay element

appearance of sustained oscillation in a system of the non-Yates-Pardee type. Assuming the time delay in the inhibition by $X_4$, we have the rate equation for Scheme 6.8:

$$
\left.
\begin{aligned}
\frac{dX_1}{dt} &= I - [k_1/(1+k_5(X_4(t-\tau))^n]X_1 \\
\frac{dX_2}{dt} &= [k_1/(1+k_5(X_4(t-\tau))^n]X_1 - k_2X_2 \\
\frac{dX_3}{dt} &= k_2X_2 - k_3X_3 \\
\frac{dX_4}{dt} &= k_3X_3 - k_4X_4,
\end{aligned}
\right\}
\tag{43}
$$

where $I$ indicates the input of $X_1$ (pool of substrate corresponding to $S_0$ in Scheme 6.7) and $k_5$ represents the feedback gain. $n$ is a parameter corresponding to $p$ in equation (41). $\tau$ denotes the time delay, and then $X_4(t-\tau)$ indicates the concentration of $X_4$ at time $(t-\tau)$, which is assumed to be 0 during $0 < t < \tau$. To exert its function, the inhibitor $X_4$ has to translate a certain distance from the produced position to the enzyme location. The time required for translation amounts to $\tau$. Hence, the inhibitor concentration acting at time $t$ is set to be $X_4(t-\tau)$. The time courses of $X_4(t)$ with $\tau = 0$ or $\tau \neq 0$ are shown in Fig. 6.8. It is clear from these results that the time delay is the main cause for sustained oscillation in a feedback system of the non-Yates-Pardee type.

The relationship between number $n$ in equation (43) and the oscillatory pattern of $X_4(t)$ is examined for Scheme 6.8 under the same conditions as in Fig. 6.8. The results are listed in Table 6.7. In the case of $n=1$, no sustained oscillation appears in $X_4(t)$, even if $\tau$ increases several fold. This indicates the close relationship between the control mode of allosteric enzyme and the mode of oscillation, as also proposed by Goldbeter and Nicolis [9]. To derive the relationship between the size of feedback loop and the mode of oscillation and to examine the effect of $\tau$ on it, some simulations are performed for Schemes 6.9 (a) and (b) which have smaller feed-

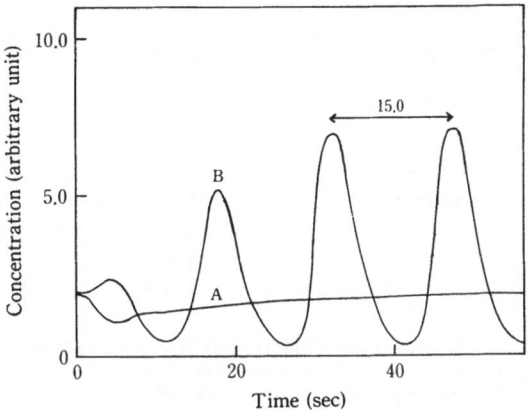

**Fig. 6. 8** Simulated time course of $X_4(t)$ in Scheme 6.8 ($n=3$). A : $\tau = 0$ ; B : $\tau = 4.0$. Initial condition : $X_{10} = X_{20} = X_{30} = 1.0$, $X_{40} = 2.0$ ; $I = 1.5$. The steady state value in $A$ is 3.0.

**Table 6. 7** Effect of $n$ value on oscillation of $X_4$

| $n$ | $\tau$ | Oscillatory mode | Period | Amplitude |
|---|---|---|---|---|
| 1 | 4. 0 | Damped | 15. 5 | — |
| 1 | 8. 0 | " | 22. 0 | — |
| 1 | 12. 0 | " | 29. 0 | — |
| 2 | 4. 0 | Sustained | 15. 0 | 2. 90 |
| 3 | 4. 0 | " | 15. 0 | 3. 44 |
| 4 | 4. 0 | " | 15. 0 | 3. 49 |

(a) $I \longrightarrow X_1 \xrightarrow{k_1} X_2 \xrightarrow[(+)]{k_2} \circ \longrightarrow X_3 \xrightarrow{k_3} X_4 \xrightarrow{k_4}$
$(-)$
$k_5$

(b) $I \longrightarrow X_1 \xrightarrow{k_1} X_2 \xrightarrow{k_2} X_3 \xrightarrow[(+)]{k_3} \circ \longrightarrow X_4 \xrightarrow{k_4}$
$(-)$
$k_5$

Scheme 6. 9. System with smaller feedback loops

back loops than Scheme 6.8. The results are summarized in Tables 6.8 and 6.9, demonstrating that Scheme 6.9 (b) with the smallest feedback loop can still cause a sustained oscillation when the value of $\tau$ is very large. It is thus concluded that relation (42) cannot be applied to the feedback

**Table 6.8**   Effect of $\tau$ on oscillation of $X_4(t)$ in Scheme 6.9 (a)

| $\tau$ | Oscillatory mode | Period | Amplitude |
|---|---|---|---|
| 0 | Undershoot | — | — |
| 0.5 | " | — | — |
| 1.0 | Damped | 5.5 | — |
| 3.0 | Sustained | 11.5 | 3.12 |
| 4.0 | " | 13.5 | 3.48 |
| 6.0 | " | 16.5 | 3.80 |

**Table 6.9**   Effect of $\tau$ on oscillation of $X_4(t)$ in Scheme 6.9 (b)

| $\tau$ | Oscillatory mode | Period | Amplitude |
|---|---|---|---|
| 0 | Undershoot | — | — |
| 0.5 | " | — | — |
| 1.0 | Damped | 3.5 | — |
| 4.0 | Sustained | 11.5 | 3.60 |
| 5.0 | " | 13.0 | 3.75 |
| 6.0 | " | 15.0 | 3.89 |

**Table 6.10**   Effect of $k_1 \sim k_4$ on oscillation of $X_4(t)$

| $k_1(=k_2=k_3)$ | $k_4$ | $\tau$ | Oscillatory mode |
|---|---|---|---|
| 1.0 | 0.5 | 0 | Damped |
| 1.0 | 0.5 | 2.0 | Sustained |
| 0.5 | 0.25 | 0.5 | Damped |
| 0.5 | 0.25 | 2.0 | " |
| 0.5 | 0.25 | 4.0 | Sustained |
| 0.1 | 0.05 | 0.5 | Damped |
| 0.1 | 0.05 | 4.0 | " |
| 0.1 | 0.05 | 10.0 | " |
| 0.1 | 0.05 | 15.0 | " |

system with time delay. Table 6.10 reveals the effect of $k_i$ ($i=1,2,3,4$) in Scheme 6.8 on the mode of oscillation. The ratio of $k_1/k_4$ ($=k_2/k_4=k_3/k_4$) is fixed at 2.0. A sustained oscillation occurs with a small value of $\tau$ when the values of $k_1 \sim k_4$ become large. No sustained oscillation is observed when the reaction rates between $X_1$ and $X_4$ are very small.

In general, the time delay in a feedback system stems from the time-consuming translation of effectors by diffusion or active transport. Hence, the factor of time delay should be taken into consideration in the analysis of most enzymatic feedback systems *in vivo*.

## 2. Frequency Conversion

We now consider the response of an oscillatory system to oscillatory input. This problem may lead to the understanding of the frequency-conversion mechanism in enzymatic reaction. The proper frequency (or natural frequency) of the system is maintained in living organisms through a certain mechanism which counters the oscillatory perturbations with various frequencies. This mechanism may be very important in metabolic regulation to synchronize the frequencies of interacting systems and to keep the proper frequency for metabolic intermediates.

Marek and Stuchl [10] first observed experimentally the synchronization in two interacting oscillatory systems using the Belousov reaction (oxidation of malonic acid by bromate in sulfuric acid with ceric/cerous ions as catalyzer [11]). In an enzymatic reaction system, Boiteux *et al.* [12] have revealed the relationship of the frequency in glycolysis between phosphofructokinase action and glucose input. By injecting glucose periodically into a phosphofructokinase reaction system, they have measured the time course of NADH absorbance to reveal the domain of entrainment of the absorbance by the fundamental, 1/2-harmonic and 1/3-harmonic frequencies in a sinusoidal source of glucose. However, clarification of the frequency-conversion mechanism by which any oscillatory input with different frequency is converted to the proper frequency of the system remains to be done.

We consider here the problem of frequency conversion using Scheme 6.10 with time delay, in which $w$ represents the constant input and the numerals indicate the values of rate constants in the corresponding steps.

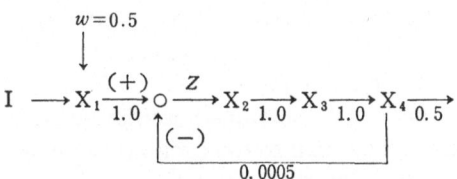

Scheme 6.10. System of frequency conversion

The activity of allosteric enzyme at the summing point is expressed by

$$Z = \frac{1.0}{1 + 0.0005(X_4(t-4))^3},\qquad(44)$$

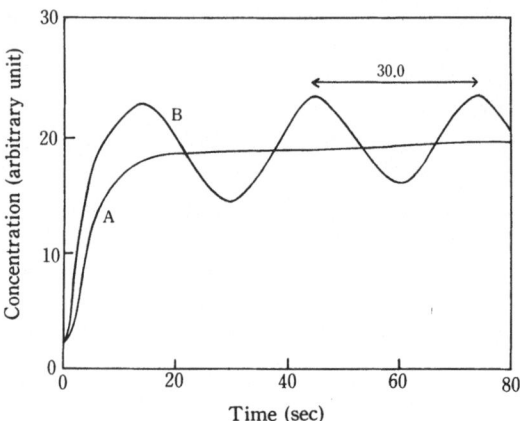

**Fig. 6. 9** Simulated time course of $X_4(t)$ in Scheme 6.10 without time delay. A : $I=10.0$; B : $I=10.0+10.0 \sin(2\pi/30) t$

where the notations are the same as those in equation (43). The oscillating mode of input (I) is assumed to be

$$I = 10.0 + 10.0 \sin\frac{2\pi}{T} t,\tag{45}$$

where $T$ indicates the period ($=1/f$, $f$ : frequency) of oscillating input. In order to compare the responses of the systems with and without time delay, we also assume that equation (44) corresponds to

$$Z' = \frac{1.0}{1+0.0005(X_4(t))^3}\tag{46}$$

for the system without time delay.

First, the possibility of frequency conversion for Scheme 6.10 without time delay is sought using equations (45) and (46). The results are shown in Fig. 6.9. This system essentially has no oscillatory property (A in Fig. 6.9). When a sinusoidal input (equation (45) with $T=30$ sec) is introduced to the system, $X_4(t)$ undergoes sustained oscillation with a period of 30 sec (B in Fig. 6.9), which is identical to the oscillatory input. Thus, the system has no capacity for converting the frequency of input to another frequency.

Secondly, with time delay ($\tau$) fixed at 4.0 sec, the time course of $X_4(t)$ is computed under the condition of a constant input ($I=0,10$, and 20 in mM sec$^{-1}$). As shown in Table 6.11, $X_4(t)$ exhibits the sustained oscillation with a period of 13.0 sec for $I=10$ mM sec$^{-1}$ and 14.0 sec for $I=20$ mM sec$^{-1}$, respectively. That is, Scheme 6.10 with $\tau=4.0$ sec has the proper period ($T_0$) of 13.0 sec or 14.0 sec corresponding to the conditions specified

**Table 6.11**  Effect of oscillatory input on oscillation of Scheme 6.10 with $\tau = 4.0$ sec.

| Input period($T$) (sec) | | Output period($T'$) (sec) | Input period($T$) (sec) | Output period($T'$) (sec) |
|---|---|---|---|---|
| ($I=$ 0. 0)— | | 0 | 26.0* | 13. 0× 2 |
| ($I=$10. 0) | Constant input | 13. 0 | 29. 0 | 13. 0×11 |
| ($I=$20. 0)— | | 14. 0 | 33. 0 | 13. 0× 5 |
| 5. 0 | | 13. 0 | 36. 0 | 13. 0×11 |
| 6. 5 | | 13. 0 | 39.0* | 13. 0× 3 |
| 13.0* | | 13. 0 | 42. 0 | 13. 0×13 |
| 16. 0 | | 13. 0× 6 | 46. 0 | 13. 0× 7 |
| 20. 0 | | 13. 0× 3 | 49. 0 | 13. 0×11 |
| 23. 0 | | 13 ·0× 7 | 52.0* | 13. 0× 4 |

* Double periodicity is observed (see text).

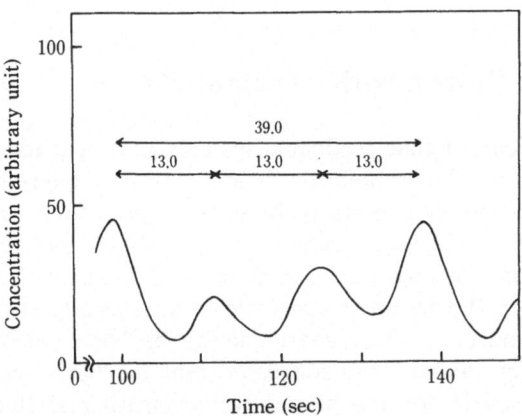

**Fig. 6.10**  Oscillatory pattern of $X_4(t)$ in Scheme 6.10 with $\tau = 4.0$ sec. Oscillatory input: $I = 10.0 + 10.0 \sin(2\pi/39)\, t$. Double periodicity is observed.

above. Next, the oscillatory patterns of $X_4(t)$ are computed for the case that the oscillatory input expressed by equation (45) is introduced to the system with $\tau = 4.0$ sec. If this system provides the mechanism of frequency conversion, the period of $X_4(t)$ must be kept near 13 sec against the oscillatory inputs with various periods. The relationship between the period $T$ of oscillatory input and the resulting period $T'$ of $X_4(t)$ is shown in Table 6.11 and Fig. 6.10. It is obvious that the resulting period $T'$ can be represented by

$$T' = mT_0, \tag{47}$$

where $T_0$ denotes the proper period of $X_4(t)$ and $m$ is a constant coefficient. Figure 6.10 illustrates the oscillation of $X_4(t)$ in the case that the period $T$ of oscillatory input is 39.0 sec, which is thrice larger than $T_0$. It follows that double periodicity is superimposed by the periods of $T$ and $T_0$, when the oscillatory input with the period of $mT_0$ is introduced to the system. Hence, it is clear that Scheme 6.10 with time delay provides the mechanism of frequency conversion, keeping the proper period against the oscillatory inputs of different frequencies.

The significance of the feedback system has been discussed in connection with the homeostatic control, especially the constant-value control. As described above, it is likely that feedback system with time delay plays a major role in the frequency conversion in addition to the homeostatic control. This section only deals with the relationship between feedback system and its oscillatory behavior. In the following section, we shall discuss the homeostatic capability of enzyme systems from the aspect of constant-value control.

## 6.4   Feedback System with Constant Output

It is widely recognized that feedback systems play a leading role in the homeostatic control of metabolic pathways. Allosteric enzymes in closed system have been studied extensively with respect to their cooperative characteristics and molecular mechanism of feedback control. It has been speculated that the constant-value control in metabolism is realized in a feedback system by the action of allosteric enzyme as a regulatory element.

On the other hand, extensive research on the feedback control in the fields of engineering and industry has suggested that feedback systems are not necessarily responsible for the homeostatic control and that many constraint conditions are required for the system to have specific regulatory characteristics. In the area of technology, the structure of optimal feedback systems, which usually yields excellent controllability, has been clarified on the basis of the optimal control theory. Since some kinds of optimal feedback systems are also thought to operate in the regulation of enzyme systems, the elucidation of the structure of such systems is an interesting and important problem.

In this section, we are concerned with the homeostatic control capability in various enzymatic feedback systems, focusing especially on the performance for constant output. Furthermore, the constraint conditions for optimal regulation imposed on the enzymatic feedback system are examined in reference to the relationship between the structure and controllability of the system. That is, this section mainly deals with how the enzymatic feedback system can keep the concentration of its end product at a constant value against external perturbations on the input or intermediate of the system. It is assumed that the external perturbation (or disturbance) in the form of impulse or step is intruded into the system at a stationary state. Then we follow the behavior of the system in the transient phase and new stationary state.

## 1. Response of Feedback System to Disturbance

Figure 6.11 presents the structures ineffective for constant-value control. Though each scheme has a structure including multiple or cascade feedback loops, each system demonstrates a large deviation from stationary value against the step perturbation. In fact, the concentration of $X_4$ at new stationary state ($\bar{X}_4$) can be represented by

$$\bar{X}_4 = \frac{I}{k_i}, \tag{48}$$

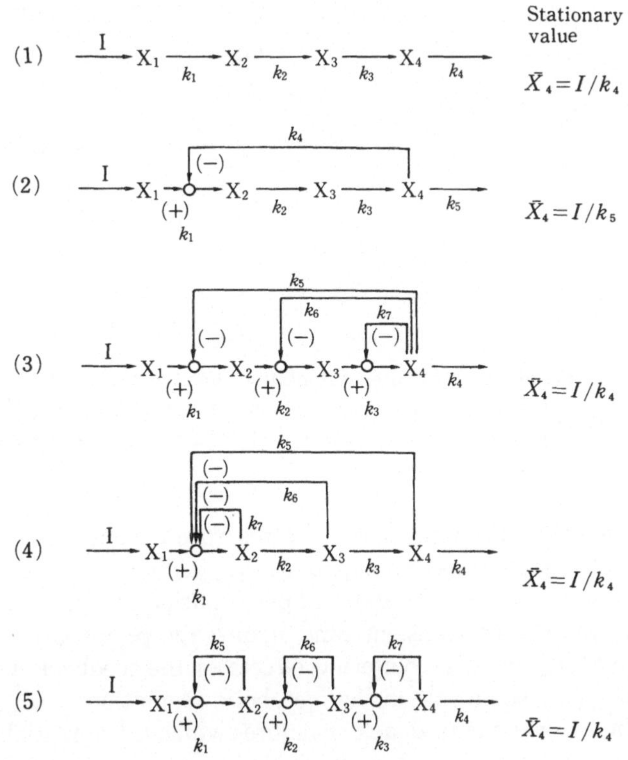

Fig. 6.11 Structures ineffective for constant-value control.

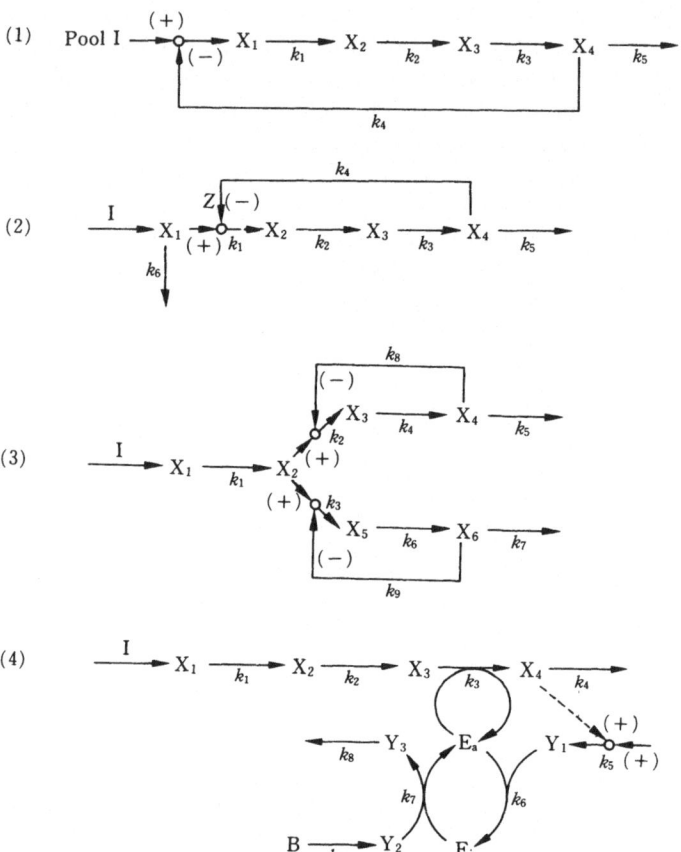

**Fig. 6.12** Structures potentially effective for constant-value control. (1) Yates-Pardee type; (2) non-Yates-Pardee type with input-consuming process; (3) branching feedback type; (4) feedback with two-factor realizing system.

where I indicates the input and $k_i$ is the rate constant for efflux of $X_4$. Therefore, the system has no capability in yielding a constant output in a stationary state against the external perturbation on the input. However, slow response or long transient time against the perturbation is observed, which can be regarded as buffering effects of the feedback system. Such effects may play some role in the metabolic regulation.

Figure 6.12 gives the feedback structures which are potentially capable of the constant-value control. Scheme 6.11 shows a typical structure in which the external step-perturbation $Y_1$ is introduced to $X_1$ in the system of Fig.

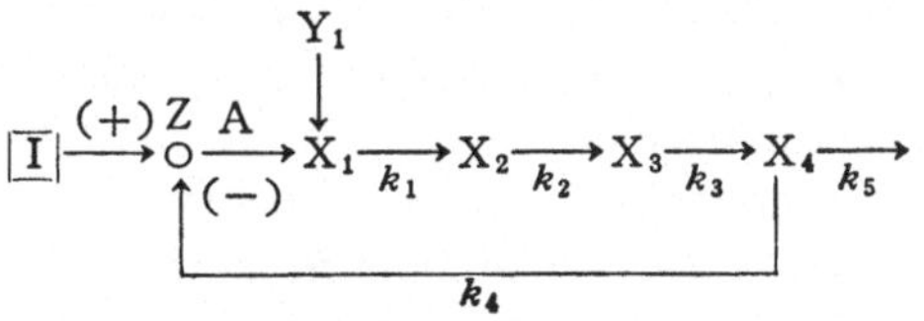

Scheme 6.11. Feedback system with external perturbation imposed on the intermediate step

6.12 (1). "I" represents a large pool which supplies the constant input, and A is the net flow of the input. The control mode $Z$ at the summing point is expressed by

$$Z = \frac{1}{k_4(X_4)^3},$$

(49)

where we assume $ZI = A$. The effect of $Y_1$ on the value of $X_4$ at the new stationary state ($\bar{X}_4$) in Scheme 6.11 is summarized in Table 6.12. The value of $A$ decreases with increase in $Y_1$, approaching almost zero with a larger $Y_1$. However, Scheme 6.11 apparently has no mechanism to exclude $Y_1$ higher than $A$; $X_4$ is affected directly by such $Y_1$. These results indicate that the feedback system with a single large loop may be relatively unstable and ineffective for the external perturbation on the inside of the loop. Furthermore, Scheme 6.11 causes sustained oscillation, which is unfavorable property for attainment of the constant-value control.

In Scheme 6.11, $X_4$ is assumed to regulate directly the input flow, though the molecular mechanism of such a direct input-control is not known. The pool size of $I$ is expected to increase with the feedback gain $k_4$, because the

**Table 6.12** Effect of $Y_1$ on $\bar{X}_4$ in Scheme 6.11

| $Y_1$ | Influx rate to $X_1$ (A) | $\bar{X}_4$ | Relative deviation* |
|---|---|---|---|
| 0 | 0.389 | 0.795 | 0 |
| 0.1 | 0.325 | 0.850 | 6.93 |
| 0.2 | 0.259 | 0.918 | 15.4 |
| 0.3 | 0.200 | 1.00 | 25.7 |
| 0.4 | 0.150 | 1.10 | 38.3 |
| 1.0 | 0.023 | 2.04 | 157 |
| 2.0 | 0.003 | 4.01 | 404 |

* Deviation of $\bar{X}_4$ from that at $Y_1 = 0$.
$I = 10.0$, $X_{10} = X_{20} = X_{30} = X_{40} = 1.0$

net input-flow becomes smaller and the input substrate is accumulated in the pool. However, the pool size is assumed to maintain a constant value. The following two postulates can generally make constant pool-size occur : 1) The accumulated amount of the input substrate by negative feedback is negligible in comparison with the amount in the pool. 2) Near the summing point there exists another process which consumes the accumulated input substrate. The schemes of (2) and (3) in Fig. 6.12 have such input-consuming processes. These schemes are also potentially effective for the constant-value control. Among the schemes similar to Fig. 6.12(2), Scheme 6.12 is found to have the most effective structure for excluding the external perturbation. This system has several short loops which do not overlap

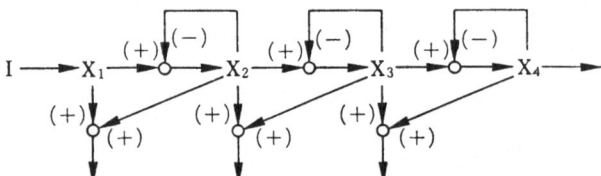

Scheme 6.12.   System effective for exclusion of external perturbation

each other and the positive feedback on the branched steps. The positive feedback can eliminate the accumulated input substrate.

In technological fields, the structure of the optimal regulator in a dynamic system has been determined by means of the dynamic control theory [13, 14]. Accordingly, it is valuable to compare the behavior of enzymatic feedback systems with that of the optimal regulator in order to characterize the regulatory mechanism in enzyme systems more precisely and to predict those effective systems which would function in the real metabolic networks. Figure 6.13(1) illustrates the structure of the optimal regulator for a linear dynamic system which is capable of rapidly excluding the effect of an impulse perturbation. $f$ in the box represents the feedback gain which is obtained from the solution of the Riccati differential equation and $u$ is a control variable (input) to $X_1$. The broken lines indicate that the concentrations of each reactant are used for calculation of the gain $f$. Figure 6.13(2) shows the optimal regulator for continuous perturbation, which includes the integration element as well as proportional elements of $K$ and $K'$ as the feedback gain. $K$ corresponds to $f$ in Fig. 6.13(1) and $K'$ is a function of the difference between $u$ and the perturbation $w(t)$.

The transient time courses of $X_4(t)$ in Scheme 6.11, Fig. 6.13 (1) and (2) are computed for the case that an impulse perturbation of $-0.2$ mM is introduced to $X_1$ at $t=0$. The parameters used are : $Y_1=0,\ I=10$ for

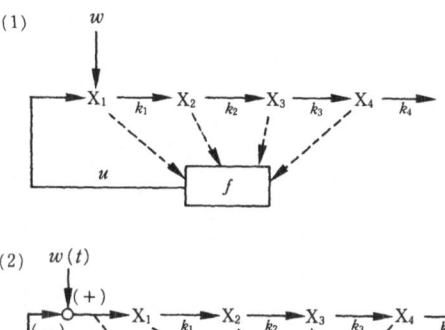

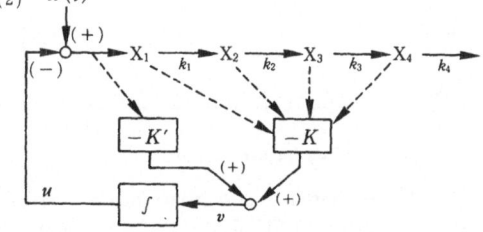

**Fig. 6.13** Structure of optimal regulators. (1) For impulse perturbation $w$ ; (2) for continuous perturbation $w(t)$ (see text).

Scheme 6.11 and $k_1 = k_2 = k_3 = 1.0$, $k_4 = 0.5$ for Fig. 6.13 (1) and (2). For computation of $u(t)$ of the latter two schemes, a cost function (see Section 9.2),

$$J = \frac{1}{2} \int_{t_0}^{t_f} [(X_4(t))^2 + (u(t))^2] dt, \tag{50}$$

is used, where $t_0$ and $t_f$ represent the initial and final times, respectively. The results are shown in Fig. 6.14, in which the ordinate indicates the deviation of $X_4(t)$ from the stationary value (or desired value) $\bar{X}_4$. Contrary to the schemes of Fig. 6.13 (1) and (2), Scheme 6.11 yields a damped oscillation induced by the external impulse perturbation. Moreover, the transient time of Scheme 6.11 is very long. Thus, it is clear that the transient characteristic of Scheme 6.11 is inferior to those of Fig. 6.13 (1) and (2) with respect to the regulatory mechanism.

Next, the time course of deviation $(X_4(t) - \bar{X}_4)$ is computed for the case that the constant step-perturbation of $1.0$ mM is introduced to $X_1$ at $t = 0$. The values of parameters and the cost function are the same as above. Results are given in Fig. 6.15. Only the scheme of Fig. 6.13(2) (curve C) is able to eliminate completely the effect of perturbation after a short transient-period. Curves A and B indicate a large deviation from the stationary state, implying that Scheme 6.11 and Fig. 6.13(1) lack the capability in excluding the continuous external perturbation. It follows

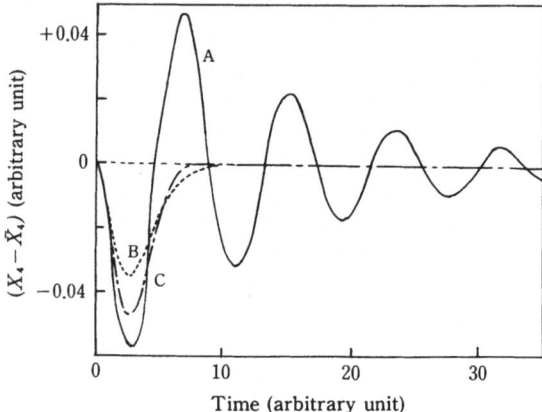

**Fig. 6. 14** Effect of impulse perturbation on feedback system. A : Scheme 6.11 ; B : Scheme of Fig. 6.13 (1) ; C : Scheme of Fig. 6.13 (2). An impulse perturbation $(-0.2\,\text{mM})$ is imposed to $X_1$ at $t=0$.

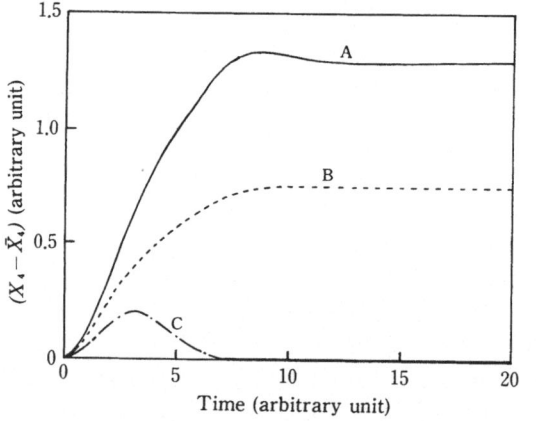

**Fig. 6. 15** Effect of step perturbation on feedback system. A : Scheme 6.11 ; B : Scheme of Fig. 6.13 (1) ; C : Scheme of Fig. 6.13 (2). A step perturbation $(1.0\,\text{mM})$ is imposed on $X_1$ at $t=0$.

from these results that Scheme 6.11 (enzymatic feedback system) is far from the optimal control and that the structure of Scheme 6.11 should be revised in order for the system to function as optimal feedback system.

## 2. Optimal Structure of Enzymatic Feedback System

The structural difference between Scheme 6.11 and the schemes of Fig.

6.13 (1) and (2) may be stated as follows: 1) In Scheme 6.11, the concentrations of all reactants have to be positive, while in the schemes of Fig. 6.13 the values of $X_i(t)$ $(i=1,2,3,4)$ and the control variable $u$ are allowed to be negative. 2) In the schemes of Fig. 6.13, $X_i(t)$ are all observable at all times and $u(t)$ is determined on the basis of all $X_i(t)$. In the enzymatic feedback system, however, the summing point is regulated only by the value of end product $X_4(t)$. 3) In the optimal regulator of Fig. 6.13, feedback gain [$f$ in (1) and $K$ and $K'$ in (2)] can be designed as time-dependent coefficient, while in the enzymatic feedback system feedback gain $k_4$ is time-invariant. The enzymatic feedback system is a kind of chemical process, requiring essentially the nonnegativity condition for reactant concentration. The regulatory element in allosteric enzyme is inhibited by a single or a couple of reactants in the feedback system, but not simultaneously by several reactants. Thus, the differences explained by 1) and 2) are considered to be inherent in the enzyme system. Therefore, we here assume that the feedback gain $k_4$ of the enzyme system is time-dependent and attempt to evaluate the optimal value of $k_4(t)$ which yields the minimum deviation in $(X_4(t)-\bar{X}_4)$.

Scheme 6.13 presents a typical enzymatic feedback system of the non-Yates-Pardee type, in which the feedback gain $k_4$ has a constant value.

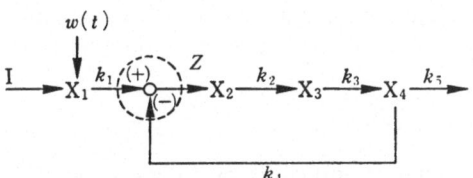

Scheme 6.13.   Non-Yates-Pardee-type system with constant feedback gain

By using a kind of Hill equation, the regulation at the summing point may be represented by

$$Z=\frac{k_1}{1+k_4(X_4(t)^3)},\qquad(51)$$

which corresponds to the activity of allosteric enzyme. In Scheme 6.13, $w(t)$ is an external step-perturbation and the other notations are the same as in Scheme 6.11. Scheme 6.14 is a counterpart of Scheme 6.13, in which the feedback gain $k_4(t)$ is assumed to be time-dependent. The control mode at the summing point may be written as

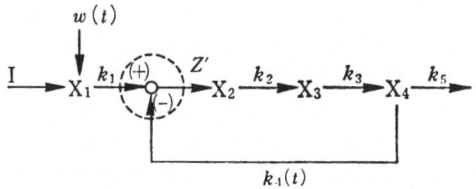

Scheme 6.14.  Non-Yates-Pardee-type system with time-
dependent feedback gain

$$Z' = \frac{k_1}{1 + k_4(t) X_4(t)^3} . \tag{52}$$

In this case, it is also assumed that the control variable $u(t)$ is equal to $Z'$:

$$u(t) = Z' . \tag{53}$$

Optimization technique can determine $u(t)$ so that the control variable $u(t)$ minimizes the cost function defined by

$$J = \frac{1}{2}(X_4(t_f) - \bar{X}_4)^2 + \frac{1}{2}\int_{t_0}^{t_f}\{(X_4(t) - \bar{X}_4)^2 + 0.01\, u^2\}\, dt, \tag{54}$$

where $t \in [t_0, t_f]$ and $\bar{X}_4$ represents the desired constant-value for $X_4$. Since the optimal trajectory (time course) of $X_4(t)$ can also be determined in the optimization of $u(t)$, the feedback gain can be evaluated from

$$k_4(t) = \frac{k_1 - u(t)}{u(t) X_4(t)^3} . \tag{55}$$

The maximum principle is applied to optimization of $u(t)$ together with the conjugate gradient method (see Section 9.2).

The effects of step perturbation $w(t)$ on the time courses of $X_4(t)$ in Schemes 6.13 and 6.14 are shown in Fig. 6.16.  Under the conditions used, the stationary value $\bar{X}_4$ is taken as 2.0 mM before $w(t)$ is introduced to the system.  If the system provides itself with a device to realize the constant-value control, $X_4(t)$ should be kept at $\bar{X}_4$ ($=2.0$ mM) against the external perturbation.  As seen in Fig. 6.16, curve 2 ($X_4(t)$ in Scheme 6.14) converges to 2.0 mM after the nonstationary state, while the other curves ($X_4(t)$ in Scheme 6.13) reach 3.0 mM ($= (I + w(t))/k_5$) at new stationary states even if a large value is given to $k_4$.  This reveals that Scheme 6.13 undergoes an elongated transient-time and has no capability in excluding the external perturbation.  The time-dependent feedback gain in Scheme 6.14 ($k_4(t)$ in equation (55)) varies as in Fig. 6.17, and the phase diagrams of the enzyme activity at the summing point ($Z$ for Scheme 6.13 and $Z'$ for

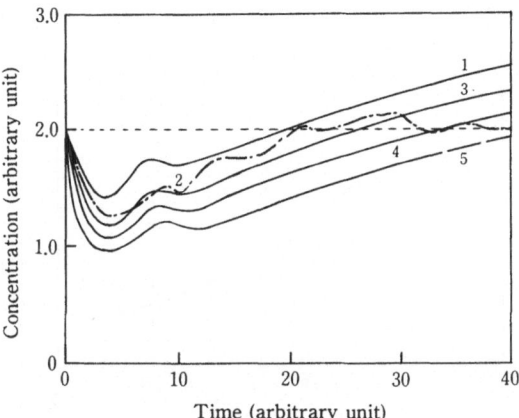

**Fig. 6. 16** Effect of step perturbation on $X_4(t)$ in Schemes 6.13 and 6.14.
1: $k_4 = 1.0$ (6.13); 2: $k_4 = 1.0$ (6.14); 3: $k_4 = 2.0$ (6.13); 4: $k_4 = 3.0$
(6.13); 5: $k_4 = 5.0$ (6.13). $k_1 = k_2 = k_3 = 1.0$, $k_5 = 0.5$, $I = 1.0$, and $w(t) = 0.5$.
Numbers in parentheses indicate the scheme number.

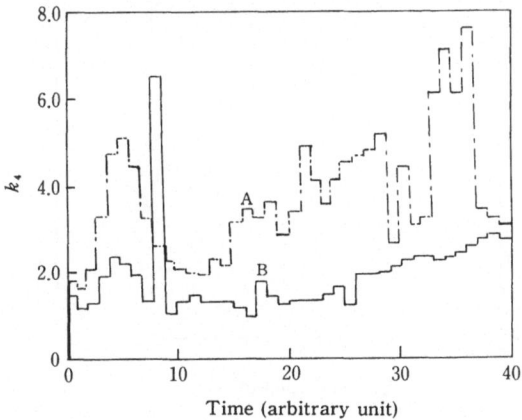

**Fig. 6. 17** Time course of $k_4(t)$ in Scheme 6.14. $k_4$ is assumed to have a
constant value in every 1.0 time unit. A: $w(t) = 1.0$; B: $w(t) = 0.5$.

Scheme 6.14) *vs.* $X_4(t)$ are given in Figs. 6.18 and 6.19. Unlike the activity
of ordinary allosteric enzyme in Scheme 6.13, the activity $Z'$ of putative
enzyme in Scheme 6.14 with the time-dependent feedback gain becomes a
multiple-valued function of $X_4(t)$ having a complicated profile. This infers
that in order to realize homeostatic control the regulatory enzyme in the
feedback system needs to have the activity accompanied by the hysteresis
with the end product (inhibitor). It has been presumed from biochemical

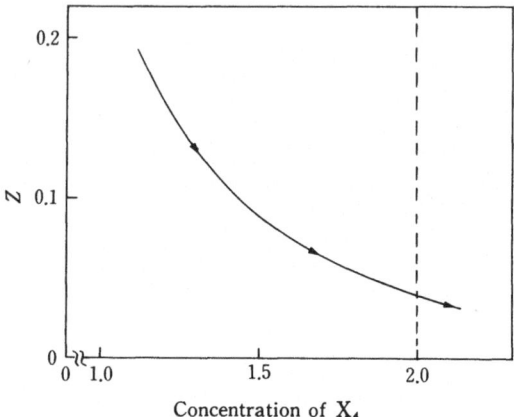

**Fig. 6. 18** Phase diagram of $Z$ (equation (51)) *vs.* $X_4(t)$ in Scheme 6.13. Both ordinate and abscissa have arbitrary units. The broken line indicates the desired concentration of $X_4$.

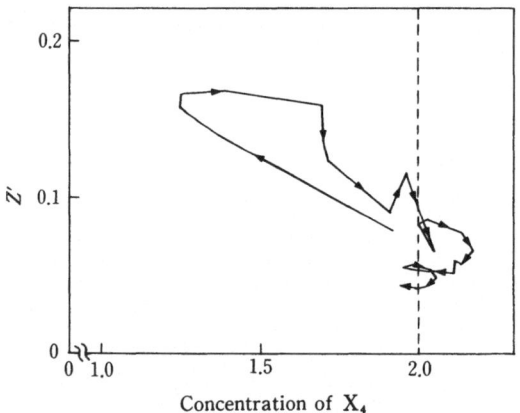

**Fig. 6. 19** Phase diagram of $Z'$ (equation (52)) *vs.* $X_4(t)$ in Scheme 6.14. Notations are the same as in Fig. 6.18.

studies that the sigmoidal feature in the activity of allosteric enzyme may lead a feedback system to have homeostatic control. The present results, however, reveal that sole sigmoidicity of allosteric enzyme does not always give rise to its capability in excluding the external perturbation.

This section is concerned with the capability of the enzymatic feedback system for constant-value control. The results are summarized as follows: (1) The existence of sigmoidicity in activity of allosteric enzyme does not always imply that the feedback system can achieve homeostatic control. (2) A regulatory enzyme providing the outstanding capability of homeostatic control must have a reminiscent feature or hysteresis in addition to the well-known property of allosteric enzyme in the feedback system. The existence of such regulatory enzymes has not yet been confirmed in living organisms. (3) There might be some auxiliary system to the feedback system to enable conjoint homeostatic control. (4) An unknown control system with an unknown regulatory enzyme might be engaged in the constant-value control. This problem will be partly considered in the following section.

## 6.5  Two-Factor System

It has been accepted that the dynamic behavior of physiological systems at various hierarchical levels is fundamentally regulated by the combined action of excitatory (activatory) and inhibitory factors. For enzyme systems, the enzymatic activity and hence the system output are generally thought to be continuously regulated by the action of activator or inhibitor, that is, the enzymatic activity changes continuously in response to the concentrations of activator and inhibitor.

On the other hand, it has been well known that some physiological systems respond to the net magnitude of competing excitatory and inhibitory factors. For instance, the neuron fires only when the excitatory input (impulse) exceeds the inhibitory input, as expressed by $Z(x,y)=1$ in Table 6.13; $Z(x, y)$ represents the output of the neuronal system when $x$ and $y$ are the magnitudes or amounts of the excitatory and inhibitory inputs, respectively.

Rosen [15] has introduced a similar idea into the biochemical systems and postulated a regulatory mechanism called the two-factor model. In his basic model, an enzyme system produces two opposing factors, activator $x$ and inhibitor $y$, and the output is controlled by a function $Z(x,y)$ which takes only values of either 1 or 0. Even though such a discrete output might offer a new type of energy or material production, the two-factor model would become more significant if the output carries information to control

**Table 6.13**  Two-factor model

| |
|---|
| $y>x : Z(x, y)=0$ |
| $y \leq x : Z(x, y)=1$ |
| $x$ : excitatory factor (activator) |
| $y$ : inhibitory factor (inhibitor) |

the other biochemical systems. Thus, it seems an interesting problem to examine the possiblity of enzyme systems functioning as the two–factor model. This section mainly deals by means of computer simulation with the two–factor system attained in enzyme systems [16].

## 1. Coupled Reaction System

A typical scheme for a coupled enzymatic reaction system is given in Scheme 6.15. The system shown in the lower part has a switching function for active $E_a$ and inactive $E_i$ enzymes. In a coupled reaction system in the upper part the active enzyme catalyzes the formation of outputs. The outputs of the whole system are taken to be $X_1$ and $X_2$. The excitatory factor $x$ and inhibitory factor $y$ produce catalytically the actual activator $Y_2$ and inhibitor $Y_1$, respectively, from the external large pools. It is assumed that $x$ and $y$ are a sort of catalyzer and not consumed during the reaction. $E_0$ is another enzyme catalyzing the step $X_2 \rightarrow X_1$, which is assumed to have a constant activity.

First, we examine the effects of ratio $f$ $(=y/x)$ on the concentrations of $X_1$, $X_2$, $E_a$ and $E_i$ in the stationary state. The ratio $f$ is changed with variation in the value of $x$ against a fixed value of $y$ at 20. The results of the simulation are shown in Fig. 6.20. The concentrations of $E_a$ and $E_i$ change stepwise at $f=1$. The concentrations of $X_2$ and $X_1$ vary in a similar manner except for the appearance of a slightly curved corner. The magnitude of the curved part apparently depends on the total concentration of $E_0$.

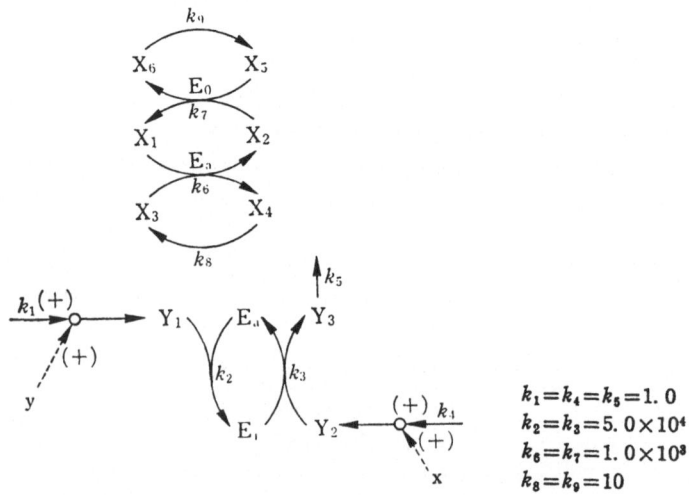

Scheme 6.15.   Coupled enzymatic reaction system

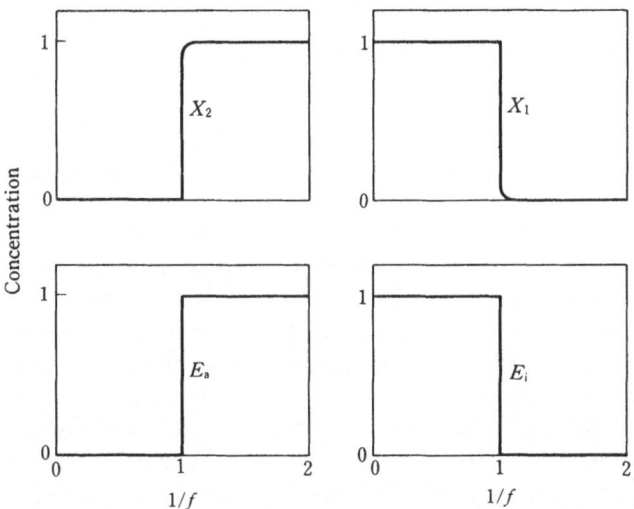

**Fig. 6. 20**  Effect of $f$ on steady-state concentrations of $E_a$, $E_i$, $X_1$, and $X_2$ in Scheme 6.15. $1/f = x/y$. Intial concentration: $E_a = 0.5$, $E_i = 0.5$, $X_1 = X_2 = 0.5$, $X_3 = X_5 = 10$, $X_4 = X_6 = 0$, $Y_1 = Y_2 = Y_3 = 1.0$.

Thus, the coupled reaction system can realize the behavior of function $Z(x, y)$. We call the system in the lower part of Scheme 6.15 a two-factor realizing system (TRS). The TRS is a kind of cyclic enzyme system.

## 2.  Regulation of Lactose Synthetase System
It is well known that the lactose synthetase system provides a typical example of interconversion of enzymatic activity. UDP-galactose reacts with $N$-acetylglucosamine to form UDP and disaccharide $N$-acetyllactosamine by the catalysis of galactosyltransferase. In the presence of lactose synthetase, UDP-galactose also reacts with glucose to produce UDP and lactose. Galactosyltransferase consists of both catalytic and regulatory subunits. The catalytic subunit by itself can produce $N$-acetyllactosamine. When $\alpha$-lactoalbumin, a form of milk protein, binds at the modifier site on the regulatory subunit, galactosyltransferase interconverts to lactose synthetase to produce nutrient lactose for a baby, as shown in Table 6.14 [16]. During pregnancy, the reaction proceeds along equation (1) in the table. At the time of birth, hormonal levels change drastically, so that $\alpha$-lactoalbumin is synthesized in a large amount. Galactosyltransferase interconverts to lactose synthetase as represented in mode (3) in the table, which leads to production of a large amount of lactose. Hence, the interconversion of

**Table 6.14**  Control mode of lactose synthetase system

| | |
|---|---|
| (1) | UDP-galactose + $N$-acetylglucosamine $\xrightarrow[\text{galactosyltransferase}]{}$ |
| | UDP + $N$-acetyllactosamine |
| (2) | UDP-galactose + glucose $\xrightarrow[\text{lactose synthetase}]{}$ UDP + lactose |
| (3) | galactosyltransferase + $\alpha$-lactalbumin $\longrightarrow$ lactose synthetase |

galactosyltransferase is considered dependent upon the amount of $\alpha$-lacto-albumin. However, if the TRS should exist in this type of reaction as a regulatory element, it would be expected that a very small increase in $\alpha$-lacto-albumin level would cause an abrupt formation of lactose synthetase.

Scheme 6.16 presents an underlying model scheme for the lactose synthetase system in which $X_1$, $X_2$ and $X_3$ denote UDP-galactose, $N$-acetyllactosamine and lactose, respectively, and $E_1$ and $E_2$ represent galactosyltransferase and lactose synthetase. $E_1$ is able to convert to $E_2$ in the presence

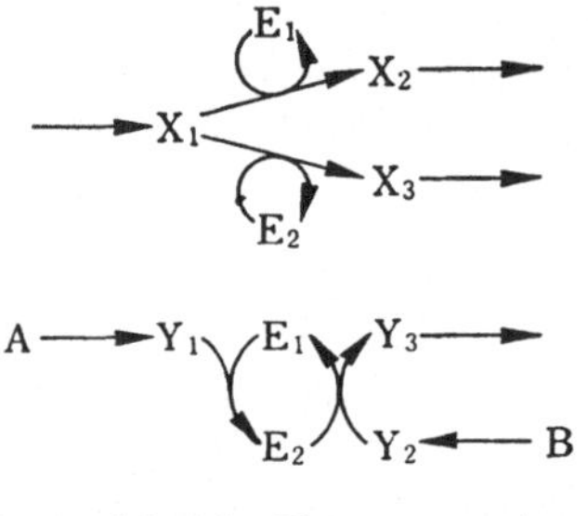

Scheme 6.16.   Model of lactose synthetase system

of $Y_1$ ($\alpha$-lactoalbumin), as shown in the lower part of the scheme. In addition to the experimental results, it is further assumed that $E_2$ can interconvert to $E_1$ accompanied by a specific reaction-step $Y_2 \rightarrow Y_3$. $A$ and $B$ represent hormonal levels which correspond to $y$ and $x$ in Scheme 6.15.

The effects of the levels of $A$ and $B$ on the time course of chemical species are followed by the simulation and the results are given in Fig. 6.21. In the figure, the arrow indicates the switching time at which $A$ becomes larger than $B$. When $A$ exceeds $B$, $E_1$ is switched off and $E_2$ is switched on. As a result, $X_3$ and $Y_1$ are synthesized in large amounts, while the production of $X_2$ decreases to zero. When $B$ exceeds $A$, the situation reverses completely. As mentioned above, it has been supposed that the interconversion of galactosyltransferase to lactose synthetase is induced by the amount of $\alpha$-lactoalbumin. However, if the TRS were involved in such a system (Scheme 6.16), the interconversion can proceed to completion

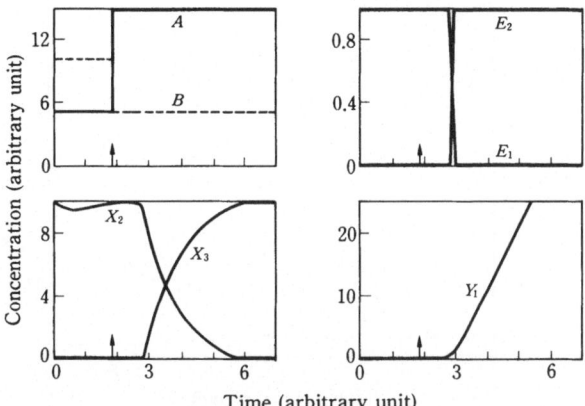

**Fig. 6. 21** Effect of levels of $A$ and $B$ on time course of Scheme 6.16. Arrows indicate the switching time when the level of $A$ exceeds that of $B$.

with $Y_1$ just exceeding $Y_2$. The TRS would thus be effective as a switching control element in the real metabolic regulation.

### 3. Feedback Control System

A typical example of enzymatic feedback system in which the TRS is installed as a control element is shown in Scheme 6.17, in which A represents the input to the system and $k_i$ denote the rate constants. $X_i$, $Y_2$ and $Y_3$ are reactants or intermediates, and $Y_1$ serves as an inhibitor. $E_a$ and $E_i$

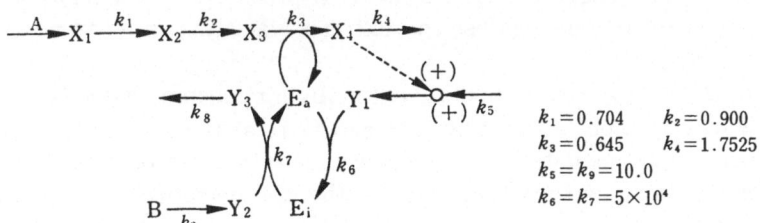

$$k_1 = 0.704 \quad k_2 = 0.900$$
$$k_3 = 0.645 \quad k_4 = 1.7525$$
$$k_5 = k_9 = 10.0$$
$$k_6 = k_7 = 5 \times 10^4$$

Scheme 6.17. Feedback system with the TRS

represent the active and inactive enzymes, respectively. B is the precursor of $Y_2$ and its concentration is assumed to be equal to the desired constant concentration of $X_4$ (end product). Under an initial condition,

$$X_1 = 0.821, \quad X_2 = 0.533, \quad X_3 = 0.915, \quad X_4 = 0.543,$$

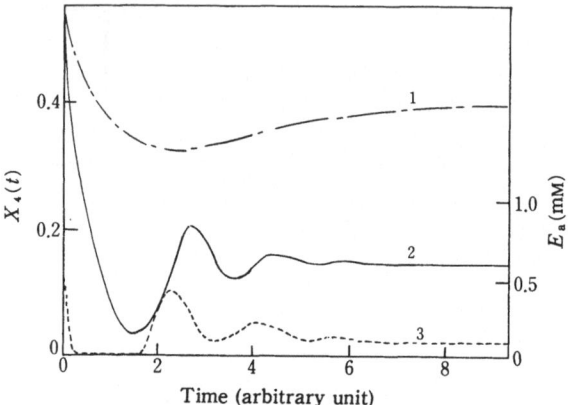

**Fig. 6.22**  Time course of Scheme 6.17.  1: $X_4(t)$ without TRS; 2: $X_4(t)$ ; 3: $E_a$. The desired value of $X_4$ is 0.15 ($B=0.15$).

$$Y_1 = Y_2 = Y_3 = 10.0, \quad E_a = E_1 = 0.5, \quad A = 0.701, \quad \text{(in mM)}, \quad (56)$$

the time course of $X_4(t)$ is simulated and the results are shown in Fig. 6.22 [16]. The concentration of precursor B is fixed at 0.15 mM. Curve 1 in the figure indicates the time course of $X_4(t)$ of linear chain ($\to X_1 \to X_2 \to X_3 \to X_4 \to$) without the TRS, which converges to 0.4 mM. In contrast, curve 2 with the TRS reaches with damped oscillation to 0.15 mM, which is equal to the given value of $B$.  The time course of $X_4(t)$ with ten-fold larger value of $A$ ($=7.01$ mM) is shown in Fig. 6.23, revealing that a ten-fold increase in the input leads to no change in the stationary value of $X_4(t)$.  This suggests that Scheme 6.17 provides an excellent constant-value-control system against the extreme variation in exogeneous substrate supply.

In order for a system to operate optimally for constant-value control, the following conditions at least need be satisfied ; a) the concentration of end product or intermediates at a stationary state does not deviate as a result of the external perturbation, and b) the concentration of end product converges to the constant value at stationary states against the input perturbation. It is obvious from Fig. 6.23 that Scheme 6.17 has a structure satisfying the above condition b), if we are concerned only with the constant value of $X_4$.

Next, we examine the effects of external perturbation on the stationary values of all chemical species.  The effect of step perturbation ($r$) imposed at $X_i$ on the stationary values of $X_i(\bar{X}_i)$ is summarized in Table 6.15.  The relative deviation $Q$ is defined by

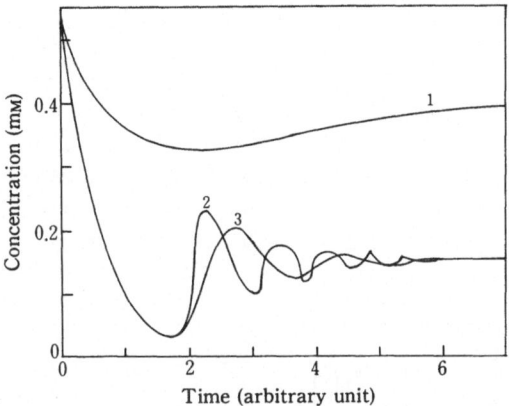

**Fig. 6. 23**   Time course of $X_4(t)$ in Scheme 6.17.   1: without the TRS;
2: input $A=7.01$; 3: input $A=0.701$.

**Table 6. 15**   Effect of step perturbation $r$ on steady states in Scheme 6.17

| $r$ | $\bar{X}_1$ | $Q$ | $\bar{X}_2$ | $Q$ | $\bar{X}_3$ | $Q$ | $\bar{X}_4$ | $Q$ |
|---|---|---|---|---|---|---|---|---|
| 0 | 0. 996 | 0 | 0. 779 | 0 | 43. 1 | 0 | 0. 150 | 0 |
| 1. 30 | 2. 84 | 185 | 2. 22 | 185 | 93. 7 | 117 | 0. 150 | 0 |
| 2. 50 | 4. 55 | 357 | 3. 56 | 357 | 128 | 197 | 0. 150 | 0 |
| 6. 00 | 9. 52 | 856 | 7. 45 | 856 | 227 | 427 | 0. 150 | 0 |
| 10. 0 | 15. 2 | 1430 | 11. 9 | 1430 | 341 | 691 | 0. 150 | 0 |

$$Q=\frac{100|\beta-\alpha|}{\alpha},\tag{57}$$

where $\alpha$ is the stationary value of $X_i$ under the condition of $r=0$ and $\beta$ is
the stationary value of $X_i$ when the step perturbation $r$ is introduced to $X_i$.
It is seen in Table 6.15 that the stationary value $\bar{X}_4$ remains constant at
0.15 mM (equal to $B$), while the stationary values of $X_1$, $X_2$ and $X_3$ increase
with $r$.

The effect of ramped perturbation imposed at $X_1$ on the stationary values
of $X_i$ is demonstrated in Table 6.16.  The perturbation is represented by

$$\epsilon=\frac{1}{2}t^2,\tag{58}$$

and time 0 indicates the starting point of the perturbation.  $Q$ is as defined
above.  $\bar{X}_4$ does not vary with the ramped perturbation.  The same results
as in Tables 6.15 and 6.16 are obtained in cases where the external

**Table 6.16**  Effect of ramped perturbation on $X_i$ in Scheme 6.17

| Time | $X_1$ | $Q$ | $X_2$ | $Q$ | $X_3$ | $Q$ | $X_4$ | $Q$ |
|---|---|---|---|---|---|---|---|---|
| 0 | 0.996 | 0 | 0.779 | 0 | 43.1 | 0 | 0.150 | 0 |
| 5.0 | 6.20 | 522 | 3.73 | 379 | 49.9 | 15.8 | 0.150 | 0 |
| 10.2 | 13.3 | 1240 | 9.16 | 1080 | 77.6 | 80.0 | 0.150 | 0 |
| 25.2 | 34.6 | 3370 | 25.8 | 3210 | 309 | 617 | 0.150 | 0 |
| 45.1 | 63.1 | 6240 | 48.1 | 6080 | 972 | 2160 | 0.150 | 0 |

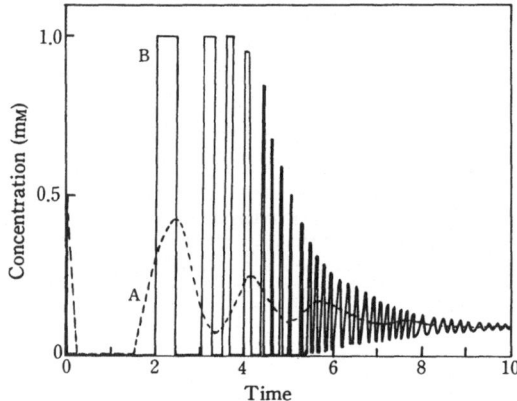

**Fig. 6.24**  Time course of $E_a(t)$ in Scheme 6.17. A : $k_5(=k_9)=10.0$ ; B : $k_5(=k_9)=1000$. $X_4(t)$ converges to the desired value of 0.15 in both cases.

perturbations are introduced to $X_2$ or $X_3$. On the other hand, when the perturbation is introduced directly on $X_4$ itself, $X_4$ increases with the values of $r$ and $\epsilon$. This infers that for the constant-value control of $X_4$ Scheme 6.17 also has a structure which can satisfy the above condition a). Scheme 6.17 is not suitable, however, for the control of the other intermediates ($X_1$, $X_2$ and $X_3$).

We are further concerned with the effect of $k_5$ ($=k_9$) in Scheme 6.17 on the time course of active enzyme $E_a$ for the case where $E_a$ and $E_i$ are initially partitioned in equal amounts. The time courses of $E_a(t)$ are shown in Fig. 6.24 in which, with increase in the value of $k_5$ ($=k_9$), $E_a(t)$ fluctuates stepwise in the early phase, followed by damped spike-oscillation (curve B). In spite of such fluctuation in $E_a(t)$, $X_4(t)$ converges to the desired value ($B=0.15$ mM), regardless of the value of $k_5$. For the initial concentration of total enzyme, $E_a(0)=E_i(0)=40\ \mu M$ (instead of 500 $\mu M$ in

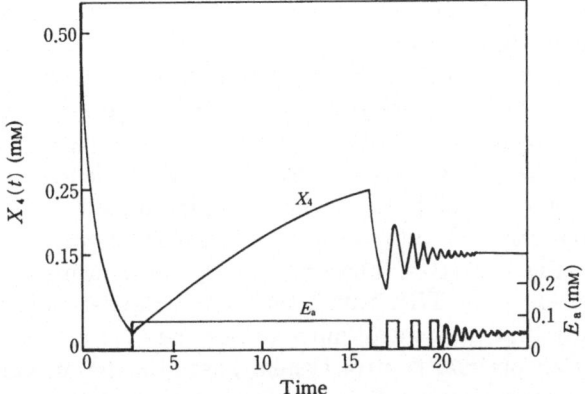

**Fig. 6. 25**  Effect of initial concentration of $E_a$ and $E_1$ on time course of $X_4(t)$ in Scheme 6.17. $k_5 = k_9 = 1000$. $E_{a0} = E_{10} = 0.04$ mM. The desired value of $X_4(t) = 0.15$ mM.

(A) Simple feedback control

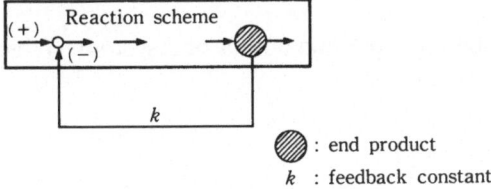

                : end product

      $k$   : feedback constant

(B) Adaptive control

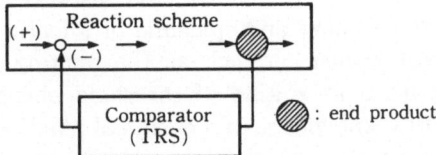

                : end product

**Fig. 6. 26**  Comparison of the mechanisms of simple feedback system and adaptive control. The adaptive control system employs the TRS as comparator.

the previous cases), the time courses of $X_4(t)$ and $E_a(t)$ display quite different profiles. As shown in Fig. 6.25, the rapidly decreased $X_4(t)$ increases again gradually during the first 16 time-units, while $E_a(t)$ keeps the maximum value (0.08 mM). After $t = 16$, $X_4(t)$ suddenly follows a kind of damped oscillation with high frequency, converging to the desired value of 0.15 mM. $E_a(t)$ exhibits the corresponding change. Thus, it is obvious that Scheme 6.17 can also maintain $X_4$ at a desired level against the endogenous parametric variation.

As described in the preceding section 6.4, a simple feedback system such as in Fig. 6.26(A) cannot keep the end product or intermediates at desired values against various perturbations. On the other hand, it is found that the feedback system with the TRS as a control element can achieve a desired value for the end product which is equal to the value of $B$ in Scheme 6.17, and that the TRS may serve as a comparator installed in the adaptive control (Fig. 6.26(B)). These regulatory mechanisms, especially catastrophic behavior of the TRS, have been discussed in detail [18]. It has been found recently that the control mode of Scheme 6.17 is in good agreement with that of an optimal control system based on the maximum principle [17]. Furthermore, by installing the TRS as control element, we can construct a rate-sensitive feedback system at the molecular level. This system is more effective for excluding perturbation than Scheme 6.17 which is regarded as a concentration-sensitive feedback system [17].

## 6.6  System with Threshold

Let $x(t)$ and $y(t)$ be the input and output of a system. When the input-output relation is represented by

$$y(t) = \begin{cases} 0 & x < \theta \\ f(x) & x \geq \theta, \end{cases} \qquad (59)$$

$\theta$ is called the threshold value of the input. The threshold phenomena are frequently observed in biological systems; for instance, they are well known in the drug-response curve and epidemic of infectious diseases. In physical chemistry, phase transition such as the melting and boiling of substances may be regarded as a kind of threshold phenomena. There seems, however, to be little knowledge of a chemical reaction system which can give rise to a distinct threshold phenomenon. If an enzyme system *in vivo* exhibited the threshold and its structure could be determined, the fundamental mechanism might be understood for the threshold on the

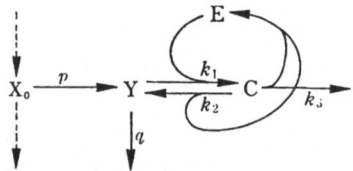

Scheme 6.18.  System with threshold

physiological level. In this section, we estimate the structure of enzyme system which is required to yield the threshold phenomenon.

Ličko [19] reported that the steady-state solution of the rate equation of the system represented by Scheme 6.18 demonstrated the threshold in the relationship between the input $(X_0)$ and the output $(Y)$. The plot of the input $X_0$ (on the abscissa) against the output $Y$ (on the coordinate) at stationary state exhibits a discontinuous point at a certain value $(\theta)$ of the input. The threshold value $\theta$ may be represented by

$$\theta = \frac{k_3 E_0}{p},\tag{60}$$

where $E_0$ is the total concentration of enzyme, and $p$ and $k_3$ are the rate constants at the indicated steps in Scheme 6.18. The results of the computer simulation with variation in the value of $K_m$ of enzyme E reveal that the threshold is only observed with $K_m \to 0$ as shown in Fig. 6.27 [20, 21]. This implies that the threshold on $Y$ appears only when the enzyme system consumes $Y$ stoichiometrically and rapidly.

In general, the input-output curve with a distinct threshold exhibits a saturation profile, and in an extreme case the curve may change its profile stepwise at the threshold value. In order to realize the saturation profile, the feedback loops are added to Scheme 6.18, leading to Scheme 6.19. However, the saturation of $Y$ is not observed in Scheme 6.19; $Y$ increases smoothly with increase in $X_0$. Consequently, a positive feedforward path is

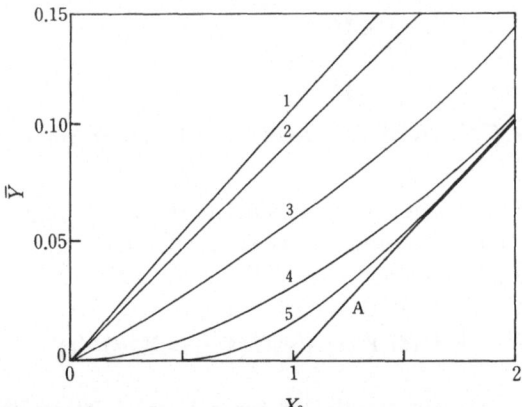

**Fig. 6.27**  Effect of $K_m$ on the relation between input $X_0$ and output $\bar{Y}$ in Scheme 6.18. $K_m$: 1: 2.0; 2: 1.0; 3: 0.1; 4: 0.01; 5: 0.001. The straight line A represents $\bar{Y} = (p/q)(X_0 - k_3 E_0/p)$, which corresponds to $K_m = 0$.

(A)

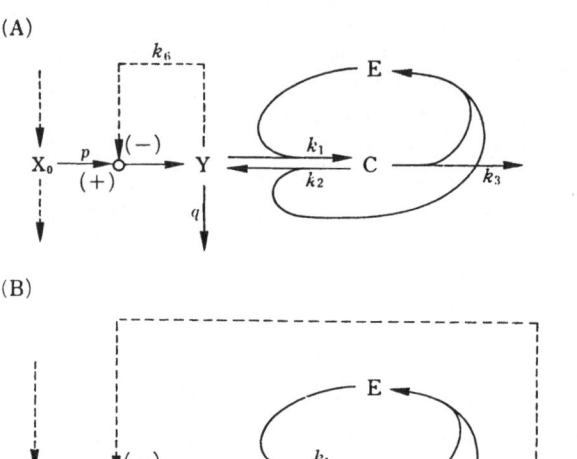

(B)

Scheme 6.19.   Addition of feedback loops to Scheme 6.18

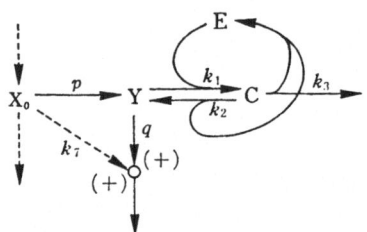

Scheme 6.20.   Addition of feedforward loop to Scheme 6.18

added to Scheme 6.18 to generate Scheme 6.20 [22]. The effect of the feedforward loop is represented by $k_7 X_0$ in the last term of the rate equation for output $Y$ :

$$\frac{dY(t)}{dt} = pX_0 - k_1 E(t) Y(t) + k_2 C(t) - k_7 X_0 q Y(t). \qquad (61)$$

The results of the computer simulation for Scheme 6.20 are shown in Fig. 6.28. The feedforward loop causes a distinct threshold and saturation in the curves. The value of $k_7$ does not change the general profile of saturation, but its increase causes $Y(t)$ to have lower values.

Representation of the feedforward effect by the term of $k_7 X_0$ does not

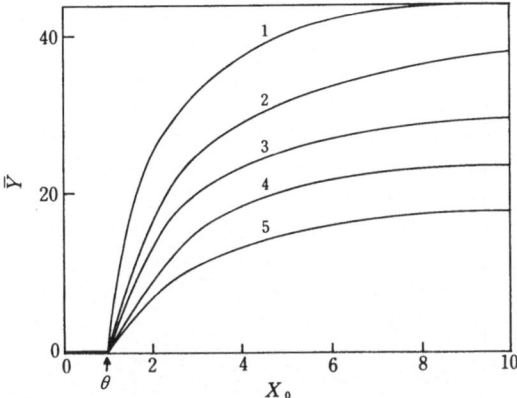

**Fig. 6. 28** Effect of $k_7$ on the input-output relation of Scheme 6.20. $k_7$: 1: 0.4; 2: 0.5; 3: 0.6; 4: 0.8; 5: 1.0. $\theta = 1.0$.

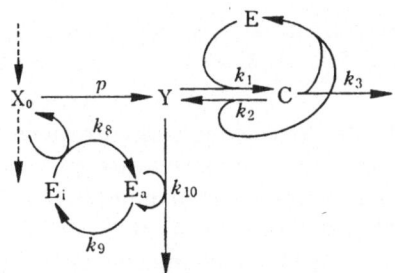

Scheme 6. 21.  Enzyme system with feedforward loop

reflect the actual enzyme systems. The feedforward loop should be performed through the action of enzymes. Thus, the molecular mechanism shown in Scheme 6.21 is chosen as the structure of the feedforward loop [22], in which $E_a$ and $E_i$ represent active and inactive enzyme species, respectively. In this scheme, the reaction processes related to the positive feedforward loop are represented by

$$\frac{dY(t)}{dt} = pX_0 - k_1 E(t) Y(t) + k_2 C(t) - k_{10} E_a(t) Y(t)$$

$$\frac{dE_i(t)}{dt} = - k_8 X_0 E_i(t) + k_9 E_a(t)$$

$$\frac{dE_a(t)}{dt} = k_8 X_0 E_i(t) - k_9 E_a(t)$$

$$K = \frac{\bar{E}_a}{X_0 \bar{E}_i} = \frac{k_8}{k_9},$$

(62)

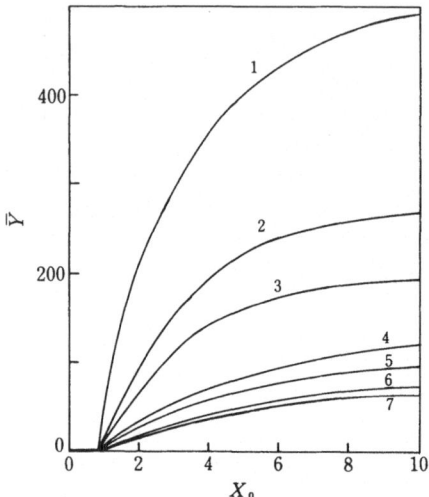

**Fig. 6. 29** Effect of $K$ on the input-output relation of Scheme 6.21. $K$:
1: $10^{-2}$; 2: $2 \times 10^{-2}$; 3: $3 \times 10^{-2}$; 4: 0.1; 5: 0.2; 6: 1.0; 7: 10.

where $\bar{E}_a$ and $\bar{E}_i$ indicate the enzyme concentrations at a stationary state. The results of the computer simulation are presented in Fig. 6.29. As expected, $K$, which may express the magnitude of feedforward effect, does not affect the general profile of saturation either. Hence, it is obvious that the system with feedforward through the conversion $E_i \leftrightarrow E_a$ can realize the threshold phenomenon with saturation.

Since the positive feedforward loop is considered to have an effect similar to that of positive feedback loop, we attempt to install a positive feedback loop in the chemical reaction system to obtain a threshold with saturation. The postulated structure is given in Scheme 6.22. The rate equation for $Y$

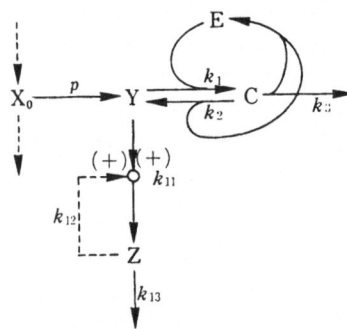

Scheme 6. 22.   Positive feedback system with threshold

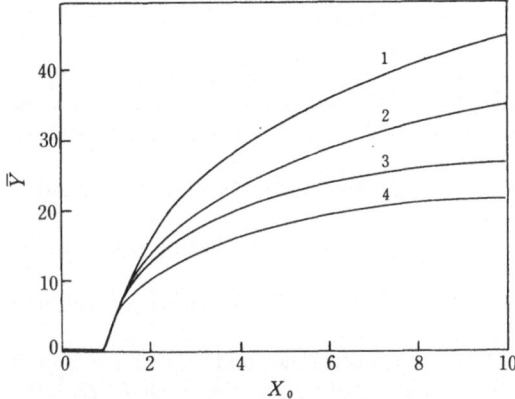

**Fig. 6. 30** Effect of $k_{13}$ on the input-output relation of Scheme 6.22. $k_{13}$: 1: 0.3; 2: 0.2; 3: 0.15; 4: 0.1. $k_{12} = 0.10$.

and $Z$ may be written as

$$\frac{dY(t)}{dt} = pX_0 - k_1 E(t) Y(t) + k_2 C(t) - k_{11}[1 + k_{12}Z(t)] Y(t)$$

$$\frac{dZ(t)}{dt} = k_{11}[1 + k_{12}Z(t)] Y(t) - k_{13}Z(t).$$ (63)

The results of the computer simulation are shown in Fig. 6.30. With a small value of $k_{13}$ the saturation curve is obtained. Therefore, it is clear that the threshold with saturation can also be realized by positive feedback loop.

The threshold in input-output curve of enzymatic reaction may be attained by an unexpectedly simple reaction-scheme. For realizing the saturation curve, however, the enzyme system should contain either positive feedforward or feedback loop which accelerates the consumption of output Y. The roles of positive feedforward and feedback loops in the enzyme system have not been discussed in connection with the specific behavior of the system. The results of the simulation that positive feedforward and feedback loops cause the saturation curve imply that these devices play a role in the homeostatic control of metabolites against the fluctuation of input flow.

## References

[ 1 ] Okamoto, M. and K. Hayashi (1985). Kinetic characteristics of biocbemical cycle reaction systems : Amplification of substrate cycle system. *Biotech. Bioeng.*, **27**, 122-136.

[ 2 ]   Newsholme, E.A. and C. Start (1973). "Regulation in Metabolism," Wiley-Interscience, New York.

[ 3 ]   Chipman, D.M. (1971). A kinetic analysis of the reaction of lysozyme with oligosaccharides from bacterial cell walls. *Biochemistry,* **10**, 1714-1722.

[ 4 ]   Koga, D. and K. Hayashi (1976). Activation process of pepsingen. *J. Biochem.,* **79**, 549-558.

[ 5 ]   Goldbeter, A. and R. Caplan (1976). Oscillatory enzymes. *Ann. Rev. Biophys. Bioeng.,* **5**, 449-476.

[ 6 ]   Higgins, J. (1967) . The theory of oscillating reactions. *Ind. Eng. Chem.,* **59**, 18-62.

[ 7 ]   Morales, M. and D. McKay (1967). Biochemical oscillation in controlled system. *Biophys. J.,* **7**, 621-625.

[ 8 ]   Walter, C. F. (1970). The occurrence and signficance of limit cycle behavior in controlled biochemical systems. *J. Theor. Biol.,* **27**, 259-272.

[ 9 ]   Goldbeter, A. and G. Nicolis (1976). An allosteric enzyme model with positive feedback applied to glycolytic oscillations. *Prog. Theor. Biol.,* **4**, 65-160.

[10]   Marek, M. and I. Stuchl (1975). Synchronization in two interacting oscillatory systems. *Biophys. Chem.,* **3**, 241-248.

[11]   Field, R. J., E. Koros and R. M. Noyes (1972). Oscillations in chemical systems. II. Thorough analysis of temporal oscillation in the bromate-cerium-malonic acid system. *J. Am. Chem. Soc.,* **94**, 8649-8664.

[12]   Boiteux, A., A. Goldbeter and B. Hess (1975). Control of oscillating glycolysis of yeast by stochastic, periodic, and steady source of substrate: a model and experimental study. *Proc. Natl. Acad. Sci. USA,* **72**, 3829-3833.

[13]   Johnson, C. D. (1968). Optimal control of the linear regulator with constant disturbances. *IEEE Trans. Automatic Control,* **AC-13**, 416-421.

[14]   Johnson, C. D. (1970). Further study of the linear regulator with disturbance —The case of vector disturbances satisfying a linear differential equation. *ibid.,* **AC-15**, 222-228.

[15]   Rosen, R. (1967). Two-factor models, neural, and biochemical automata. *J. Theor. Biol.,* **15**, 282-279.

[16]   Okamoto, M., A. Katsurayama, M. Tsukiji, Y. Aso and K. Hayashi (1980). Dynamic behavior of enzymatic system realizing two-factor model. *ibid.,* **83**, 1-16.

[17]   Okamoto, M. and K. Hayashi (1984). Optimal control mode of a biochemical feedback system. *BioSystems,* **16**, 315-321.
        Okamoto, M. and K.Hayashi (1984). Homeostatic capability of rate sensitive feedback system: Mathematical model. *Am. J. Physiol.,* **247**, R927-R931.
        Okamoto, M. and K.Hayashi (1984). Frequency conversion mechanism in enzymatic feedback systems. *J. Theor. Biol.,* **108**, 529-537.

[18]   Okamoto, M. and K.Hayashi (1983). Dynamic behavior of cyclic enzyme system. *ibid.,* **104**, 591-598.
        Okamoto, M. and K.Hayashi (1985). Control mechanism for a bacterial sugar-transport system: Theoretical hypothesis. *ibid.,* **113**, 785-790.

[19] Ličko, V. (1972). Some biochemical threshold mechanisms. *Bull. Math. Biophys.*, **34**, 103-112.

[20] Okuyama, K., M. Okamoto, Y. Aso and K. Hayashi (1976). Studies on the threshold mechanism of enzymatic reaction. *J. Fac. Agr. Kyushu Univ.*, **20**, 87-95.

[21] Okuyama, K., M. Okamoto and K. Hayashi (1977). On the saturation of the threshold-type response in enzymatic reaction system. *ibid.*, **21**, 67-77.

[22] Okuyama, K., M. Okamoto, Y. Aso and K. Hayashi (1978). Threshold of enzymatic reaction system: Further studies with a simplified model. *ibid.*, **22**, 153-160.

# CHAPTER 7

# Analysis of Reaction–Diffusion Systems

Enzyme systems actually operate in inhomogeneous cellular environments in which chemical species are spatially distributed to constitute most commonly reaction-diffusion systems. In the case of *in vitro* and engineering systems, likewise, many enzyme systems need to be treated as reaction-diffusion systems. In fact, the enzyme systems in homogeneous system analyzed in the preceding chapters result from the approximation by ignoring the spatial distribution of chemical species in systems. The rate equation for such inhomogeneous systems is expressed by a set of nonlinear partial differential equations, as described in Chapter 1. The methods of linear and steady-state approximations are explained in Chapter 2, and the numerical methods are introduced in Chapter 3. The rate equation can be analyzed by these methods to reveal and characterize the dynamic behavior of the objective inhomogeneous system, which is carried out in the following.

In this chapter we are concerned with a reaction-diffusion system with feedback loop [1] and a partition-chromatographic column of a reacting system [2]. The dynamic behavior of these systems is examined by computer simulation, that is, the numerical solution of a boundary value problem of a system of partial differential equations. In Section 7.1 the behavior of a sequential metabolic pathway with negative feedback loop working as reaction-diffusion system is analyzed with respect to the effectiveness of enzyme feedback system in the homeostatic control of metabolism. In Section 7.2 the action of partition-column chromatography is studied for $N$-acetylglucosamine which undergoes a mutarotation reaction and diffusion during the flow through the column.

---

This chapter was written by Katsuya Hayashi.

## 7.1   Reaction-Diffusion System with Feedback Loop

In order to survey the effectiveness of enzyme feedback system in the homeostatic control of metabolism, we consider the enzyme system shown
in Scheme 7.1, in which S, $E_n$, $C_n$ and $P_n$ represent the substrate, enzyme,
enzyme-substrate complex and product, respectively, and $k_{ij}$ are the rate
constants. The first enzyme $E_1$ is assumed to be an allosteric enzyme which

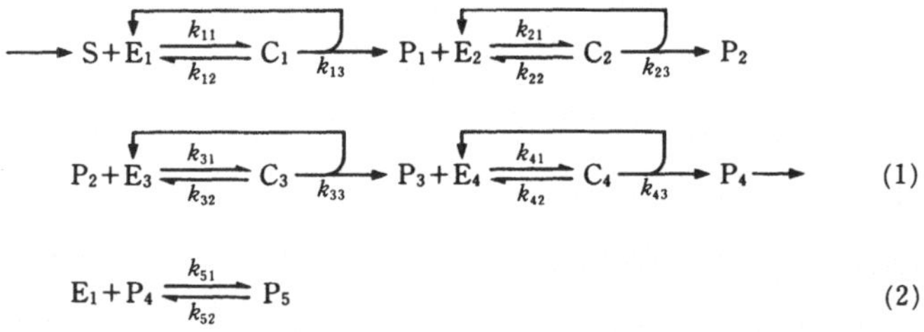

$$\longrightarrow S + E_1 \underset{k_{12}}{\overset{k_{11}}{\rightleftarrows}} C_1 \xrightarrow{k_{13}} P_1 + E_2 \underset{k_{22}}{\overset{k_{21}}{\rightleftarrows}} C_2 \xrightarrow{k_{23}} P_2$$

$$P_2 + E_3 \underset{k_{32}}{\overset{k_{31}}{\rightleftarrows}} C_3 \xrightarrow{k_{33}} P_3 + E_4 \underset{k_{42}}{\overset{k_{41}}{\rightleftarrows}} C_4 \xrightarrow{k_{43}} P_4 \longrightarrow \qquad (1)$$

$$E_1 + P_4 \underset{k_{52}}{\overset{k_{51}}{\rightleftarrows}} P_5 \qquad\qquad\qquad (2)$$

Scheme 7.1.   Linear enzyme system with negative feedback
loop

interacts negatively with product $P_4$. The complex of product $P_4$ and
enzyme $E_1$ is denoted by $P_5$. It is further assumed that enzymes $E_1 \sim E_4$ are
located at the fixed positions $L_1 \sim L_4$ in a region of the one-dimensional
space and that substrate S enters into the region through boundary (0) as
shown in Fig. 7.1. All chemical species except enzymes and enzyme-re-
actant complexes can move in the region by free diffusion, but the behavior of chemical species at the boundaries is specified by the boundary conditions.

### 1.   Rate Equation

The activity of allosteric enzyme $E_1$ in the presence of the inhibitor $P_4$ is
postulated to be represented by a simple Hill equation: the rate constant
$k_{13}$ for the activity of $E_1$ is replaced by

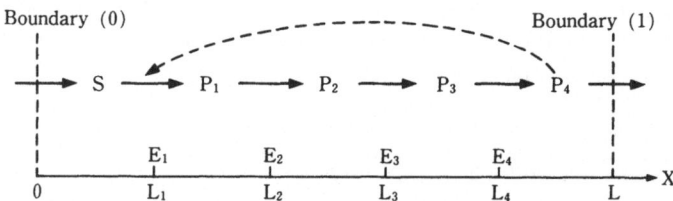

**Fig. 7.1**   Reaction-diffusion system with feedback loop.

$$\frac{k_{13}}{1+\alpha P_5} \quad (\alpha: \text{constant}).\tag{1}$$

The rate equation for the fixed enzyme system may be given by

$$\left.\begin{aligned}
\frac{\partial [E_1]}{\partial T} &= -k_{11}[S][E_1]+k_{12}[C_1]+\frac{k_{13}}{1+\alpha [P_5]}[C_1]\\
\frac{\partial [E_2]}{\partial T} &= -k_{21}[P_1][E_2]+(k_{22}+k_{23})[C_2]\\
\frac{\partial [E_3]}{\partial T} &= -k_{31}[P_2][E_3]+(k_{32}+k_{33})[C_3]\\
\frac{\partial [E_4]}{\partial T} &= -k_{41}[P_3][E_4]+(k_{42}+k_{43})[C_4]\\
\frac{\partial [C_n]}{\partial T} &= -\frac{\partial [E_n]}{\partial T},\quad n=1,2,3,4\\
\frac{\partial [P_5]}{\partial T} &= k_{51}[P_4][E_1]-k_{52}[P_5]
\end{aligned}\right\}\tag{2}$$

$$\left.\begin{aligned}
\frac{\partial [S]}{\partial T} &= D_s\frac{\partial^2 [S]}{\partial X^2}-k_{11}[S][E_1]+k_{12}[C_1]\\
\frac{\partial [P_1]}{\partial T} &= D_1\frac{\partial^2 [P_1]}{\partial X^2}-k_{21}[P_1][E_2]+k_{22}[C_2]+\frac{k_{13}}{1+\alpha [P_5]}[C_1]\\
\frac{\partial [P_2]}{\partial T} &= D_2\frac{\partial^2 [P_2]}{\partial X^2}-k_{31}[P_2][E_3]+k_{32}[C_3]+k_{23}[C_2]\\
\frac{\partial [P_3]}{\partial T} &= D_3\frac{\partial^2 [P_3]}{\partial X^2}-k_{41}[P_3][E_4]+k_{42}[C_4]+k_{33}[C_3]\\
\frac{\partial [P_4]}{\partial T} &= D_4\frac{\partial^2 [P_4]}{\partial X^2}+k_{43}[C_4]-k_{51}[P_4][E_1]+k_{52}[P_5],
\end{aligned}\right\}\tag{3}$$

where brackets represent the molar concentration, and $D_s$ and $D_n$ ($n=1,2,3,4$) are diffusion coefficients (in cm$^2$ sec$^{-1}$) of the substrate and product $P_n$ ($n=1,2,3,4$), respectively. The conservation equation is written as

$$[E_n]+[C_n]=E_{n0}\delta_{XL_n}\quad (n=1,2,3,4),\tag{4}$$

where $\delta_{XL_n}$ is a Kronecker's delta and $E_{n0}$ ($n=1,2,3,4$) are the total concentrations of enzymes.

The initial conditions are set as

$$\begin{aligned}
&[S(0,0)]=S_0\quad [S(X,0)]=0,\quad 0<X\le L\\
&[P_n(X,0)]=0\quad (n=1,2,\cdots,5)\quad 0<X\le L\\
&[C_n(X,0)]=0\quad (n=1,2,3,4)\quad 0<X\le L\\
&[E_n(L_n,0)]=E_{n0}\quad (n=1,2,3,4)\\
&[E_n(X,0)]=0\quad X\ne L_n\quad (n=1,2,3,4),
\end{aligned}\tag{5}$$

and the boundary conditions as

$$[S(0,T)] = S_0$$

$$\frac{\partial[P_n(0,T)]}{\partial X} = a[P_n(0,T)] \quad (n=1,2,3,4)$$

$$\frac{\partial[S(L,T)]}{\partial X} = a[S(L,T)] \tag{6}$$

$$\frac{\partial[P_n(L,T)]}{\partial X} = a[P_n(L,T)] \quad (n=1,2,3,4)$$

$$(a : \text{positive constant}).$$

The boundary conditions indicate that the products at boundary (0) and the substrate and products at boundary (1) can permeate to the outside at rates in proportion to their concentrations at the boundaries.

For dimensionless formulation of the rate equation, new variables and parameters are defined as follows:

$$x = X/L, \quad t = T/\theta, \quad \theta = L^2/D_S,$$

$$\eta_n = D_n/D_S \quad (n=1,2,3,4)$$

$$S = [S]k_{11}/k_{12}$$

$$P_n = [P_n]k_{n+11}/k_{n+12} \quad (n=1,2,3)$$

$$P_4 = [P_4]k_{51}/k_{52}$$

$$E_n = [E_n]/E_{n0} \quad (n=1,2,3,4)$$

$$P_5 = [P_5]/E_{10} \tag{7}$$

$$\sigma_n = k_{n1}E_{10}\theta \quad (n=1,2,3,4)$$

$$\sigma_5 = k_{51}E_{10}$$

$$\varepsilon_1 = \frac{k_{13}k_{21}\theta}{k_{22}(1+E_{10}P_5)}, \quad \varepsilon_2 = \frac{k_{31}k_{23}E_{20}\theta}{k_{32}}$$

$$\varepsilon_3 = \frac{k_{41}k_{33}E_{30}\theta}{k_{42}}, \quad \varepsilon_4 = \frac{k_{51}k_{43}E_{40}\theta}{k_{52}}$$

$$\zeta_{n2} = k_{n2}\theta, \quad \zeta_{n3} = k_{n3}\theta, \quad (n=1,2,\cdots,5).$$

Hence, the dimensionless rate equation may be written as

$$\frac{\partial S}{\partial t} - \frac{\partial^2 S}{\partial x^2} + \sigma_1(SE_1 + E_1 - 1) = 0 \tag{8}$$

$$\frac{\partial P_1}{\partial t} - \eta_1\frac{\partial^2 P_1}{\partial x^2} + \sigma_2(P_1E_2 + E_2 - 1) + \varepsilon_1(E_1 - 1) = 0$$

$$\frac{\partial P_2}{\partial t} - \eta_2\frac{\partial^2 P_2}{\partial x^2} + \sigma_3(P_2E_3 + E_3 - 1) + \varepsilon_2(E_2 - 1) = 0 \tag{9}$$

$$\frac{\partial P_3}{\partial t} - \eta_3\frac{\partial^2 P_3}{\partial x^2} + \sigma_4(P_3E_4 + E_4 - 1) + \varepsilon_3(E_3 - 1) = 0$$

$$\frac{\partial P_4}{\partial t} - \eta_4 \frac{\partial^2 P_4}{\partial x^2} + \sigma_5 (P_4 E_1 - P_5) + \varepsilon_4 (E_4 - 1) = 0 \tag{10}$$

$$\frac{\partial E_1}{\partial t} + \zeta_{12} (SE_1 + E_1 - 1) + \zeta_{13} (E_1 - 1) / (1 + \alpha E_{10} P_5) = 0 \tag{11}$$

$$\frac{\partial E_2}{\partial t} + \zeta_{22} (P_1 E_{22} + E_2 - 1) + \zeta_{23} (E_2 - 1) = 0$$

$$\frac{\partial E_3}{\partial t} + \zeta_{32} (P_2 E_3 + E_3 - 1) + \zeta_{33} (E_3 - 1) = 0 \tag{12}$$

$$\frac{\partial E_4}{\partial t} + \zeta_{42} (P_3 E_4 + E_4 - 1) + \zeta_{43} (E_4 - 1) = 0$$

$$\frac{\partial P_5}{\partial t} + \zeta_{52} (P_5 - P_4 E_1) = 0. \tag{13}$$

## 2. Difference Equation

The time interval $[0, t]$ and the space region $[0, 1]$ are divided into equal steps with small width $\Delta t = k$ and small length $\Delta x = h$, respectively. Let $S(i, j)$ and $P(i, j)$ represent the concentrations at a lattice point $(i, j)$. The difference equations are derived from the dimensionless rate equation by the Crank-Nicolson method (see Section 3.4) ; from equation (8)

$$-\frac{r}{2} S(i-1, j+1) + \left\{1 + r + \frac{k}{2} \sigma_1 E_1 (i, j)\right\} S(i, j+1)$$

$$-\frac{r}{2} S(i+1, j+1) = \frac{r}{2} \{S(i-1, j) + S(i+1, j)\}$$

$$+ (1-r) S(i, j) - \frac{k}{2} \sigma_1 \{S(i, j) E_1 (i, j) + 2 E_1 (i, j) - 2\}; \tag{14}$$

from equation (9)

$$-\frac{r}{2} \eta_n P_n (i-1, j+1) + \left\{1 + r + \frac{k}{2} \sigma_{n+1} E_{n+1} (i, j)\right\} P_n (i, j+1)$$

$$-\frac{r}{2} \eta_n P_n (i+1, j+1)$$

$$= \frac{r}{2} \eta_n \{P_n (i-1, j) + P_n (i+1, j)\} + (1 - r\eta_n) P_n (i, j)$$

$$-\frac{k}{2} \sigma_{n+1} \{P_n (i, j) E_{n+1} (i, j) + 2 E_{n+1} (i, j) - 2\}$$

$$- k\varepsilon_n \{E_n (i, j) - 1\} \qquad (n = 1, 2, 3) ; \tag{15}$$

from equation (10)

$$-\frac{r}{2} \eta_4 P_4 (i-1, j+1) + \left\{1 + r\eta_4 + \frac{k}{2} \sigma_5 E_1 (i, j)\right\} P_4 (i, j)$$

$$-\frac{r}{2}\,\eta_4 P_4\,(i+1,j+1) = \frac{r}{2}\,\eta_4\{P_4\,(i-1,j)+P_4\,(i+1,j)\}$$
$$+ (1-r\eta_4)\,P_4\,(i,j) - \frac{k}{2}\,\sigma_5\{P_4\,(i,j)\,E_1\,(i,j) - 2\,P_5\,(i,j)\}$$
$$- k\varepsilon_4\{E_4\,(i,j)-1\}\,;$$

(16)

from equation (11)

$$\left[\frac{1}{k}+\frac{\varsigma_{12}}{2}\{S(i,j+1)+1\}+\frac{\varepsilon_5}{2}\right]E_1\,(i,j+1)$$
$$=\left[\frac{1}{k}+\frac{\varsigma_{12}}{2}\{S(i,j)+1\}-\frac{\varepsilon_5}{2}\right]E_1\,(i,j)$$
$$+\varsigma_{12}+\varsigma_{13}\Big/\left[1+\frac{\alpha E_{10}}{2}\{P_5\,(i,j+1)+P_5\,(i,j)\}\right];$$

(17)

from equation (12)

$$\left[\frac{1}{k}+\frac{1}{2}\{\varsigma_{n2}P_{n-1}\,(i,j+1)+\varsigma_{n3}\}\right]E_n\,(i,j+1)$$
$$=\left[\frac{1}{k}+\frac{1}{2}\{\varsigma_{n2}P_{n-1}\,(i,j+1)+\varsigma_{n3}\}\right]E_n\,(i,j)$$
$$+\varsigma_{n2}+\varsigma_{n3} \qquad (n=2,3,4)\,; \qquad (18)$$

(18)

from equation (13)

$$\left(\frac{1}{k}+\frac{\varsigma_{52}}{2}\right)P_5\,(i,j+1) = \left(\frac{1}{k}-\frac{\varsigma_{52}}{2}\right)P_5\,(i,j)$$
$$+\frac{1}{2}\,\varsigma_{52}\{P_4\,(i,j+1)\,E_1\,(i,j)+P_4\,(i,j)\,E_1\,(i,j)\}\,,$$

(19)

where $r=k/h^2$.

## 3.   Time Course from Simulation

All the difference equations $(14)\sim(19)$ are numerically solved by the Gauss elimination method with $\Delta t=k=0.001$ and $\Delta x=h=0.01$. The boundary conditions are represented by equations $(3.81)\sim(3.83)$ and used directly for the computation. The numerical solutions are obtained with the values of $E_{n0}=1.0\,(n=1,2,3,4)$, $S_0=0.1$, $L_1=0.2$, $L_2=0.4$, $L_3=0.6$, $L_4=0.8$, $D_s=10^{-5}\,\mathrm{cm^2\,sec^{-1}}$, $D_n=10^{-4}\,\mathrm{cm^2\,sec^{-1}}$, $(n=1,2,3,4)$, $k_{11}=k_{21}=k_{31}=k_{41}=10^3$ $\mathrm{M^{-1}\,sec^{-1}}$, $k_{12}=k_{22}=k_{32}=k_{42}=50\,\mathrm{sec^{-1}}$, $k_{13}=k_{23}=k_{33}=k_{43}=30\,\mathrm{sec^{-1}}$, $k_{51}=10^3$ $\mathrm{M^{-1}\,sec^{-1}}$, $k_{52}=10\,\mathrm{sec^{-1}}$, $\alpha=30$, $a=30$. The spatial distribution of reactant concentrations at $t=4$ is shown in Fig. 7.2, in which the concentrations of each reactant change linearly with the inflections at $L_n$ $(n=1,2,3,4)$. The time course of the system at $x=0.9$ is given in Fig. 7.3. When $S$ is increased

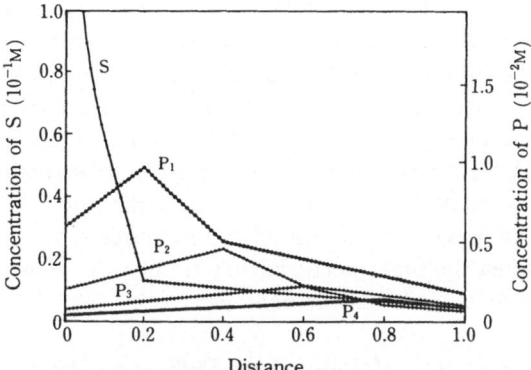

**Fig. 7. 2**  Spatial distribution of reactant concentrations at $t=4$.

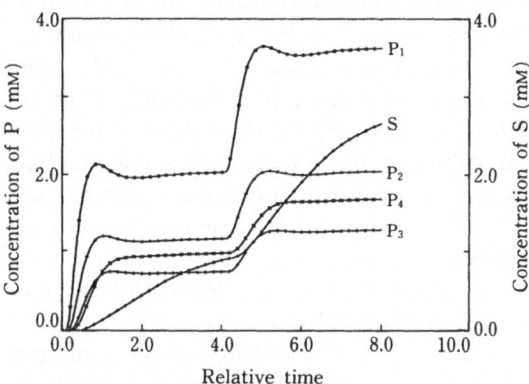

**Fig. 7. 3**  Time courses of reactant concentrations at $x=0.9$.

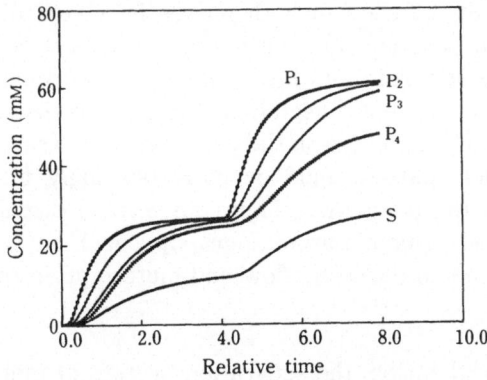

**Fig. 7. 4**  Time courses of reactant concentrations at $x=0.9$. $P_1$, $P_2$ and $P_3$ are not permeable through boundaries.

stepwise to $0.2$ at $t=4$, the new stationary state of the products appears after the overshoots which are caused by the time-delayed regulation of $E_1$.

Next, we simulate the time courses assuming that products $P_1$, $P_2$ and $P_3$ cannot permeate to the outside through boundaries (0) and (1). As shown in Fig. 7.4, the attainment of stationary state is quite slow. The increment in stationary level is exactly proportional to that in substrate concentration which can be regarded as a kind of external perturbation. This implies that the enzyme feedback system simulated here cannot perform the homeostatic control against the input perturbation, though the system contains an allosteric enzyme as a feedback element.

## 7.2  Chromatography of Reaction-Diffusion System

We consider here the partition-chromatographic separation of $\alpha$- and $\beta$-anomers of a sugar, 2-acetamido-2-deoxy-D-glucopyranose (GlcNAc, $N$-acetylglucosamine) [2], which mutarotate or interconvert at a very high rate [3]. Partition-column chromatography of the reacting system involves three basic processes: a) vertical flow of the eluent and solute along the column, b) diffusion of the solute in vertical and lateral directions, and c) reaction of the solute in the liquid and support phases. In the column, the solute diffuses in and from the support, staying for a certain period in the support in which neither lateral nor vertical flows occur. Such a process decelerates flow rate of the solute depending on the partition coefficient of the solute between the liquid and support phases.

For this system, the vertical distribution of the reactants, $\alpha$- and $\beta$-anomers of GlcNAc, is estimated, assuming that it is mainly governed by the diffusion process of reactants in the liquid and support phases. The vertical component of diffusion in both phases is related directly to the reactant distribution. The lateral diffusion in the support phase results in the slowdown of the flow rate and hence affects the vertical distribution, while that in the liquid phase has negligible effect on the distribution. This suggests that the diffusion process on the partition of the reactants between the liquid and support phases could be correspondingly included in the overall flow rates of the reactants. Thus, the vertical distribution of the reactants in partition-column chromatography could be formulated by means of one-dimensional diffusion, flow and mutarotation reaction.

### 1.  Rate Equation
From the above consideration, the separation process of mutarotating $\alpha$- and $\beta$-anomers of GlcNAc in the partition column may be represented by

$$\frac{\partial c_\alpha}{\partial T} = D_\alpha \frac{\partial^2 c_\alpha}{\partial X^2} - v_\alpha \frac{\partial c_\alpha}{\partial X} - k_1 c_\alpha + k_2 c_\beta$$

$$\frac{\partial c_\beta}{\partial T} = D_\beta \frac{\partial^2 c_\beta}{\partial X^2} - v_\beta \frac{\partial c_\beta}{\partial X} + k_1 c_\alpha - k_2 c_\beta,$$

(20)

where $c_\alpha$ and $c_\beta$ are the respective molar concentrations of $\alpha$- and $\beta$-anomers of GlcNAc at time $T$ and position $X$ ($X=0$ at the column top). $D$ and $v$ denote the diffusion coefficient and the flow rate, respectively, and $k_1$ and $k_2$ are the first-order rate constants for mutarotation of

$$\alpha\text{-GlcNAc} \underset{k_2}{\overset{k_1}{\rightleftharpoons}} \beta\text{-GlcNAc}.$$

The initial and boundary conditions are given as

$$c_\alpha(X,0) = c_\beta(X,0) = 0, \quad 0 < X$$
$$c_\alpha(0,0) = c_{\alpha 0}, \quad c_\beta(0,0) = c_{\beta 0}$$
$$c_\alpha(0,T) = c_\beta(0,T) = 0, \quad 0 < T.$$

(21)

The new variables and parameters are defined by

$$x = X/L, \quad t = v_\alpha T/L, \quad C_\alpha = \frac{c_\alpha}{c_{\alpha 0} + c_{\beta 0}},$$

$$C_\beta = \frac{c_\beta}{c_{\alpha 0} + c_{\beta 0}}, \quad K_1 = L k_1/v_\alpha, \quad K_2 = L k_2/v_\alpha,$$

$$\theta_\alpha = D_\alpha/v_\alpha L, \quad \theta_\beta = D_\beta/v_\alpha L, \quad V = \frac{v_\beta}{v_\alpha},$$

(22)

where $L$ is the column length. The dimensionless rate equation is derived such that

$$\frac{\partial C_\alpha}{\partial t} = \theta_\alpha \frac{\partial^2 C_\alpha}{\partial x^2} - \frac{\partial C_\alpha}{\partial x} - K_1 C_\alpha + K_2 C_\beta$$

$$\frac{\partial C_\beta}{\partial t} = \theta_\beta \frac{\partial^2 C_\beta}{\partial x^2} - V \frac{\partial C_\beta}{\partial x} + K_1 C_\alpha - K_2 C_\beta$$

$$C_\alpha(x,0) = C_\beta(x,0) = 0, \quad 0 < x \le 1$$

$$C_\alpha(0,0) = C_{\alpha 0}, \quad C_\beta(0,0) = C_{\beta 0}$$

$$C_\alpha(0,t) = C_\beta(0,t) = 0, \quad 0 < t$$

(23)

## 2.  Difference Equation

We can make the state of chromatographic column correspond to a lattice of $t$ and $x$ with the respective step-length of $k$ and $h$. According to the flow rate of $\alpha$- and $\beta$-anomers, $C_\alpha$ and $C_\beta$ at the same lattice point move to different lattice points after $\Delta t$ ($=k$) (see Fig. 7.5). The lattice is constructed so that after $\Delta t$, $C_\alpha$ and $C_\beta$ locate exactly on the lattice points, that is, the ratio of $v_\alpha$ and $v_\beta$ is represented by a pair of integers, for instance

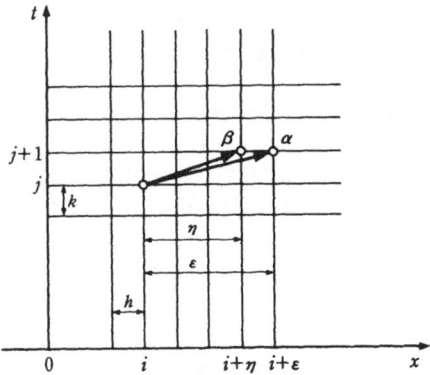

**Fig. 7.5** Lattice for numerical solution.

4 : 3, as illustrated in Fig. 7.5. The mutarotation is assumed to occur only at a lattice point. The difference expression of the flow term may be given by

$$\frac{\partial C_\alpha}{\partial x} = \frac{C_\alpha(i,j) - C_\alpha(i-\varepsilon,j)}{\varepsilon h}$$
$$\frac{\partial C_\beta}{\partial x} = \frac{C_\beta(i,j) - C_\beta(i-\eta,j)}{\eta h} \tag{24}$$

where $\varepsilon$ and $\eta$ are jumping lengths of $\alpha$- and $\beta$-anomers, respectively. The difference expressions of $\partial C/\partial t$ and $\partial^2 C/\partial x^2$ are the same as those given in Section 3.4. The difference rate-equation may be written as

$$\begin{aligned}
C_\alpha(i,j+1) = & \frac{k\theta_\alpha}{2\,\zeta_1}\{C_\alpha(i+1,j+1) + C_\alpha(i-1,j+1) \\
& + C_\alpha(i-\varepsilon+1,j) + C_\alpha(i-\varepsilon-1,j)\} \\
& + \frac{k(h-\varepsilon h^2 K_1 - \varepsilon\theta_\alpha)}{\varepsilon\zeta_1}\,C_\alpha(i-\varepsilon,j) \\
& + \frac{h(\varepsilon h-k)}{\varepsilon\zeta_1}\,C_\alpha(i,j) + \frac{kh^2 K_2}{\zeta_1}\,C_\beta(i-\eta,j) \\
C_\beta(i,j+1) = & \frac{k\theta_\beta}{2\,\zeta_2}\{C_\beta(i+1,j+1) + C_\beta(i-1,j+1) \\
& + C_\beta(i-\eta+1,j) + C_\beta(i-\eta-1,j)\} \\
& + \frac{k(hV-\eta h^2 K_2 - \eta\theta_\beta)}{\eta\zeta_2}\,C_\beta(i-\eta,j) \\
& + \frac{h(\eta h-kV)}{\eta\zeta_2}\,C_\beta(i,j) + \frac{kh^2 K_1}{\zeta_2}\,C_\alpha(i-\varepsilon,j),
\end{aligned} \tag{25}$$

where $\zeta_1 = (h^2 + k\theta_\alpha)$ and $\zeta_2 = (h^2 + k\theta_\beta)$. Since $C_\alpha(i,j)$ and $C_\beta(i,j)$ move by $\varepsilon h$ and $\eta h$, respectively, owing to the flow in time interval $k$, they do not affect the terms $C_\alpha(i,j+1)$ and $C_\beta(i,j+1)$. Then $h$ and $k$ should satisfy the equation,

$$\varepsilon h - k = 0 \qquad (26)$$
$$\eta h - kV = 0,$$

which is solved to provide both values.

As described in Section 3.4, equation (25) can be arranged with unknown variables, $C_\alpha(i+1,j+1)$, $C_\alpha(i,j+1)$ and $C_\alpha(i-1,j+1)$, and known variables represented by $b(i)$. The same manipulation is also applied to $\beta$-anomer. Thus, we have

$$C_\alpha(i,j+1) = A\{C_\alpha(i+1,j+1) + C_\alpha(i-1,j+1)\} + b(i)$$
$$C_\beta(i,j+1) = E\{C_\beta(i+1,j+1) + C_\beta(i-1,j+1)\} + g(i)$$
$$b(i) = A\{C_\alpha(i-\varepsilon+1,j) + C_\alpha(i-\varepsilon-1,j)\} + BC_\alpha(i-\varepsilon,j)$$
$$\qquad + CC_\alpha(i,j) + DC_\beta(i-\eta,j) \qquad (27)$$
$$g(i) = E\{C_\beta(i-\eta+1,j) + C_\beta(i-\eta-1,j)\} + FC_\beta(i-\eta,j)$$
$$\qquad + GC_\beta(i,j) + HC_\alpha(i-\varepsilon,j),$$

where

$$A = \frac{k\theta_\alpha}{2\,\zeta_1}, \quad B = \frac{k(h - \varepsilon h^2 K_1 - \varepsilon\theta_\alpha)}{\varepsilon\zeta_1}, \quad C = \frac{h(\varepsilon h - k)}{\varepsilon\zeta_1}, \quad D = \frac{kh^2 K_2}{\zeta_1}$$

$$E = \frac{k\theta_\beta}{2\,\zeta_2}, \quad F = \frac{k(hV - \eta h^2 K_2 - \eta\theta_\beta)}{\eta\zeta_2}, \quad G = \frac{h(\eta h - kV)}{\eta\zeta_2}, \quad H = \frac{kh^2 K_1}{\zeta_2}$$

## 3. Numerical Solution of Difference Equation

The difference equation (27) is solved by the over-relaxation method (see Section 3.4). The simulated chromatographic pattern of $\alpha$- and $\beta$-anomers of GlcNAc is shown in Fig. 7.6. The values of parameters used for the computation are as follows: $C_{\alpha 0}/C_{\beta 0} = 1.45$, $v_\alpha = 4.0$ cm min$^{-1}$, $v_\beta = 3.0$

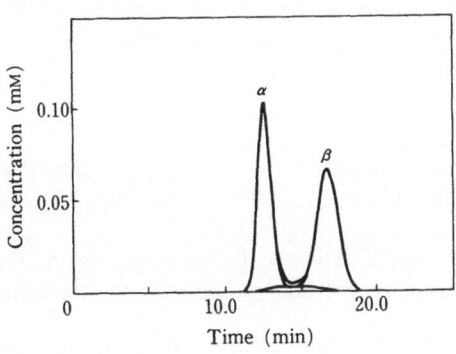

**Fig. 7.6** Simulated elution pattern of $\alpha$- and $\beta$-anomers of GlcNAc.

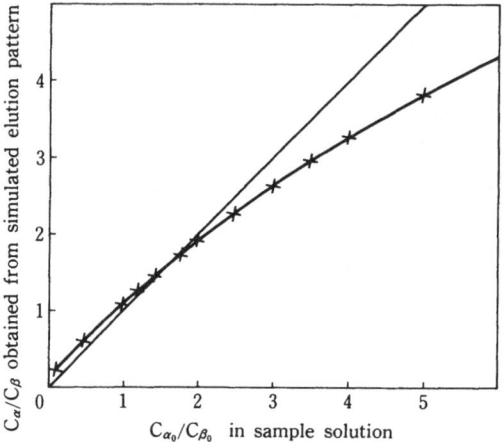

**Fig. 7.7** Comparison of $C_{\alpha_0}/C_{\beta_0}$ in the sample solution and $C_\alpha/C_\beta$ in the simulated elution pattern. Straight line indicates the case of identical ratios.

cm min$^{-1}$, $D_\alpha = D_\beta = 2.4 \times 10^{-2}$ cm$^2$ min$^{-1}$, $k_1 = 5.3 \times 10^{-3}$ min$^{-1}$ and $k_2 = 6.0 \times 10^{-3}$ min$^{-1}$. The simulated pattern is in good agreement with that obtained by experimental runs.

When the simulation is performed with $C_{\alpha 0}/C_{\beta 0} = 1.45$, which corresponds to the completely equilibrated mixture of $\alpha$- and $\beta$- anomers, the ratio $C_\alpha/C_\beta$ has completely the same value. This implies that the ratio $C_{\alpha 0}/C_{\beta 0}$ in a sample solution could be determined experimentally by partition-column chromatography. On the other hand, if $C_{\alpha 0}/C_{\beta 0}$ is larger than 3.0, which corresponds to a mixture highly rich in $\alpha$-anomer, the ratio $C_\alpha/C_\beta$ from the simulation is far underestimated as shown in Fig. 7.7. Hence, the direct experimental estimation of sample solution with such a large $C_{\alpha 0}/C_{\beta 0}$ is not possible. To obtain a correct estimation, the experimental data should be analyzed by further computer simulation.

# References

[ 1 ] Kuhara, S., S. Iwamoto, A. Fujita and K. Hayashi (1984). Feedback control of enzymatic reaction-diffusion system. *J. Fac. Agr., Kyushu Univ.*, **29**, 1-21.

[ 2 ] Kuhara, S., S. Iwamoto, Y. Yanase, T. Fukamizo and K. Hayashi (1982). Note on the separation of reacting system by chromatography. *ibid.*, **27**, 33-45.

[ 3 ] Fukamizo, T. and K. Hayashi (1982). Separation and mutarotation of anomers of chitooligosaccharides. *J. Biochem.*, **91**, 619-626.

CHAPTER 8

# Determination of Reaction Scheme and Kinetic Parameters

As demonstrated by many examples in the preceding chapters, an enzymatic reaction has its own specific mechanism and kinetic parameters with which its kinetic behavior can be characteristically generated. Therefore, enzyme kinetics is concerned principally with the determination of the mechanism and parameters of a particular enzyme. The scheme of an enzymatic reaction can be estimated if the experiment is performed under a quasi-steady-state condition and the data are treated in the usual way based on the approximations of the rate equation. The next important problem is to evaluate the kinetic parameters for the estimated reaction scheme.

Actually, the reactions for which we could determine the scheme and kinetic parameters using these usual procedures would be the exceptions among all enzymatic reactions. It is acknowledged that even for simple reactions strong nonlinearity often prevents the experimental conditions from allowing the quasi-steady-state approximation. Moreover, the common methods of kinetic analysis cannot be directly applied to metabolic systems, which operate as complex and large-scale networks comprised of many enzymatic reactions and their regulatory mechanisms. Hence, new analytical procedures are needed to overcome these shortcomings. The enzyme dynamics procedure described in this book is regarded as one such new method.

In this chapter we introduce the analytical procedure of combined computer simulation and optimization as an extension of the method employed and discussed in the preceding chapters. This is capable of yielding more exact answers for problems not yet completely solved. In Section

This chapter was written by Satoru Kuhara, Kiyokazu Nemoto, Yukihiro Eguchi and Naoto Sakamoto.

8.1 the procedure for estimation of a reaction scheme is explained and applied to some enzyme systems. The reaction schemes most suitable for generation of the observed or desired behavior are chosen by this procedure for the activation reaction of pepsinogen and spike-type oscillatory reaction. Another kind of application is attempted for estimation of the reaction mechanism of $p$-hydroxybenzoate hydroxylase. Section 8.2 is concerned with the procedure for evaluation of kinetic parameters; gradient and direct-search methods are introduced as optimization techniques. Evaluation of the binding free energy in lysozyme is done as an example of application of this procedure.

## 8.1   Procedure for Estimation of Reaction Scheme

The estimation of reaction schemes (or mechanisms) is normally done on the basis of experimental data. Simple mechanisms can be determined by direct treatment of the data. For complex reactions, however, a certain anlaysis is required because it becomes difficult to correlate the data to the mechanism. One such procedure is an algorithm for estimating a mechanism by analysis of the dynamic characteristics of reaction.

The first step in this method is to formulate a rate equation based exactly on a model scheme for the reaction mechanism. The dynamic characteristics of the model scheme are revealed from the numerical solution of the rate equation. We then examine whether the dynamic behavior thus obtained is consistent with the experimental data. If the comparison should not agree, the scheme is revised and the procedure is iterated until satisfactory agreement is attained.

In addition to the estimation of a reaction scheme reproducing the experimental data, we sometimes need to deal with selection of a scheme of enzymatic reactions generating a time course with a specified feature. A knowledge of the characteristics of basic schemes with specific behavior provides good information to use in the analysis of unknown reactions and the design of new experiments. The procedure for estimation of a reaction scheme is again very useful for this type of problem.

### 1.   Algorithm for Estimation of Reaction Scheme

We here describe an iterative algorithm for estimation of a reaction scheme from the dynamic behavior. Figure 8.1 demonstrates a flow chart of this algorithm, each step of which is explained in the following.

(1) Accumulation of experimental data: In this step we collect experimental data on the objective reaction system. The data generally include the results of both static and dynamic examinations of the reaction. For the

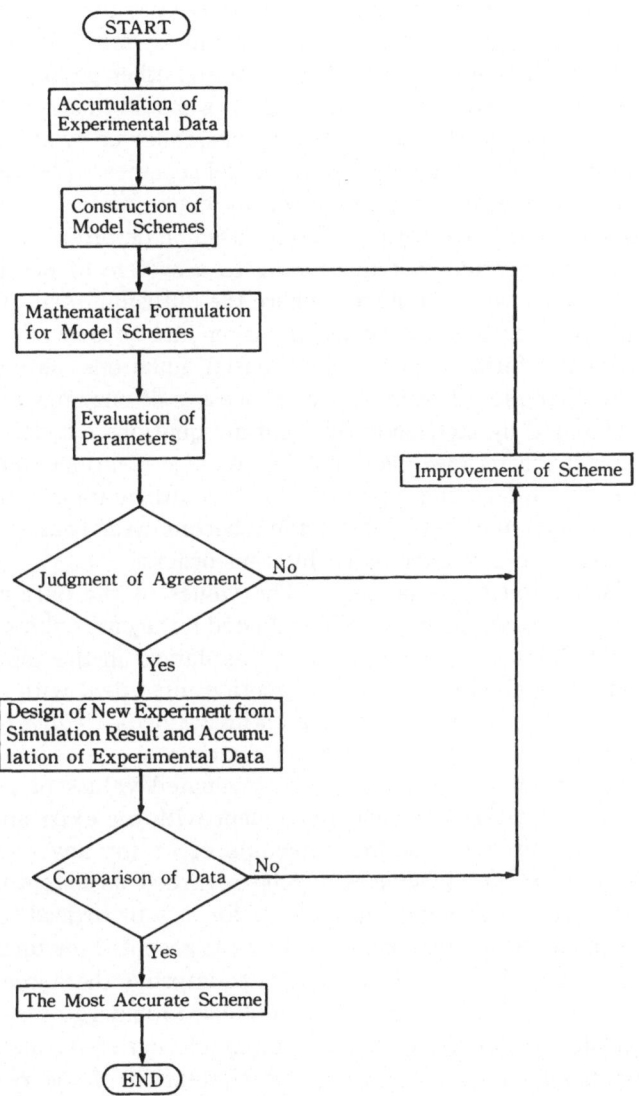

**Fig. 8. 1**   Flow chart for estimation algorithm of reaction scheme.

present purpose the data on dynamic behavior are more helpful in grasping
the temporal progress in the reaction.

(2) Construction of model schemes : The model schemes for the reaction are constructed on the basis of the experimental data obtained in step

(1). In the model construction the characteristics of the reaction like time delay, induction time and oscillation should be taken into consideration. If one feature is overemphasized in modeling, other possible schemes could be carelessly ignored. All probable schemes should thus be considered carefully. It is best to start with simple schemes, followed by more complex ones.

(3) Mathematical formulation for model schemes : The various schemes constructed in step (2) are expressed by mathematical models. In many cases the mathematical formulation is based on deterministic approach and homogeneous reaction system, leading to a system of nonlinear ordinary differential equations. In a case where the diffusion of substrates and products and the location of enzymes are taken into account, the mathematical model has the form of partial differential equations, as discussed in the preceding chapters. The molecular processes of enzymes and metabolites are represented by stochastic differential equations exactly reflecting molecular mechanisms (see Section 9.1). As also mentioned in the preceding chapters, the differential equations in the mathematical models for enzymatic reactions and metabolic systems become stiff because the rate constants have values of extremely different orders.

(4) Evaluation of parameters : The values of the parameters employed in the mathematical model are evaluated if they are unknown. Many evaluative methods have been devised, as explained in the following section. Any method applied to enzymatic reaction must deal with the stiffness in the model. An incorrect selection of optimization procedure results in waste of computation time or inaccurate evaluation.

(5) Judgment of agreement : The evaluated values of parameters are examined with respect to their agreement with the experimental data. If they do not agree within the allowable error for any value of the parameters, the model scheme would be regarded as inappropriate. Then, the procedure is resumed from step (3) for new or revised schemes. When the agreement is satisfactory, the scheme is accepted for further treatment. It is modified and returned to step (3) to improve the agreement. All the schemes constructed in step (2) are obliged to undergo these steps.

(6) Design of new experiment from simulation result and accumulation of experimental data : The computer simulation of the reaction process under various conditions is performed for schemes successful through step (5); several probable schemes reach this step successfully. Therefore, the simulation aims particularly at the establishment of an experimental condition for distinguishing these schemes. The experiments are carried out under that condition to obtain new data.

(7) Comparison and improvement : The results of the simulation and the experiment in step (6) are compared to keep the scheme in good agree-

ment. If the comparison is poor, schemes are revised based on the experimental data from step (6) and returned to step (3) for iteration of the procedure. It is comparatively easy to decide how to improve the schemes
by using information on the basic schemes with specific dynamic characteristics. When we have a few satisfactory schemes, or it is difficult to determine whether one exactly represents the reaction mechanism, we simulate the schemes to design new experiments. The experimental data thus
obtained are compared with the results of simulation to further improve
the scheme.

(8) Selection of the most accurate scheme: The scheme yielding the
best result in step (7) is selected as the most accurate.

Examples of application of this algorithm are described in the following.

## 2. Estimation of Activating Mechanism of Pepsinogen

Pepsinogen is the proenzyme of pepsin, a digestive enzyme in the gastric
juices, and is activated to pepsin by the release of about 40 amino acid
residues in acidic solution. This process of activation has been investigated
extensively to reveal the following characteristics:

(1) The time course of the activation displays an induction phase.
(2) The absorption spectrum of the process continues to vary after the
completion of activation, indicating further change in structure of
pepsin.
(3) The induction phase disappears in the activation at higher temperature.
(4) The activation is essentially an autonomous reaction, although pepsin, a product of the process, enhances the activation of pepsinogen.

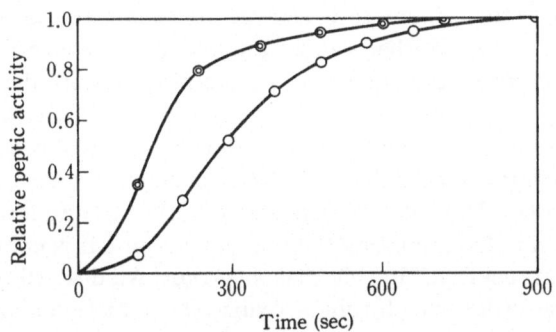

**Fig. 8.2** Time courses of activation of pepsinogen (experiment). ○:
0°C; ◎: 15°C.

**Table 8.1**  Postulated activation mechanisms of pepsinogen

| | |
|---|---|
| $X_1 \longrightarrow X_2 \longrightarrow X_3$ | $X_1 \longrightarrow X_2 \longrightarrow X_3 \longrightarrow X_4$ |
| $X_1 \longrightarrow X_2 \longrightarrow X_3$ | $X_1 \longrightarrow X_2 \longrightarrow X_3 \longrightarrow X_1$ |
| $X_1 \longrightarrow X_2 \longrightarrow X_3$ | $X_1 \longrightarrow X_2 \longrightarrow X_3 \longrightarrow X_4$ |
| $X_1 \longrightarrow X_2 \longrightarrow X_3$ | $X_1 \longrightarrow X_2 \longrightarrow X_3 \longrightarrow X_4$ |
| $X_1 \longrightarrow X_2 \longrightarrow X_3$ | $X_1 \longrightarrow X_2 \longrightarrow X_3 \longrightarrow X_4$ |
| $X_1 \longrightarrow X_2 \longrightarrow X_3$ | $X_1 \longrightarrow X_2 \longrightarrow X_3 \longrightarrow X_4$ |
| $X_1 \longrightarrow X_2 \longrightarrow X_3 \longrightarrow X_4$ | $X_1 \longrightarrow X_2 \longrightarrow X_3 \longrightarrow X_4$ |
| $X_1 \longrightarrow X_2 \longrightarrow X_3 \longrightarrow X_1$ | $X_1 \longrightarrow X_2 \longrightarrow X_3 \longrightarrow X_4$ |
| $X_1 \longrightarrow X_2 \longrightarrow X_3 \longrightarrow X_1$ | $X_1 \longrightarrow X_2 \longrightarrow X_3 \longrightarrow X_4$ |
| $X_1 \longrightarrow X_2 \longrightarrow X_3 \longrightarrow X_4$ | $X_1 \longrightarrow X_2 \longrightarrow X_3 \longrightarrow X_4$ |
| $X_1 \longrightarrow X_2 \longrightarrow X_3 \longrightarrow X_4$ | $X_1 \longrightarrow X_2 \longrightarrow X_3 \longrightarrow X_4$ |

Many problems still remain to be solved regarding the activating mechanism of pepsinogen. We here apply the above procedure to this process of activation in order to estimate its mechanism in more detail [1].

First, we are concerned with the induction phase occurring in the time course of reaction. As shown in Fig. 8.2, the activation of pepsinogen at 0°C and pH 2.5 undergoes an induction phase of about 2 min, after which the activity of protein degradation incraeses rapidly to completion of the activation in 10 min. In estimating the reaction scheme it is common to begin with a one-step reaction. In this case, however, we discard one-step reactions of the first order, which cannot realize the induction phase. It is demonstrated in Chapter 6 that the multi-step reactions can display the induction phase. Hence, we examine the two- and three-step reactions given in Table 8.1. From the simulation of these schemes the six reactions of

$$X_1 \xrightarrow{k_1} X_2^* \xrightarrow{k_2} X_3^* \qquad (k_4)$$

8.1

$$X_1 \xrightarrow{k_1} X_2 \xrightarrow{k_2} X_3^* \xrightarrow{k_3} X_4^* \qquad (k_4)$$

8.4

$$X_1 \xrightarrow{k_1} X_2 \xrightarrow{k_2} X_3^* \xrightarrow{k_3} X_4^*$$

8.2

$$X_1 \xrightarrow{k_1} X_2 \xrightarrow{k_2} X_3^* \xrightarrow{k_3} X_4^* \qquad (k_4)$$

8.5

$$X_1 \xrightarrow{k_1} X_2 \xrightarrow{k_2} X_3^* \xrightarrow{k_3} X_4^* \qquad (k_4)$$

8.3

$$X_1 \xrightarrow{k_1} X_2 \xrightarrow{k_2} X_3^* \xrightarrow{k_3} X_4^* \qquad (k_1)$$

8.6

Schemes 8.1 - 8.6. Probable schemes of activation reaction of pepsinogen

$X_1$ : pepsinogen ; final species ($X_3$ or $X_4$) : pepsin ; other $X_i$ : intermediates ; species with asterisk : active species. $k_i$ represent the rate constants.

**Table 8.2**  Values of rate constants for Schemes 8.1-8.6

| Scheme | $k_1(10^3 \text{sec}^{-1})$ | $k_2(10^5 \text{sec}^{-1})$ | $k_3(10^5 \text{sec}^{-1})$ | $k_4(10^3 \text{M}^{-1}\text{sec}^{-1})$ |
|---|---|---|---|---|
| 8.1 | 0.70 | 4.7 | — | 8.0 |
| 8.2 | 8.9 | 4.7 | 1.7 | — |
| 8.3 | 1.2 | 6.9 | 1.7 | 2.15 |
| 8.4 | 1.4 | 2.3 | 1.6 | 2.54 |
| 8.5 | 2.5 | 5.9 | 1.7 | 5.10 |
| 8.6 | 0.1 | 0 | 1.6 | 29.0 |

Schemes 8.1-8.6 are found to follow time courses almost similar to the experimental one. In these schemes the chemical species with an asterisk (∗) have the pepsin activity. The rate constants listed in Table 8.2 are determined so that they lead to the best agreement with the experimental data. The first procedure thus chooses six schemes from among many.

The next procedure is to select more accurate schemes by performing the simulation and experiment for these six. Simulation yields the typical time courses of the activation shown in Fig. 8.3. It is seen that these schemes follow quite different time courses in the early stage of reaction. Figure 8.4 demonstrates the temporal change in difference spectrum during the ear-

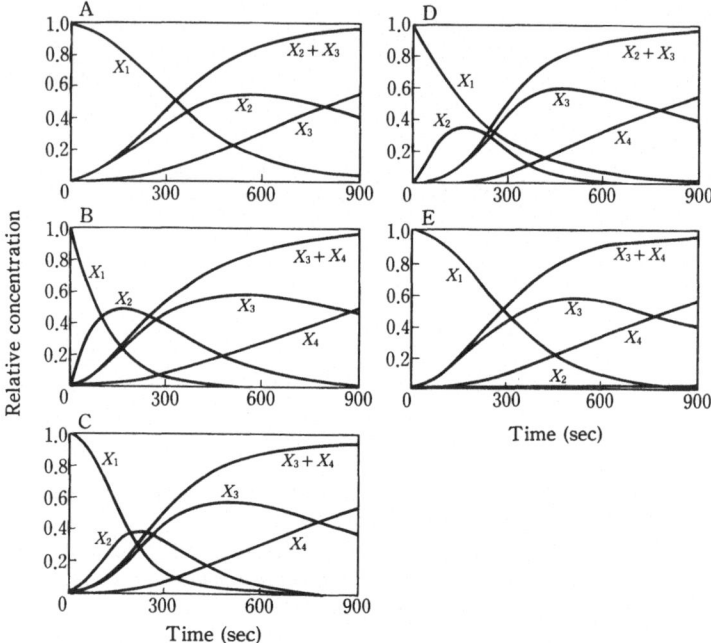

**Fig. 8. 3** Time courses of probable schemes (simulation). A : Scheme 8.1 ; B : Scheme 8.2 ; C : Scheme 8.3 ; D : Scheme 8.4 ; E : Scheme 8.5.

ly stage of reaction experimentally measured by the stopped-flow method. The early stage of activation (production of $X_2$) clearly exhibits an exponential progress. It follows from a comparison of the simulation results of Fig. 8.3 with Fig. 8.4 that Schemes 8.2 and 8.4 provide the desired behavior. Scheme 8.2 is a linear sequential reaction, while Scheme 8.4 is a nonlinear reaction including a bimolecular process of $X_2$ and $X_3$.

We now need to decide which is the more accurate of the two schemes. The examination is made with respect to the dependency of the activation reaction on the initial concentration. The results of simulation in Fig. 8.5 indicate that the dependency of Scheme 8.4 nearly coincides with the experimental result. We can thus estimate the primary scheme of the activation reaction by the experiment and simulation.

The next step aims at the improvement of Scheme 8.4 to more realistic one. One of the characteristics in the activation reaction is observed in the electrophoretic pattern tracking the change in molecular weight of the protein during the reaction (Fig. 8.6). The formation of a dimeric intermediate of 3~4% of the total protein is detected (denoted by the arrow in

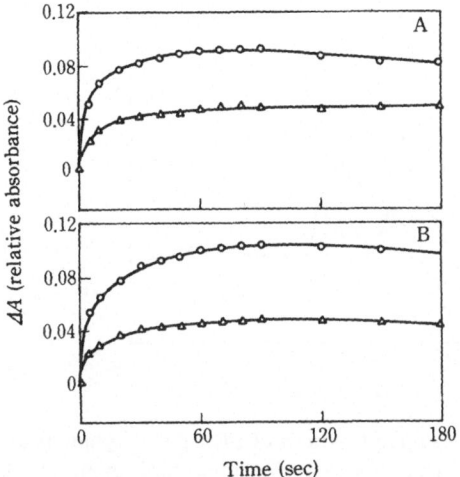

**Fig. 8. 4** Time course of difference spectrum in early stage of pepsinogen activation (experiment). A : pH 3.0 ; B : pH 2.5. ○ : $A_{260nm} - A_{220nm}$ ; △ : $A_{298nm} - A_{220nm}$.

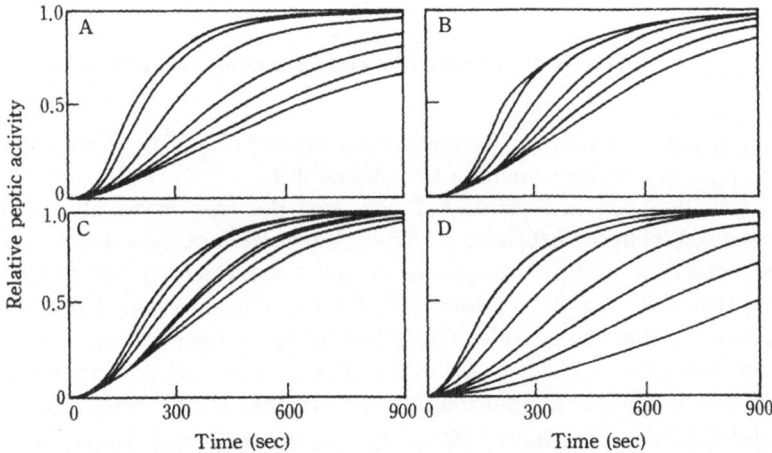

**Fig. 8. 5** Effect of initial concentration on time course in early stage (simulation). The four segments of the figure correspond to A : Scheme 8.3 ; B : Scheme 8.4 ; C : Scheme 8.5 ; D : Scheme 8.6 ; and the curves (from top to bottom) to initial concentrations (in %) of 0.6, 0.4, 0.2, 0.1, 0.0666, 0.04, and 0.02, respectively.

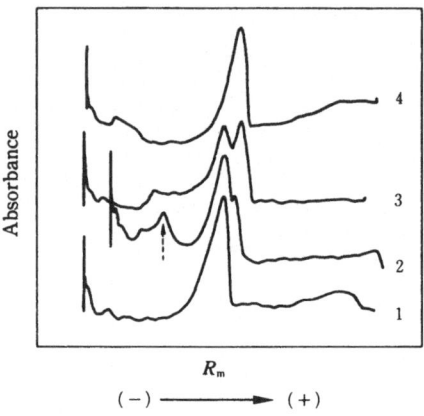

**Fig. 8.6**  Temporal pattern of SDS-polyacrylamide gel electrophoresis in the activation reaction. Sampling time (min) : 1 : 0.0, 2 : 6.0, 3 : 8.0, 4 : 15.0. The arrow indicates a dimeric intermediate.

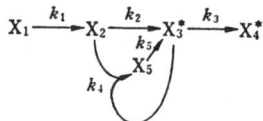

Scheme 8.7.   Addition of a dimeric intermediate to Scheme 8.4

the figure). Therefore, we consider Scheme 8.7, which arises from addition of the dimeric intermediate to Scheme 8.4.

One problem in Scheme 8.7 concerns the step in which the peptide is released from the pepsinogen. The following characteristics are noted. The structure of protein changes after the completion of activation, as demonstrated in Fig. 8.7. Moreover, $X_3^*$ has the activity. These lead to the postulate that the peptide is detached before the generation of active pepsin and that the path $X_3^* \rightarrow X_4^*$ is the slow isomerization (structural change) of active pepsin. In addition, it is known that the activation of pepsinogen and the isomerization of $X_3^* \rightleftarrows X_4^*$ are strongly dependent on the hydrogen ion concentration. Hence, the intake and release of hydrogen ion are supposed to occur in these processes. Taking these experimental results into consideration, we modify Scheme 8.7 to the final scheme (Scheme 8.8). This scheme can explain the known experimental data without contradiction, even if it might be further improved with new knowledge on the activation process of pepsinogen.

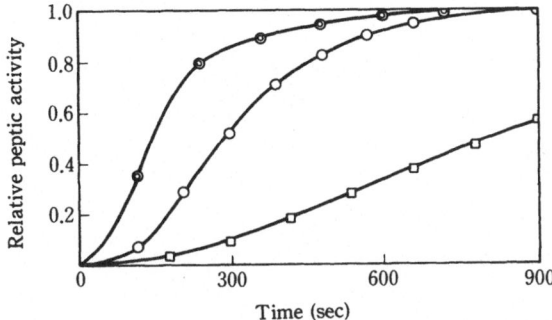

**Fig. 8.7** Time courses of activation of pepsinogen (experiment). ○: activated at 0°C ; ◎ : at 15°C ; □ : difference absorbance at 293 nm at 4°C.

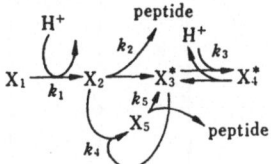

Scheme 8.8.   Activation process of pepsinogen

## 3.  Estimation of a Basic Scheme for Generation of Spike-Type Oscillation

Some biological systems are known to generate a particular spike-type oscillation in their *in vivo* behavior, for instance, in the humoral secretion of sex, growth and pituitary hormones. The underlying enzyme systems would naturally be expected to operate, but their basic schemes are little known. Only the reaction with a single enzyme and an inhibitor of Scheme 8.9 has been analyzed by Karfunkel and Seeling [2].

The quasi-steady-state approximation for Scheme 8.9 yields a time course with definite spike-type oscillation, as shown in Fig. 8.8 [2]. The numerical solution of the rate equation of Scheme 8.9 also reveals a sort of spike-type oscillation (Fig. 8.9). The reaction in Scheme 8.9 takes place in an open system and has a step working as a pool of the product in the form of $P_nI$. The spike-type oscillation apparently stems from these characteristics.

We are here concerned with the possibility of spike-type oscillation in the end product of a sequential reaction system found in the metabolic processes. It would be interesting and valuable to estimate the basic scheme

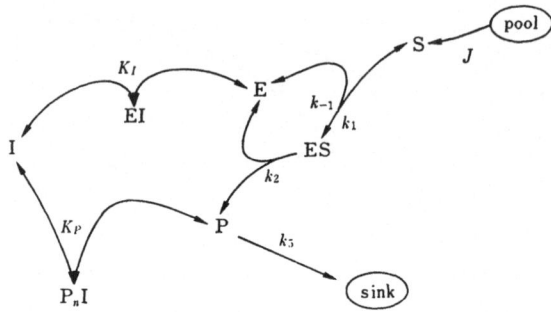

Scheme 8.9. System generating a spike-type oscillation

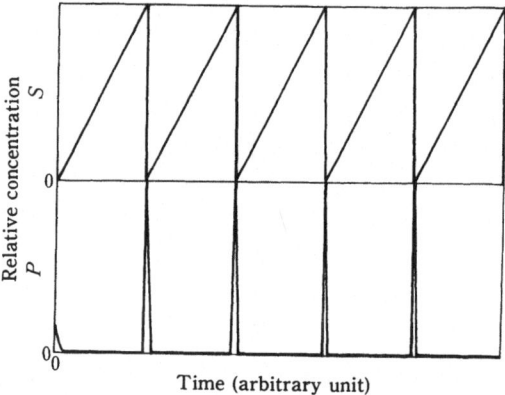

**Fig. 8.8** Time course of Scheme 8.9 (quasi-steady-state approximation).

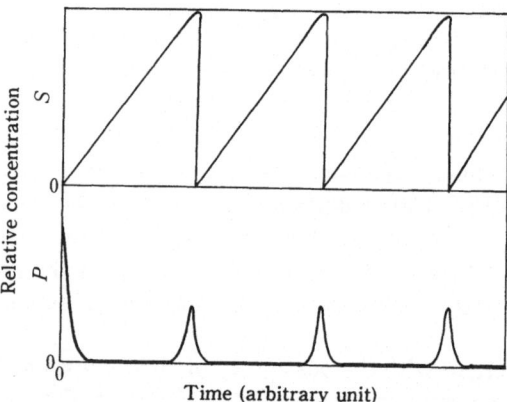

**Fig. 8.9** Time course of Scheme 8.9 (simulation).

with such a capacity and the following properties:
  (1) The system contains a structure for accumulation of reaction inter-
      mediates.
  (2) The accumulated species is released promptly within a certain time
      period.
  (3) The oscillation is sustained.
  For estimation of a basic scheme, let us begin with the linear-chain reaction system of Scheme 8.10. As discussed in Chapter 6, the end product,
$X_5$, of this scheme follows a sigmoidal time course without oscillation
because no pool is provided in the system to store the intermediates. Hence,
an equilibrating step to serve as a pool is added to Scheme 8.10, leading
to Scheme 8.11, for example. Nevertheless, simulation reveals that this
scheme still yields only a sigmoidal time course exhibiting no oscillation, as
shown in Fig. 8.10.
  We then consider Scheme 8.12 which is derived by introduction of

$$\xrightarrow{k_1} X_1 \xrightarrow{k_2} X_2 \xrightarrow{k_3} X_3 \xrightarrow{k_4} X_4 \xrightarrow{k_5} X_5 \xrightarrow{k_6}$$

Scheme 8.10.  Linear-chain reaction system
$X_i$: reactant, $k_i$: rate constant

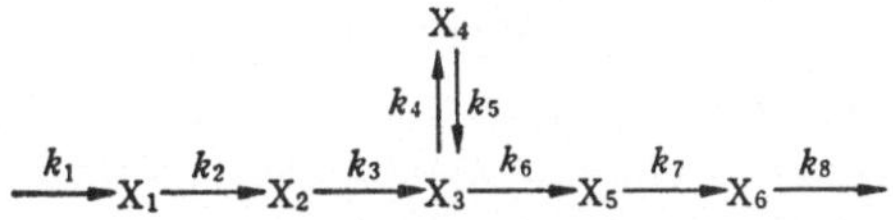

$$X_4$$
$$k_4 \Big\Updownarrow k_5$$
$$\xrightarrow{k_1} X_1 \xrightarrow{k_2} X_2 \xrightarrow{k_3} X_3 \xrightarrow{k_6} X_5 \xrightarrow{k_7} X_6 \xrightarrow{k_8}$$

Scheme 8.11.  System with a pool of intermediate

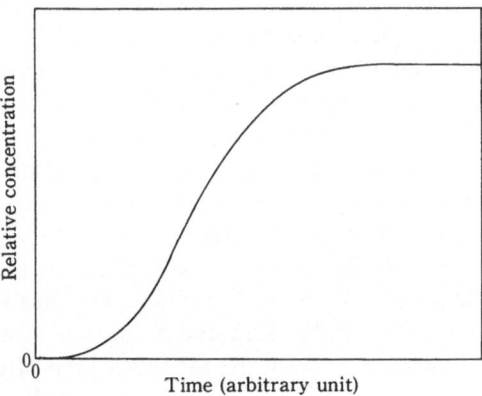

Fig. 8.10  Time course of $X_6$ in Scheme 8.11.

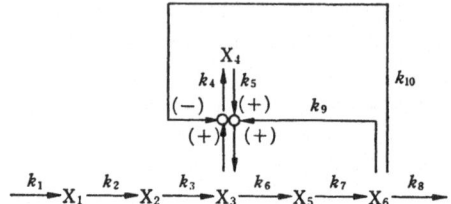

Scheme 8. 12.   Addition of feedback loops to Scheme 8.11

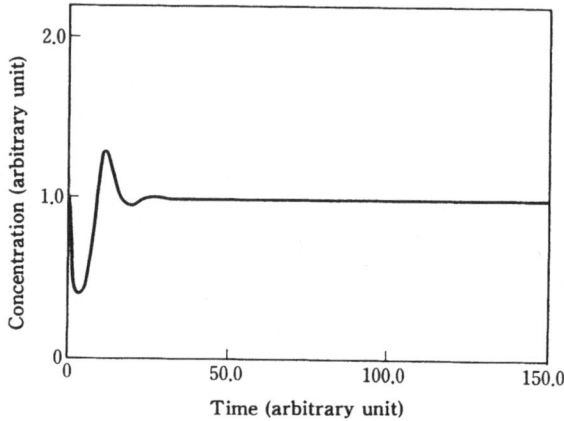

**Fig. 8. 11**   Time course of $X_6$ in Scheme 8.12.  Rate constant : $k_1 = k_8 = 2.0$, $k_2 = k_3 = k_4 = k_6 = k_7 = k_9 = k_{10} = 1.0$ (arbitrary unit).

feedback loops to Scheme 8.11.  The feedback loops work as regulatory devices providing the pool with more of accumulation and release capacity. Figure 8.11 illustrates a typical time course of Scheme 8.12 obtained from the simulation.  Wide range of rate-constant values cannot lead to spike-type oscillation, although the overshoot is recognized in the time course.  In the steady state of Scheme 8.12 we have $\bar{X}_6 = k_1/k_6$, which is constant regardless of the feedback loops and opposing to oscillation.

None of the schemes considered above can generate sustained oscillations, and hence not spike-type oscillations.  As described in Chapter 6, a sustained oscillation emerges in a linear-chain reaction system with introduction of a feedback loop of the end product.  We thus examine three reaction systems of Schemes 8.13, 8.14 and 8.15, which are essentially similar to Scheme 8.12 appended with the feedback loop of species $X_6$.  It is found from the simulation of these schemes that only Scheme 8.13 generates a sustained oscillation of the spike type as in Fig. 8.12.  The

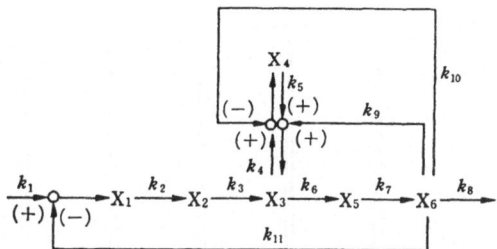

Scheme 8. 13.   Linear system with pool and feedback loops ( I )

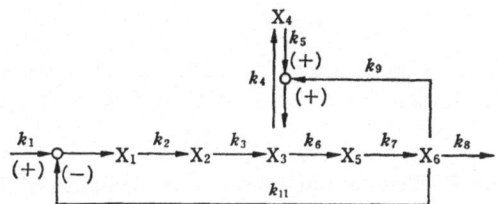

Scheme 8. 14.   Linear system with pool and feedback loops ( II )

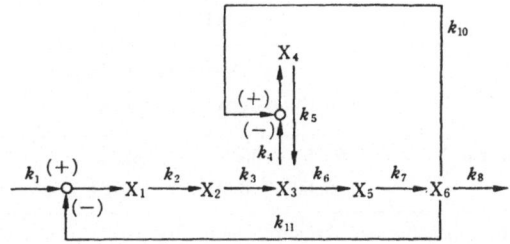

Scheme 8. 15.   Linear system with pool and feedback loops (III)

effects of the rate constants on the behavior of Scheme 8.13 are then studied with the simulation and shown in Table 8.3. Obviously, the values of $k_9$ and $k_{10}$ strongly affect the appearance of spike-type oscillation.

On the other hand, the change in the position of feedback input of $X_6$ to the path $X_1 \rightarrow X_2$ causes the oscillation to disappear. We now consider Scheme 8.16 with a branch from $X_1$. The characteristics of a scheme with a branch from the position preceding the summing point of feedback are described in detail in Chapter 6. Figure 8.13 presents the result of simulation

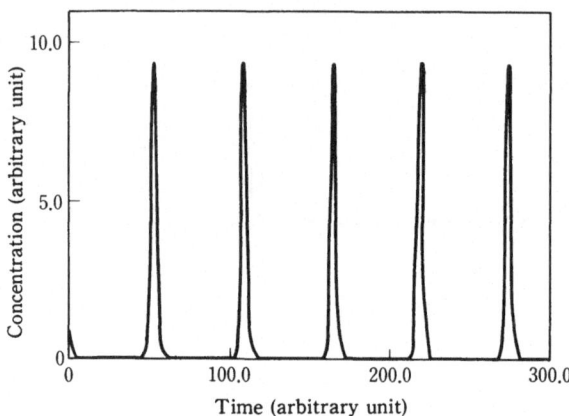

**Fig. 8.12** Time course of $X_6$ in Scheme 8.13. Rate constant : $k_1 = k_8 = 2.0$, $k_2 = k_3 = k_4 = k_5 = k_6 = k_7 = k_9 = k_{10} = 1.0$, $k_{11} = 100.0$ (arbitrary unit).

**Table 8.3** Effect of rate constants on oscillation of $X_6$

|  | Value of $k_9$ $(=k_{10})$ | | | |
|---|---|---|---|---|
|  | 0.1 | 1.0 | 10.0 | 100.0 |
| Period of oscillation | — | 55.5 | Damped | Damped |
| Peak height | — | 9.34 | | |
|  | Value of $k_1$ $(=k_8)$ | | | |
|  | 0.5 | 1.0 | 2.0 | 4.0 |
| Period of oscillation | 69.5 | 60.0 | 55.5 | 51.0 |
| Peak height | 6.60 | 3.08 | 9.34 | 9.80 |

Other rate constants are the same as in Fig. 8.12.

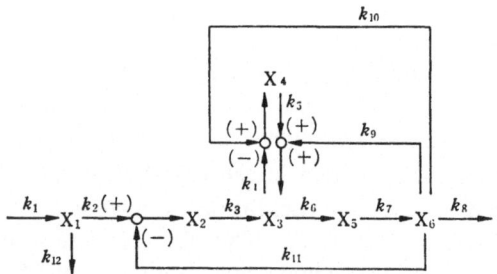

**Scheme 8.16.** Addition of a branch to the linear system

for Scheme 8.16, indicating that a scheme with such a branch follows the time course of spike-type oscillation.

The following conditions for generation of spike-type oscillation are derived from the above results of improvement in the schemes by simulation:

(1) A branched equilibrating reaction must operate to store intermediates.

(2) Feedback loops are required for the branched reaction.

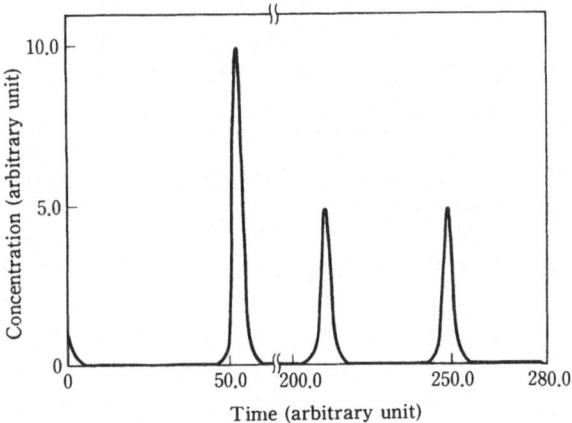

**Fig. 8. 13** Time course of $X_6$ in Scheme 8.16. Rate constant: $k_{11} = 400.0$, $k_{12} = 0.05$. Other rate constants are the same as in Fig. 8.12.

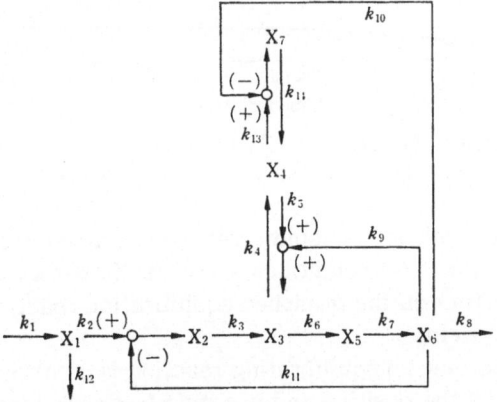

Scheme 8. 17. Addition of two-step equilibrating reaction to Scheme 8.16 ( I )

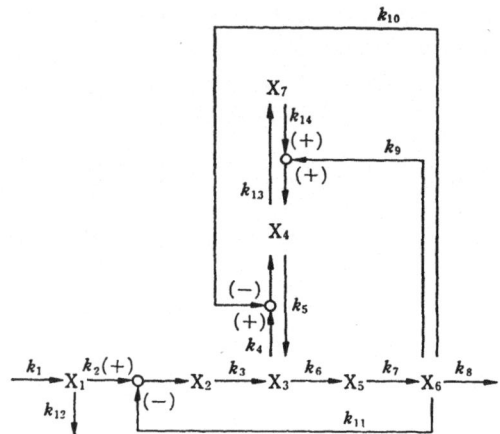

Scheme 8.18.  Addition of two-step equilibrating reaction to Scheme 8.16 (II)

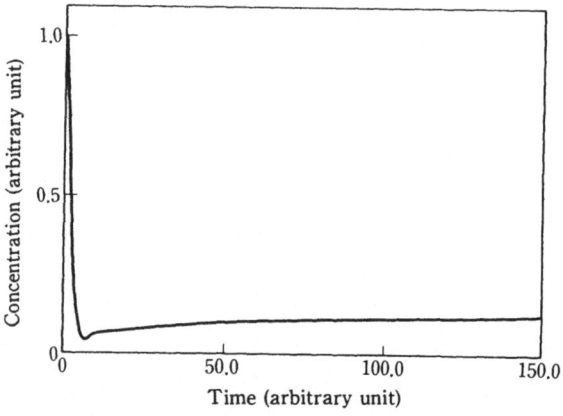

**Fig. 8.14**  Time course of $X_6$ in Scheme 8.17. Rate constant : $k_{13} = k_{14} = 1.0$. Other rate constants are the same as in Fig. 8.13.

(3) A basic feedback loop should exist for sustained oscillation.
In Scheme 8.16 with these conditions satisfied, $X_6$ works as both positive and negative effectors on the branched equilibrating reaction, which seems unlikely in real systems.

We then assume that an equilibrating reaction is comprised of two steps on which $X_6$ exerts the positive and negative feedback loops, respectively. Schemes 8.17 and 8.18 are probable for this mechanism. It follows from Figs. 8.14 and 8.15 that only Scheme 8.18 generates a spike-type oscil-

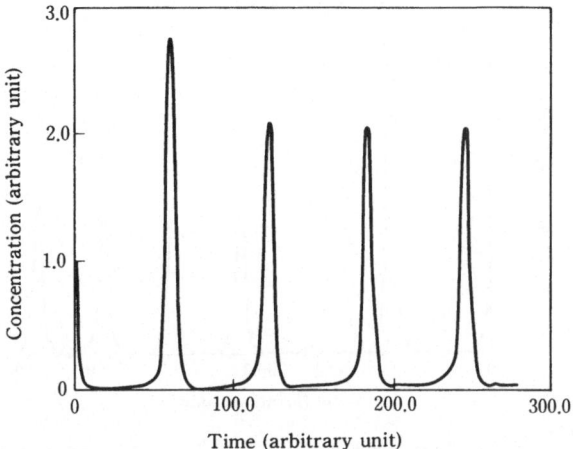

**Fig. 8. 15** Time course of $X_6$ in Scheme 8.18. Rate constants are the same as in Fig. 8.14.

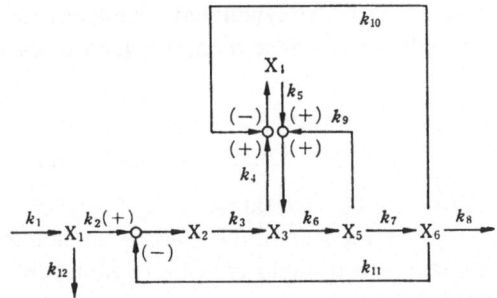

Scheme 8. 19. A linear system generating a spike-type oscillation

lation. The conditions would also be satisfied with Scheme 8.19 which has a one-step equilibrating reaction affected by two different species. Simulation reveals, as in Fig. 8.16, that the scheme can produce a typical spike-type oscillation.

These results imply that any scheme similar to Scheme 8.19 satisfies the three conditions and generates a basic spike-type oscillation. It is demonstrated here how the simulation can solve the problem of estimation of basic schemes which generate a spike-type oscillation in enzyme systems. Other probable schemes might have been overlooked in this example because of the estimation beginning with a limited number of scheme types.

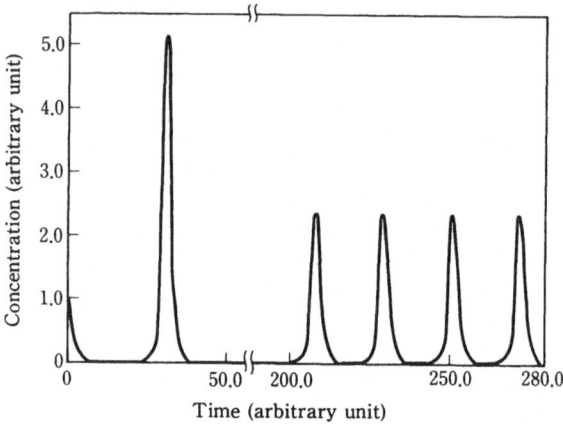

**Fig. 8.16** Time course of $X_6$ in Scheme 8.19. Rate constants are the same as in Fig. 8.13.

## 4. Reaction Scheme of $p$-Hydroxybenzoate Hydroxylase

$p$-Hydroxybenzoate hydroxylase is a monooxygenase, catalyzing the following reaction:

$p$-hydroxybenzoate + NADPH + H$^+$ + O$_2$
$$\longrightarrow \text{protocatechuate} + \text{NADP} + \text{H}_2\text{O}.$$

This reaction requires three reactants, $p$-hydroxybenzoate (PHBA), oxygen (O$_2$) and reduced nicotinamide-adenine dinucleotide phosphate (NADPH), and the enzyme is a flavin enzyme of molecular weight 68,000 with FAD as chromophoric prosthetic group. In the absence of NADPH or O$_2$ a partial reaction can be observed spectrophotometrically. Following crystallization of the complex of substrate PHBA and holoenzyme from *Pseudomonas*, the kinetic study of the reaction was performed to lead to a reasonable mechanism of three partial reactions:

$$E_{ox} + S \cdots\cdots\cdots\cdots\cdots\cdots\longrightarrow E_{ox}S$$
$$E_{ox}S + NADPH + H^+ \cdots\cdots\longrightarrow E_{red}S + NADP$$
$$E_{red}S + O_2 \cdots\cdots\cdots\cdots\cdots\longrightarrow E_{ox} + P + H_2O,$$

where the reduced and oxidized forms of the enzyme are denoted by $E_{red}$ and $E_{ox}$, respectively, S represents the substrate PHBA, and P the product protocatechuate.

**Measurement by Stopped-Flow Method**

The measurement of change in absorbance of FAD by the stopped-flow

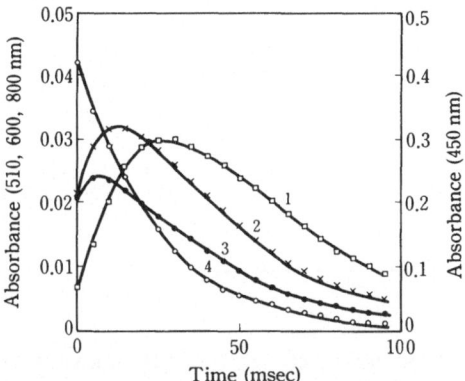

**Fig. 8.17** Time course of absorbance in reduction process. 1 : 800 nm ; 2 : 600 nm ; 3 : 510 nm ; 4 : 450 nm. Initial condition : $E_{ox}S = 43.6\mu$M, $S = 373\mu$M, $NADPH = 207\mu$M. Experimental condition : pH 8.0, 50°C, anaerobic, about 1.5 msec dead period.

method makes it possible to analyze the number of intermediate steps and the structure of intermediates involved in each partial reaction. We here analyze the second partial reaction (reduction process) as an example. Figure 8.17 illustrates the time course of absorbance change in the reduction process of PHBA hydroxylase observed at 10 nm wavelength intervals between 450 nm and 850 nm by the stopped-flow method. The characteristics in the intervals are recognized in the figure as :

( i )  450–500 nm      monotonous decrease
( ii )  510 nm          peak at about 5 msec
(iii)  520–650 nm      peak at about 15 msec
(iv)  700–850 nm      peak at about 25 msec,

where the time is measured after the dead period unless stated otherwise. In addition, the peak between 660 nm and 700 nm varies continuously from 15 msec to 25 msec. This observation suggests the presence of three intermediates corresponding to the peaks at 5, 15 and 25 (in msec).

**Construction of Model Scheme**

It is shown in Fig. 8.18 that the rate constant $k_{red}$ of the apparent first-order for the reduction reaction depends on the concentration in NADPH. This reaction thus seems to bear the intermediate step instead of a simple bimolecular mechanism. Hence, we assume the following linear multi-step scheme :

$$E_{ox}S + NADPH \rightleftharpoons X \rightleftharpoons Y \cdots\cdots Z \rightleftharpoons E_{red}S + NADP.$$

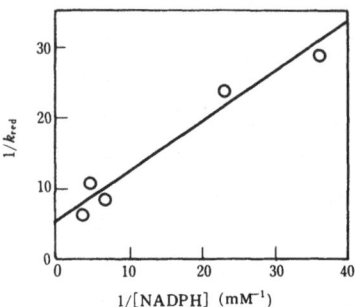

**Fig. 8.18** Relationship between NADPH concentration and rate constant $k_{red}$ of reduction process. Experimental condition: $E_0 = 4.6 \mu M$, *PHBA* $= 0.22$ mM, pH 8.2, 25°C.

Furthermore, for the first approximation the first and final steps can be treated as unimolecular reaction rather than bimolecular reaction, because the amount of NADPH is substantially more than that of $E_{ox}S$ and because the reaction tends sufficiently toward the production of $E_{red}S$. These lead to Scheme 8.20 of linear $n$-step reaction as a model for the reduction reaction. In the scheme $X_1$ and $X_{n+1}$ represent $E_{ox}S$ and $E_{red}S$, respectively,

$$X_1 \underset{k_{-1}}{\overset{k_1}{\rightleftarrows}} X_2 \underset{k_{-2}}{\overset{k_2}{\rightleftarrows}} \cdots\cdots \rightleftarrows X_n \underset{k_{-n}}{\overset{k_n}{\rightleftarrows}} X_{n+1}$$

Scheme 8.20. Linear $n$-step reaction

while $k_1$ is dependent on the concentration in NADPH. We now attempt to estimate the number of intermediate steps, rate constants and spectra of the intermediates for this scheme.

**Formulation to Estimate the Reaction Scheme and Parameters**

The observed data indicate the temporal change in total absorbance of $X_1$ $\sim X_{n+1}$ in unknown distribution. Obviously, the extinction coefficients of intermediates are also unknown. On the other hand, simplification of the reaction into Scheme 8.20 allows the simultaneous estimation of rate constants and number of intermediate steps using the following procedure.

The reaction of Scheme 8.20 is linear and its mathematical model is expressed by

$$\frac{dX}{dt} = KX,$$

$$(1)$$

where

$$X = \begin{bmatrix} X_1 \\ X_2 \\ \vdots \\ X_{n+1} \end{bmatrix}$$

$$K = \begin{bmatrix} -k_1 & k_{-1} & 0 & \cdots\cdots\cdots & 0 \\ k_1 & -(k_2+k_{-1}) & k_{-2} & \cdots\cdots\cdots & 0 \\ 0 & k_2 & -(k_3+k_{-2}) & \cdots\cdots & 0 \\ \vdots & & & & \vdots \\ 0 & & \cdots\cdots & -(k_n+k_{-(n-1)}) & k_{-n} \\ 0 & & \cdots\cdots\cdots & k_n & -k_{-n} \end{bmatrix}$$

Let us denote the absorbance experimentally observed and the extinction coefficients of enzyme species at each measuring wavelength by $A_0$ and $F_0$, respectively. Then, we have a relationship:

$$A_0 = F_0 X, \tag{2}$$

where $A_0$ is a vector of the dimension equal to the number of measuring wavelengths, and component $F^0_{ij}$ of the extinction coefficient matrix $F_0$ represents the extinction coefficient of enzyme species $X_j$ at the $i$th measuring wavelength $\lambda_i$. Using the conservation law of enzyme,

$$X_1 + X_2 + \cdots + X_{n+1} = E, \tag{3}$$

we can eliminate $X_{n+1}$ in equation (2) and get

$$A = A_0 - E \begin{bmatrix} F^0_{1,n+1} \\ F^0_{2,n+1} \\ \vdots \\ F^0_{m,n+1} \end{bmatrix}$$

$$= \begin{bmatrix} (F^0_{1,1} - F^0_{1,n+1}) & \cdots\cdots & (F^0_{1,n} - F^0_{1,n+1}) \\ \vdots & \cdots\cdots\cdots & \vdots \\ (F^0_{m,1} - F^0_{m,n+1}) & \cdots\cdots & (F^0_{m,n} - F^0_{m,n+1}) \end{bmatrix} \begin{bmatrix} X_1 \\ X_2 \\ \vdots \\ X_n \end{bmatrix} = FX(n), \tag{4}$$

where $E$ indicates the total concentration of all enzyme species, *i.e.*, the initial concentration in $E_{ox}S$. Equation (4) represents the relationship in reference to $X_{n+1}$, since $A$ is the absorbance measured in reference to the state in which all enzyme species are changed into $X_{n+1}$ in the system, that is, the state of completion of the reaction, and $F$ is the extinction coefficient relative to $X_{n+1}$.

In preparation for later analysis, we derive the quantity which is contained in the observed value $A$ and independent of the unknown quantity $F$. Let the number of intermediate steps be $n$, and $n$ wavelengths be chosen arbitrarily from $m$ measuring wavelengths. From equation (4) we have

$$A(n) = \begin{bmatrix} A_1 \\ A_2 \\ \vdots \\ A_n \end{bmatrix} = \begin{bmatrix} F_{1,1} & \cdots & F_{1,n} \\ \vdots & \cdots & \vdots \\ F_{n,1} & \cdots & F_{n,n} \end{bmatrix} \begin{bmatrix} X_1 \\ \vdots \\ X_n \end{bmatrix} = F(n) X(n), \tag{5}$$

where notations like $A(n)$ and $F(n)$ are used to express $n$-dimensional vector and $n \times n$ matrix exclusively in this discussion. If the reaction at the final step can be assumed to be irreversible, equation (1) is reduced to a linearly independent differential equation:

$$\frac{dX(n)}{dt} = K(n) X(n), \tag{6}$$

where $K(n)$ is the $n \times n$ matrix derived by eliminating the $(n+1)$th row and $(n+1)$th column of $K$ in equation (1). Actually, the reduction reaction considered here yields a spectrum at the end of reaction which coincides well with the known spectrum of $E_{red}S$.

Differentiation of both sides of equation (5) leads to

$$\frac{dA(n)}{dt} = \{F(n) K(n) F(n)^{-1}\} A(n), \tag{7}$$

where equation (6) and $X(n)$ in equation (5) are used for the derivation. Existence of the inverse matrix $F(n)^{-1}$ requires $n$ wavelengths for which the corresponding spectra have different profiles in their temporal change with each other. The eigenvalues for the matrix $F(n) K(n) F(n)^{-1}$ are equal to those for $K(n)$ and independent of $F(n)$. Solution of $f_n(\rho) = 0$, the characteristic equation for $K(n)$, yields the eigenvalues $\rho$ as functions of the rate constants $k_i$, where $f_n(\rho)$ is given as follows:

$$\left. \begin{aligned} f_0(\rho) &= 1 \\ f_1(\rho) &= -(\rho+k_1) \\ &\vdots \\ f_n(\rho) &= -\{\rho+k_n+k_{-(n-1)}\} f_{n-1} - k_{n-1} k_{-(n-1)} f_{n-2}. \end{aligned} \right\} \tag{8}$$

The temporal change in $A(n)$ and its derivative is obtained by experimental measurement, and then equation (7) gives $F(n) K(n) F(n)^{-1}$. The characteristic equation is solved to get the eigenvalues which provide information on the rate constants (unknown parameters) in conjunction with equation (8). For instance, when all reverse reactions can be ignored, equation (8) becomes

$$f_n(\rho) = (-1)^n (\rho+k_1)(\rho+k_2) \cdots (\rho+k_n), \tag{9}$$

yielding the eigenvalues of $\rho = -k_i \ (i=1,2,\cdots,n)$.

## Estimation of the Number of Intermediate Steps and Rate Constants

Starting from $n=1$ with a single step-increase, the number of intermediate steps is estimated by examining the agreement with the experimental data. With $n=1$, corresponding to equation (7),

$$\frac{dA_1}{dt} = (F_1 k_1 F_1^{-1}) A_1 \tag{10}$$

is solved for one of the measuring wavelengths between 450 nm and 850 nm. Since $F_1 k_1 F_1^{-1} = k_1$ and the eigenvalue is $-k_1$, the rate constant is given by

$$k_1 = -\left(\frac{dA_1}{dt}\right) \Big/ A_1. \tag{11}$$

The term $-d(\ln A_1)/dt$ on the right-hand side is equal to the slope of the plot in $\ln A_1$ against time $t$ (plot of the first-order reaction). $A_1$ and $dA_1/dt$ are measured at 5 msec intervals for 44 wavelengths between 450 nm and 850 nm. The distribution of the eigenvalues thus obtained is shown in Fig. 8.19. If the assumption of $n=1$ is correct, all the eigenvalues in Fig. 8.19 must focus at a single point. However, two peaks are found around 24 sec$^{-1}$ and 1240 sec$^{-1}$, implying that $n \neq 1$. This is also anticipated from Fig. 8.17. Nevertheless, the rate constant for the rate-limiting step may be about 24 sec$^{-1}$.

Let us now assume that $n=2$. We determine $F(2) K(2) F(2)^{-1}$ from the experimental data at an arbitrary two times for an arbitrary two wavelengths:

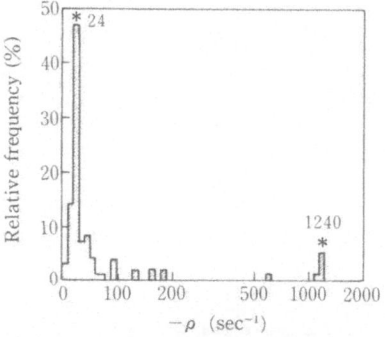

**Fig. 8.19** Corrected distribution of eigenvalues (estimated step number $=1$). The ordinate indicates the relative number of eigenvalue $f(k)$ between $k$ and $k+\Delta k$ corrected by $g(k) = f(k) k / \Delta k$. Numerals are the weighted mean value around the peak.

$$F(2) K(2) F(2)^{-1} = \begin{bmatrix} dA_1(t_1)/dt & dA_1(t_2)/dt \\ dA_2(t_1)/dt & dA_2(t_2)/dt \end{bmatrix} \begin{bmatrix} A_1(t_1) & A_1(t_2) \\ A_2(t_1) & A_2(t_2) \end{bmatrix}^{-1}. \quad (12)$$

A similar treatment leads to the distribution of eigenvalues in Fig. 8.20, in which three peaks are located around 25, 55 and 1240 (in $sec^{-1}$). This result might be explained by a situation in which two reacting steps advance most at any time during the chosen times $t_1$ and $t_2$. The first and second steps proceeding with the respective rate constants of $1240\ sec^{-1}$ and $55\ sec^{-1}$ together contribute to the measured absorbance, followed by the combined contribution from the second and third steps with the respective rate constants of $55\ sec^{-1}$ and $25\ sec^{-1}$. Hence, Fig. 8.20 implies that $n=3$.

The distribution of eigenvalues similarly obtained for $n$ equal to 3, 4 and 5 is given in Figs. 8.21, 8.22 and 8.23, respectively. It follows from these results that the number of intermediate steps must be less than 5, for only four peaks are detected in Fig. 8.23 ($n=5$). On the other hand, it is unclear whether the peaks corresponding to $125\ sec^{-1}$ in Fig. 8.23 or $105\ sec^{-1}$ and $145\ sec^{-1}$ in Fig. 8.22 are evidence that $n=4$ or a result of experimental and computational errors. This decision can be made from estimation of the intermediate spectrum and reproduction of the experimental data in the following.

**Estimation of Spectra for the Intermediates**

If all the reverse reactions are assumed negligible, the eigenvalue peaks in Figs. 8.21 and 8.22 correspond to the rate constants for $n$ equal to 3 and 4, respectively. Solution of equation (1) with these values yields the time course of each chemical species. Consequently, application of the least

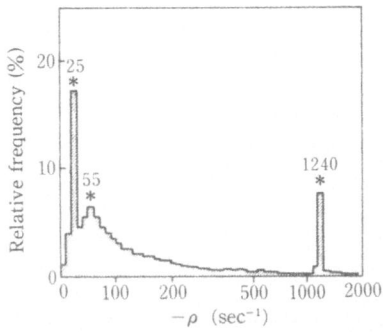

**Fig. 8. 20** Corrected distribution of eigenvalues (estimated step number=2). Notations are the same as in Fig. 8.19.

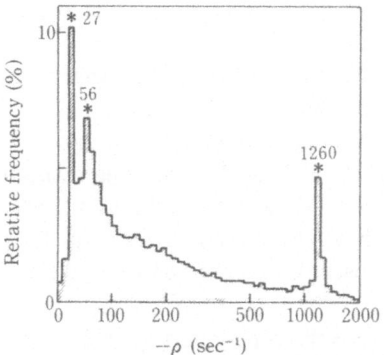

**Fig. 8. 21** Corrected distribution of eigenvalues (estimated step number = 3). Notations are the same as in Fig. 8.19.

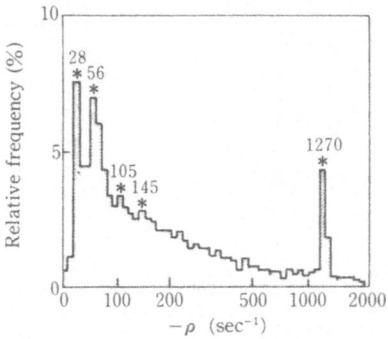

**Fig. 8. 22** Corrected distribution of eigenvalues (estimated step number = 4). Notations are the same as in Fig. 8.19.

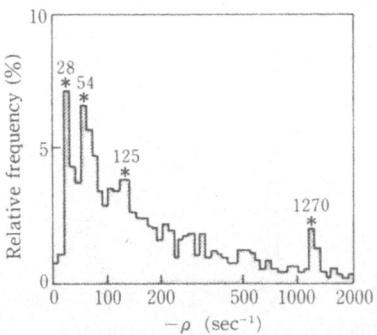

**Fig. 8. 23** Corrected distribution of eigenvalues (estimated step number = 5). Notations are the same as in Fig. 8.19.

square method to equation (4) leads to evaluation of the extinction coefficient matrix $F$.

Employing this procedure, we now estimate the spectra of the intermediates for the following two cases: $k_1 = 1260$, $k_2 = 56$ and $k_3 = 27$ (in sec$^{-1}$) for $n = 3$, and $k_1 = 1270$, $k_2 = 125$, $k_3 = 56$ and $k_4 = 28$ (in sec$^{-1}$) for $n = 4$. The results are demonstrated in Figs. 8.24 and 8.25. The estimation with $k_2 = 105$ sec$^{-1}$ or 145 sec$^{-1}$ (instead of 125 sec$^{-1}$) results in a spectrum similar to that in Fig. 8.25. The spectra of the intermediates can thus be estimated with the approximate rate constants except for those of the rate-limiting step.

Figures 8.26 and 8.27 illustrate the reproduction of the experimental data based on the spectra estimated above. In case of either $n$ equal to 3 or 4, the absorbance peaks reproduced almost coincide with the experimental

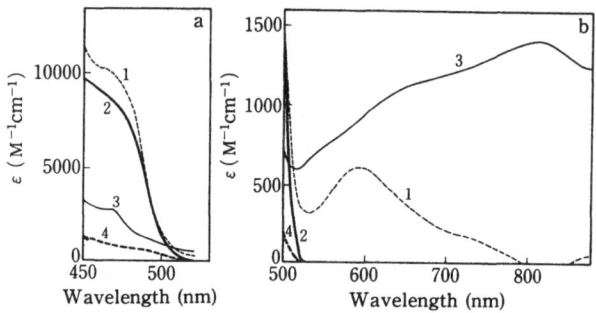

**Fig. 8. 24** Molecular extinction coefficients of chemical species in three-step reaction. 1: $X_1$; 2: $E_{ox}S$; 3: $X_2$; 4: $E_{red}S$.

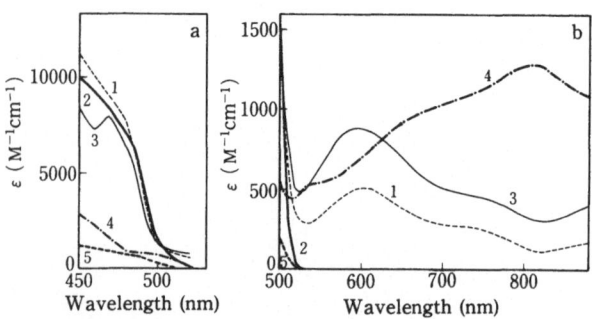

**Fig. 8. 25** Molecular extinction coefficients of chemical species in four-step reaction. 1: $X_1$; 2: $E_{ox}S$; 3: $X_2$; 4: $X_3$; 5: $E_{red}S$.

ones in Fig. 8.17. It is concluded, however, that the reduction process has four intermediate steps, from consideration of the following defects for $n = 3$: the absorbance just after the dead period $(t=0)$ is not reproduced and the peak of the absorbance at 600 nm deviates from the experimental values so that the agreement as a whole is not consistent.

**Summary of Analytical Procedure**

The procedure for estimation of the reaction scheme from the experiment is summarized in Fig. 8.28. It is desirable that experimental data be col-

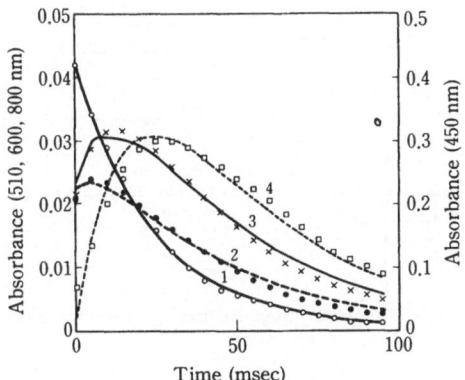

**Fig. 8.26** Time courses of absorbance in three-step reaction. Observed value: 450 nm (○), 510 nm (●), 600 nm (×), 800 nm (□). Simulated value: 450 nm (1), 510 nm (2), 600 nm (3), 800 nm (4).

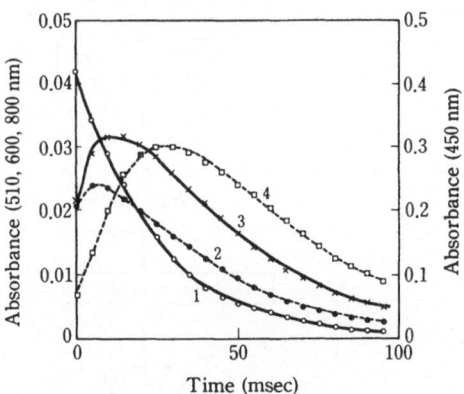

**Fig. 8.27** Time courses of absorbance in four-step reaction. Notations are the same as in Fig. 8.26.

lected for many measuring times and wavelengths more than the number of
intermediate steps. It is also necessary that not only the absorbance but its
derivative be measured accurately. Moreover, the experimental condition
should be arranged so that a bimolecular reaction can be treated as a unimolecular reaction, for example, by an excess of substrate in the reaction
system.

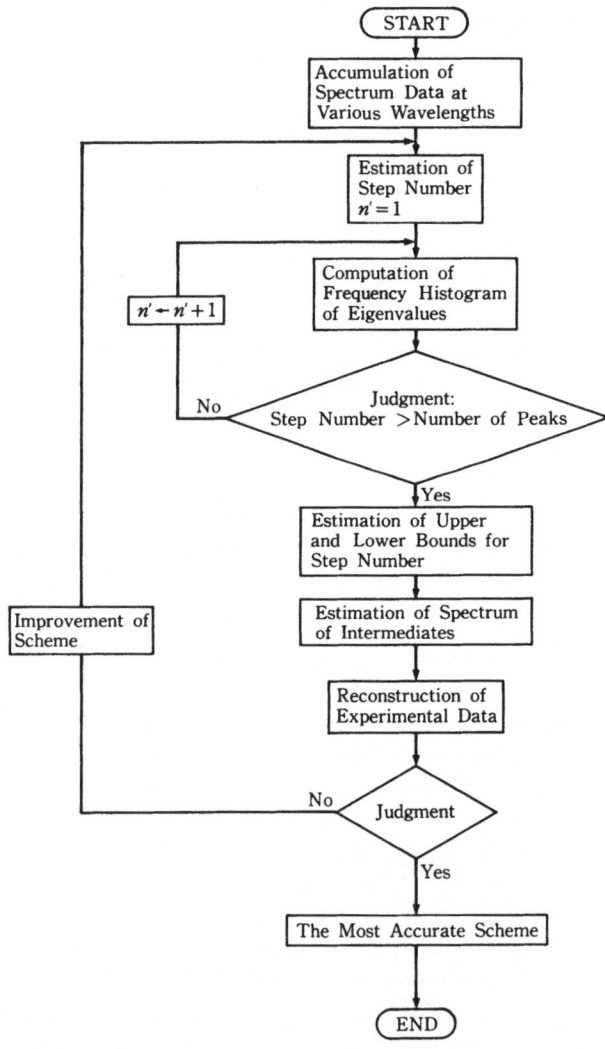

**Fig. 8. 28** Method for estimatoin of scheme of multi-step unimolecular
reaction.

## 8.2  Procedure for Evaluation of Kinetic Parameters

Determination of reaction scheme and various kinetic parameters like rate constants from the temporal change in concentrations of substrate, product and other chemical species in an enzymatic reaction involves two kinds of evaluation. One is construction of a model for the reaction scheme, which would entail a very difficult problem requiring profound biochemical insight into the functional and structural entity of the reaction. The other is evaluation of unknown kinetic parameters from a given model scheme and observed concentration changes of chemical species, which can be treated by a purely mathematical procedure called optimization.

If the results from optimization should not agree well with the observed data, the model scheme under consideration would be discarded and a revised model constructed to resume the optimization. Methods for estimation of a reaction scheme are described in the preceding section with reference to the mechanism for activation of pepsinogen, spike-type oscillation, and reaction of $p$-hydroxybenzoate hydroxylase. In this section formulation and solution of optimization problems are introduced to explain the fundamentals to the methods for evaluating kinetic parameters from observed data. The evaluation of binding free energy in lysozyme is also described as an example of optimization.

### 1.  Optimization Problem

The optimization problem we are interested in is formulated as follows (see also Section 9.2): to find the parameter $P$ ($m$-dimensional vector) such that the cost function,

$$J(P) = \sum_{i=1}^{k} \sum_{j} (y_j(t_i, P) - y_{ij})^2 , \tag{13}$$

is minimized by the use of the observed values $y_{ij}$ (of $j$th chemical species observed at time $t_i$) and postulated mathematical model,

$$\frac{d}{dt} y(t, P) = f(t, y, P), \quad y(t_0, P) = y_0 . \tag{14}$$

In equation (13) $y_j(t_i, P)$ represents a solution of equation (14) for a given $P$, and $k$ denotes the number of observation points (time). The summation with respect to $j$ is done only for the chemical species observed.

The common procedure for determination of optimal parameter $P^*$ such as $[P^*; \min J(P)]$ is the iterative evaluation of $P_{i+1}$ satisfying

$$J(P_{i+1}) \leq J(P_i), \quad i=0,1,2,\cdots \tag{15}$$

until the convergence condition meets the constraints specific to the problem. The commonly used condition for convergence is expressed as

$$J(P_{i+1}) \leq a$$

or

$$J(P_i) - J(P_{i+1}) \leq a$$

with a given positive number $a$. The methods for obtaining convergence sequence $\{P_0, P_1, \cdots\}$ have been devised and classified into the gradient method, which makes use of the derivatives of $J$, and the direct search method, which can proceed without use of the derivatives. At present, no single method applicable to any problem is available for obtaining an optimal solution $P^*$. Thus, it is desirable to provide several methods for specific optimization problems and employ those appropriate to a given problem.

In this section, we discuss a gradient method [3] and the Powell method [4,7], which is one of the direct search methods. These are described with respect to their application to the system of Scheme 8.21, which represents a model for the effect of pH on enzymatic reaction. Figure 8.29 illustrates the time course of the system with the rate constants given in Scheme 8.21 and the initial condition of $S=1.0$ mM and $EH=10\ \mu$M. The concentration in substrate S reduces rapidly in the early stage of reaction, and then decreases linearly with time.

$$S+EH_2 \underset{k_{-1}'}{\overset{k_{+1}'}{\rightleftharpoons}} EH_2S \xrightarrow{k_{+2}'} EH_2+P$$

$$k_{-01} \big\updownarrow k_{+01} \qquad k_{-01}' \big\updownarrow k_{+01}'$$

$$S+EH \underset{k_{-1}}{\overset{k_{+1}}{\rightleftharpoons}} EHS \xrightarrow{k_{+2}} EH+P$$

$$k_{+02} \big\updownarrow k_{-02} \qquad k_{+02}' \big\updownarrow k_{-02}'$$

$$S+E \underset{k_{-1}''}{\overset{k_{+1}''}{\rightleftharpoons}} ES \xrightarrow{k_{+2}''} E+P$$

| | | |
|---|---|---|
| $k_{+01} = 0.1 \times 10^7$ | $k_{+02}' = 0.1 \times 10$ | $k_{+1}'' = 0.1 \times 10$ |
| $k_{-01} = 0.1 \times 10^4$ | $k_{-02}' = 0.1 \times 10^7$ | $k_{-1}'' = 0.25 \times 10^{-2}$ |
| $k_{+01}' = 0.1 \times 10^7$ | $k_{+1} = 0.1 \times 10^6$ | $k_{+2} = 0.1 \times 10^2$ |
| $k_{-01}' = 0.1 \times 10^5$ | $k_{-1} = 0.25 \times 10^3$ | $k_{+2}' = 0.1 \times 10$ |
| $k_{+02} = 0.1 \times 10$ | $k_{+1}' = 0.1 \times 10^3$ | $k_{+2}'' = 0.1$ |
| $k_{-02} = 0.1 \times 10^7$ | $k_{-1}' = 0.25 \times 10$ | $[H] = 10^{-7}$M |

Scheme 8.21. Effect of hydrogen-ion concentration on enzymatic reaction

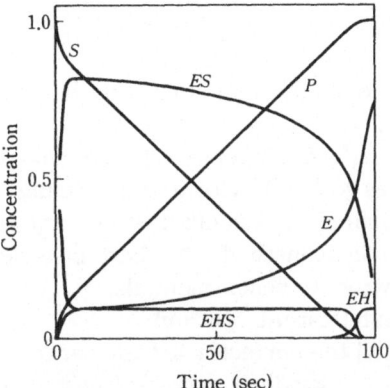

**Fig. 8.29** Time course of Scheme 8.21. Initial condition: $S = 1.0$ mM, $EH = 10\mu$M. Full scale (1.0) on the ordinate: $S = P = 1.0$ mM, $E = EH = EHS = 10\mu$M, $EH_2 = EH_2S = 50$ nM.

The optimization problem to be solved here is described as follows. The temporal change in substrate concentration obtained from the simulation is regarded as the observed values $y_i$, that is, the simulation with a given initial condition yields the observed data $y_i$. The rate constants $k_{+2}$ and $k_{+2}'$ are assumed to be unknown parameter $P$. The other rate constants and reaction scheme are as given in Scheme 8.21. We attempt to evaluate the true (optimal) values for $k_{+2}$ and $k_{+2}''$ employing the optimization methods.

## 2. Optimization by Gradient Method

In the gradient method, the corrector $\delta P$ is obtained from $J(P)$ at an evaluated value $P$ and partial derivatives of $J(P)$ with respect to $P$, where $\delta P$ is defined as the difference of $P$ from the optimal solution $P^*$, that is,

$$P^* = P + \delta P.$$

A Taylor series expansion of $J(P^*) = J(P + \delta P)$ with respect to $\delta P$ leads to the expression,

$$J(P^*) = J(P + \delta P)$$

$$= \sum_{i=1}^{k} \{y(t_i, P + \delta P) - y_i\}^2$$

$$= \sum_{i=1}^{k} \left[ \{y(t_i, P) - y_i\}^2 + 2\{y(t_i, P) - y_i\} \cdot \sum_{j=1}^{m} \frac{\partial}{\partial P_j} y(t_i, P) \delta P_j \cdot \right.$$

$$+ \sum_{j,k=1}^{m} \left\{ \frac{\partial}{\partial P_i} y(t_i, P) \right\}^t \left\{ \frac{\partial}{\partial P_k} y(t_i, P) \right\} \delta P_j \delta P_k \right], \quad (16)$$

where the expansion keeps only the first-order derivatives of $y$, and t denotes transpose of matrix.

The matrix $\partial y(t_i, P)/\partial P = Y_P(t_i, P)$ represents the concentration variation of each chemical species $y(t_i, P)$ with a small change in parameter $P$. If $Y_P(t_i, P)$ could be evaluated at each observation time $t_i$, optimization would be possible by the gradient method. We here consider the steepest descent and generalized Newton-Raphson methods.

**Derivation of $\delta P$ by Steepest Descent Method**

In the steepest descent method the corrector $\delta P$ for parameter $P$ is determined in the direction along the gradient vector of cost function $J$, which is defined as its inner product with $\delta P$ equal to the first variation $\delta J$ of the cost function $J$, i.e., the second term on the right-hand side of equation (16). Defining a vector $d$ by

$$d = \{y(t_i, P) - y_i\}^t \frac{\partial}{\partial P} y(t_i, P)$$

$$\equiv \{y(t_i, P) - y_i\}^t Y_P(t_i, P), \quad (17)$$

we thus perform the optimization process with the corrector $\delta P = \alpha d$ in the steepest descent method.

Various selection-methods for $\alpha$ are devised such as the fixed value and linear (univariable) search methods. Setting $\alpha$ at a fixed constant is the simplest method, which causes difficulty in very slow convergence around the optimal point. The linear search method proceeds with finding $\alpha$ such that

$$J(P^*) = \min_{\alpha} J(P + \alpha d).$$

Many algorithms for this method are known, such as the Fibonacci and golden division methods; these generally require much computation time. In the optimization for the system of equations (13) and (14), the computation is more loaded to solve the differential equation (14) at each step.

We here employ a simplified method in which the corrector $\delta P$ is derived from the necessary condition that $P + \delta P$ is the optimal solution, that is,

$$\frac{\partial J(P + \delta P)}{\partial \alpha} = 0,$$

where

$$J(P + \delta P) = J(P) + 2\alpha d^t d + \alpha^2 d^t \left\{ \sum_i Y_P(t_i, P)^t Y_P(t_i, P) \right\} d. \quad (18)$$

Hence, we have

$$\delta P = \frac{d^t d}{d^t \{\sum_i Y_P(t_i, P)^t Y_P(t_i, P)\} d} d .$$

(19)

**Derivation of $\delta P$ by Generalized Newton-Raphson Method**

The corrector $\delta P$ is derived from equation (16) so that

$$\frac{\partial J(P^*)}{\partial \delta P_i} = 0,$$

for every $i$; from equation (16)

$$\sum_{i=1}^{k} \sum_{j=1}^{m} \left[ \{y(t_i, P) - y_i\}^t \frac{\partial}{\partial P_j} y(t_i, P) \right.$$

$$\left. + \sum_{k=1}^{m} \left\{ \frac{\partial}{\partial P_j} y(t_i, P) \right\}^t \left\{ \frac{\partial}{\partial P_k} y(t_i, P) \right\} \delta P_k \right] = 0 .$$

(20)

Rearrangement leads to

$$\delta P = -\left[ \sum_{i=1}^{k} Y_P(t_i, P) Y_P(t_i, P)^t \right]^{-1} \left[ \sum_{i=1}^{k} Y_P(t_i, P) \Delta(t_i, P) \right],$$

(21)

where $\Delta(t_i, P) = y(t_i, P) - y_i$. This method to obtain $\delta P$ is called the generalized Newton-Raphson method.

**Evaluation of $Y_P(t_i, P)$**

Derivation of $\delta P$ by either the steepest descent method or the generalized Newton-Raphson method requires the evaluation of $Y_P(t_i, P)$ at the observation time $t_i$. If the evaluation is difficult, a direct search method such as the Powell method described later can be employed for optimization rather than the gradient method. According to the procedures of Gear [5] and Hemker [6], the system of differential equations to evaluate $Y_P(t_i, P)$ is derived and solved as follows. Differentiation of equation (14) with respect to parameter $P$ leads to a system of differential equations for $Y_P(t, P)$:

$$\frac{d}{dt} Y_P(t, P) = Y_P(t, P) F_Y(t, y, P) + F_P(t, y, P) ,$$

(22)

where

$$Y_P(0, P) = 0$$

$$[Y_P(t, P)]_{ij} = \partial y_j(t, P) / \partial P_i,$$

$$[F_Y(t, y, P)]_{kj} = \partial f_j(t, y, P) / \partial y_k,$$

$$[F_P(t, y, P)]_{ij} = \partial f_j(t, y, P)/\partial P_i ,$$

and the order of differentiation is changed in this derivation. Equation (22) is a linear system with respect to $Y_P$, which can readily be solved by the Gear method (an implicit linear multi-step method) in association with numerical integration of equation (14) for evaluation of $y(t, P)$. Application of the corrector formula in the Gear method [equation (38) in Chapter 3] to equation (22) yields the corrector formula for $Y_P$ as

$$Y_P{}^{(k+1)} = \sum_{j=1}^{q} \alpha_j * Y_P{}^{(k+1-j)} + \eta_0 * h [ Y_P{}^{(k+1)} F_Y + F_P] , \tag{23}$$

which is a system of linear algebraic equations. The solution is derived by

$$Y_P{}^{(k+1)} = (I - \eta_0 h F_Y)^{-1} \left( \sum_{j=1}^{q} \alpha_j * Y_P{}^{(k+1-j)} + \eta_0 * h F_P \right), \tag{24}$$

without resorting to the successive substitution. Now that the inverse matrix $(I - \eta_0 * h F_Y)^{-1}$ in equation (24) is the same one used to solve equation (14), $Y_P$ can be obtained simply by multiplication of the terms.

It should be mentioned that $Y_P$ can be utilized not only in optimization as above but also in sensitivity analysis, which quantitatively describes the

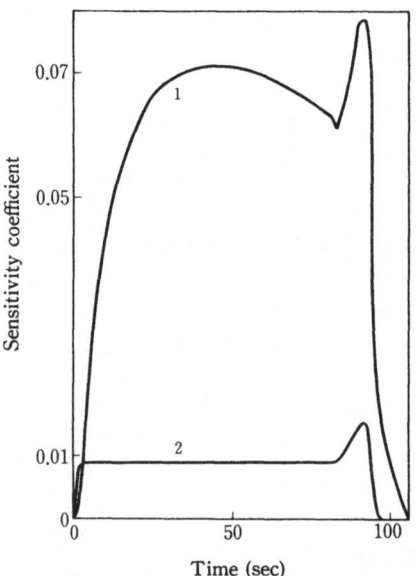

**Fig. 8. 30**  Sensitivity analysis of Scheme 8.21. 1 : $\partial J/\partial k_{+2}''$ ; 2 : $\partial J/\partial k_{+2}$.

effects of parameter $P$ on the reaction system. Figure 8.30 presents the temporal changes in sensitivity coefficients $|\partial J/\partial k_{+2}|$ and $|\partial J/\partial k_{+2}''|$ computed with $Y_P$ for the system following the time course of Fig. 8.29. Both sensitivity coefficients are virtually constant during the steady state, reaching peaks just before the end of reaction. Except for the early stage of reaction, $|\partial J/\partial k_{+2}''|$ is several times larger than $|\partial J/\partial k_{+2}|$, due to setting $k_{+2}=10\ \mathrm{sec}^{-1}$ and $k_{+2}''=0.1\ \mathrm{sec}^{-1}$, respectively. With $k_{+2}=50\ \mathrm{sec}^{-1}$ and $k_{+2}''=10\ \mathrm{sec}^{-1}$, $|\partial J/\partial k_{+2}|$ becomes about 100 times larger than $|\partial J/\partial k_{+2}''|$. Both coefficients reach zero at the end of reaction because the final state in a closed system is unaffected by $k_{+2}$ and $k_{+2}''$.

### 3. Examples of Optimization by Gradient Method

Optimization by the gradient method is done to evaluate the rate constants $k_{+2}$ and $k_{+2}''$ in Scheme 8.21. The optimization must employ a certain combination of methods specific to a given problem. It is concluded that the steepest descent method with equation (19) should start the optimization of a single parameter, followed by the correction process with $(J(P))^{1/2}\delta P$ when $J(P)$ becomes less than 1.0. On the other hand, it appears suitable for the optimization of two parameters that the steepest descent method be applied until $J(P)$ becomes less than 0.1, thence to be replaced by the generalized Newton-Raphson method of equation (21). The nonnegative condition of $P+\delta P$ should further be taken into account because no rate constants are negative in enzymatic reactions.

Hemker [6] has proposed an optimization procedure wherein the generalized Newton-Raphson method of equation (21) is applied normally unless the improvement of cost function is unlikely, and then the steepest descent method is employed. This procedure, however, is found to have extremely slow convergence.

### Optimization of a Single Parameter

We consider an optimization problem in Scheme 8.21 to evaluate the value of rate constant $k_{+2}$ with a starting value of $100\ \mathrm{sec}^{-1}$ and the experimental data of temporal change in the concentration of substrate S. In the case of a single parameter the steepest descent method of equation (19) and the generalized Newton-Raphson method of equation (21) lead to identical results, so the former method is applied in this example.

The results of optimization are shown in Table 8.4 and Fig. 8.31. The value of $k_{+2}$ converges monotonously toward the optimal solution of 10.0 $\mathrm{sec}^{-1}$ until it reaches a value of $11.4\ \mathrm{sec}^{-1}$, that is, $J(P)$ becomes less than 1.0. From then on, it oscillates around the optimal solution, as seen with points 5, 6 and 7 in Fig. 8.31; this is due to the derivation of $\delta P$ from equation (19). The exact but time-consuming method of linear search

**Table 8.4** Optimization of a single parameter by gradient method

| Step | Starting value | 1 | 2 | 3 | 4 | 5 | 6 |
|---|---|---|---|---|---|---|---|
| $k_{+2}$ | 100 | 53.5 | 28.9 | 18.9 | 3.7 | 11.4 | 7.55 |
| $J(P)$ | 12.2 | 10.5 | 6.71 | 3.38 | 1.06 | 0.211 | 0.960 |

| Step | 7 | 8 | 9 | 10 | 11 | 12 | 13 | 14 |
|---|---|---|---|---|---|---|---|---|
| $k_{+2}$ | 13.1 | 11.2 | 7.86 | 12.9 | 11.1 | 8.00 | 12.7 | 11.1 |
| $J(P)$ | 0.801 | 0.164 | 0.726 | 0.702 | 0.146 | 0.632 | 0.653 | 0.136 |

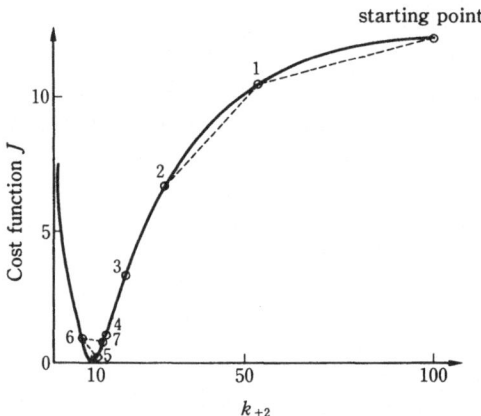

**Fig. 8.31** Convergence of $k_{+2}$ to the optimal value. Numbers indicate the step of optimization.

**Table 8.5** Improvement of optimization by introducing a factor $(J(P))^{1/2}$

| Step | Starting value | 1 | 2 | 3 | 4 | 5 | 6 | 7 |
|---|---|---|---|---|---|---|---|---|
| $k_{+2}$ | 100 | 53.5 | 28.9 | 18.9 | 13.7 | 11.4 | 9.62 | 9.75 |
| $J(P)$ | 12.2 | 10.5 | 6.71 | 3.38 | 1.06 | 0.211 | $2.05 \times 10^{-2}$ | $9.05 \times 10^{-3}$ |

| Step | 8 | 9 | 10 | 11 | 12 | 13 | 14 |
|---|---|---|---|---|---|---|---|
| $k_{+2}$ | 9.81 | 9.84 | 9.87 | 9.88 | 9.90 | 9.91 | 9.92 |
| $J(P)$ | $5.14 \times 10^{-3}$ | $3.43 \times 10^{-3}$ | $2.48 \times 10^{-3}$ | $1.87 \times 10^{-3}$ | $1.46 \times 10^{-3}$ | $1.17 \times 10^{-3}$ | $9.65 \times 10^{-4}$ |

The optimal value of $k_{+2}$ equals 10.0.

could certainly overcome the difficulty, but instead, the multiplication of $\delta P$ by a factor $(J(P))^{1/2}$ is attempted to improve the behavior when $J(P)$ is less than 1.0. The results in Table 8.5 indicate that 14 steps yield a value of $9.92\,\mathrm{sec}^{-1}$ for $k_{+2}$, which is a good approximation to the optimal solution of $10.0\,\mathrm{sec}^{-1}$. Thus, the factor $(J(P))^{1/2}$ introduced here depresses the oscillation in the optimization procedure.

## Optimization of Two Parameters

Table 8.6 demonstrates the results of optimization by the steepest descent method for rate constants $k_{+2}$ and $k_{+2}''$ with the respective starting values of 50 and 10 (in $\mathrm{sec}^{-1}$). Convergence to the optimal solution ($10.0\,\mathrm{sec}^{-1}$ for $k_{+2}$ and $0.1\,\mathrm{sec}^{-1}$ for $k_{+2}''$) gets extremely bad when $J(P)$ becomes less than $5.52\times10^{-2}$. This is not due to the choice of factor $(J(P))^{1/2}$, but is peculiar to the steepest descent method, which generally tends to have a small maximum-gradient-vector around the optimal solution.

The appropriate procedure would thus be a combination of the steepest descent method and generalized Newton-Raphson method such that the former is employed to have an optimization approach in the neighborhood

**Table 8.6**  Optimization of two parameters by steepest descent method

| Step | Start-ing value | 1 | 2 | 3 | 4 | 5 | 6 | 7 | 8 | 9 | 10 |
|------|------|------|------|------|------|------|------|------|------|------|------|
| $k_{+2}$ | 50.0 | 17.6 | 1.76 | 3.35 | 5.03 | 5.13 | 5.27 | 5.33 | 5.37 | 5.40 | 5.42 |
| $k_{+2}''$ | 10.0 | 9.38 | 8.65 | 8.78 | 8.89 | 8.90 | 8.90 | 8.90 | 8.90 | 8.90 | 8.90 |
| $J(P)$ | 10.3 | 4.28 | 1.40 | 0.487 | $5.52 \times 10^{-2}$ | $4.30 \times 10^{-2}$ | $3.75 \times 10^{-2}$ | $3.49 \times 10^{-2}$ | $3.34 \times 10^{-2}$ | $3.26 \times 10^{-2}$ | $3.20 \times 10^{-2}$ |

The optimal values of $k_{+2}$ and $k_{+2}''$ are 10.0 and 0.1, respectively. A factor $(J(P)^{1/2})$ is multiplied by $\delta P$ when the cost function becomes less than 1.0.

**Table 8.7**  Optimization of two parameters by combination of steepest descent method and generalized Newton-Raphson method

| Step | Start-ing value | 1 | 2 | 3 | 4 | 5 | 6 | 7 | 8 | 9 | 10 |
|------|------|------|------|------|------|------|------|------|------|------|------|
| $k_{+2}$ | 50.0 | 17.6 | 1.76 | 3.35 | 5.03 | 5.37 | 5.97 | 6.88 | 8.07 | 9.73 | 9.78 |
| $k_{+2}''$ | 10.0 | 9.38 | 8.65 | 8.78 | 8.89 | 4.44 | 2.22 | 1.11 | 0.556 | 0.152 | 0.136 |
| $J(P)$ | 10.3 | 4.28 | 1.40 | 0.487 | $5.52 \times 10^{-2}$ | $5.07 \times 10^{-2}$ | $4.26 \times 10^{-2}$ | $2.66 \times 10^{-2}$ | $6.00 \times 10^{-3}$ | $1.85 \times 10^{-4}$ | $5.23 \times 10^{-5}$ |

The generalized Newton-Raphson method is employed when the cost function becomes less than 0.1. The steepest descent method adopting a factor $(J(P))^{1/2}$ is used when the cost function is more than 0.1.

of the solution within which the latter works effectively. The results of optimization by this procedure are given in Table 8.7 where $J(P)$, being less than 0.1, is taken as the criterion for the neighborhood of the solution. Compared with Table 8.6 it is obvious that the optimization converges to the solution much more rapidly.

### 4. Optimization by Direct Search Method

Optimization by the gradient method must compute $(m+1)$ times to obtain the partial derivatives of cost function $J$ with respect to parameter $P$ ($m$-dimensional vector). If these computations should be complex, it is recommended that some efficient algorithms be used in the direct search method, which pursues the optimal values in independent $m$ directions. The Powell method [4] is one of such algorithms in which optimization proceeds without use of derivatives; this is discussed in the following.

### Algorithm in the Powell Method

The Powell method performs optimization by deriving a new conjugate[*] search-direction $\{d_1, d_2, \cdots, d_m\}$ from every $(m+1)$th evaluation of $J$. Starting from a value $P_0$, the linear search in every direction $d_i$ yields a sequence of evaluated values to approach the optimal solution $P^*$. The procedure is iterated until a given condition of convergence is met. The algorithm is given as follows:

Initial direction of linear search:
     0. Give $m$ linearly independent vectors $d_1{}^0, d_2{}^0, \cdots, d_m{}^0$.

Iterative linear search in $m$ times:
     1. Give a starting value $P_0{}^0$ of parameter $P$.
     2. Perform a linear search in direction $d_1{}^0$ to determine $\lambda_1{}^0$ minimizing $J(P_0{}^0 + \lambda d_1{}^0)$ and take $P_1{}^0 = P_0{}^0 + \lambda_1{}^0 d_1{}^0$.
     3. Iterate the following linear search for $i$ from 2 to $m$ until the cost function $J$ meets the convergence condition; i) perform a linear search in direction $d_i{}^0$ to determine $\lambda_i{}^0$ minimizing $J(P_{i+1}{}^0 + \lambda d_i{}^0)$, and ii) take $P_i{}^0 = P_{i-1}{}^0 + \lambda_i{}^0 d_i{}^0$.
     4. Define new directions, $d_1{}^n, d_2{}^n, \cdots, d_m{}^n$ ($n=1,2,\cdots$).
     5. Set a new direction as $d_m{}^1 = P_m{}^0 - P_0{}^0$ and take $P_{m+1}{}^0 = P_m{}^0 + \lambda_m{}^1 d_m{}^1$ and $P_0{}^1 = P_{m+1}{}^0$.

New setting for search direction:
     6. Set new directions as $d_1{}^1 = d_2{}^0, \cdots, d_{m-1}{}^1 = d_m{}^0, d_m{}^1 = P_m{}^0 - P_0{}^0$. Return to step 1 and repeat the steps for $d_i{}^n, P_i{}^n, \lambda_i{}^n; i=1,2,\cdots,m$; $n=2,3,\cdots$.

[*] The cost function $J$ is assumed to be expressed by a quadratic form as $J = 1/2\,P^t Q P$, where $Q$ is a symmetric matrix. Vectors $d_i$ and $d_j$ are conjugate if $m$ vectors $\{d_1, d_2, \cdots, d_m\}$ satisfy
     $d_i{}^t Q dj = 0, \quad i \neq j$.

The Powell method is quite efficient among the various direct-search methods, especially where $J$ is approximated well by a quadratic form of $P$, in which convergence is quick in the neighborhood of the optimal solution. Figure 8.32 illustrates the search steps for evaluation of two parameters. Linearly independent direction-vectors are usually set to orthogonal coordinate axes for the initial step. We assume $d_1{}^0 = (0,1)$ and $d_2{}^0 = (1,0)$ in Fig. 8.32. New directions are formed in steps 5 and 6 of the algorithm, so that we have $d_1{}^1 = (1,0)$ and $d_2{}^1 = P_2{}^0 - P_0{}^1$ for the second iteration.

In programming of the algorithm some techniques are required for the linear search in steps 2 and 3 as well as the nonnegative condition of parameters (rate constants). In the Powell method accuracy in optimization by linear search affects the efficiency in new search directions. Various methods for linear search are devised to cope with the situation. In any case, optimization subject to a system of differential equations leads to the increase in computation time in direct association with the iteration number of linear searches. A simplified method for linear search might suffice in some cases as follows. Taking five values, $i.e.$, $\lambda_2 = 9$, $\lambda_1 = 1$, $\lambda_0 = 0$, $\lambda_{-1} = -0.5$ and $\lambda_{-2} = -0.9$, as trial values for $\lambda$, we choose $\lambda_{-M}$ to minimize $J(\lambda)$. The coefficients in $J(\lambda) = a\lambda^2 + b\lambda + c$ are determined from the values of $J(\lambda)$ at $[\lambda_M, \lambda_0, \lambda_{-M}]$ to obtain the optimal value $\lambda^* = -b/(2a)$. This leads to a good approximation to the optimal value since $J(\lambda)$ is approximately expressed by a quadratic form in the neighborhood of the optimal solution.

It would occur that the $\lambda^*$ obtained above is in violation of the nonnegative condition for parameters. In this case, $\lambda^*$ is reevaluted so that the parameters in violation reduce their values to one-tenth of current values. Another problem concerns the linear independency of direction vectors.

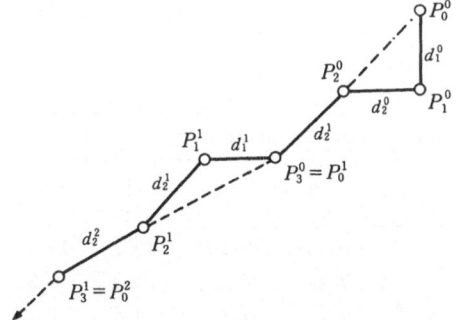

**Fig. 8.32** Diagram of iterative steps in the Powell method $(m = 2)$.

It is proved for the Powell method that at most $m$ iterations yield the optimal solution to such problems with cost function $J$ of quadratic form, if the direction vectors $\{d_1, d_2, \cdots, d_m\}$ are linearly independent at every step. Some problems, however, are not always assured of optimal solutions because the direction vectors are not always linearly independent. In order to overcome this, Powell [7] revised the algorithm by introducing criteria for formation of new direction vectors at steps 4, 5 and 6. This is called the modified Powell method, for which various procedures have been proposed [8].

## Optimization by the Powell Method

The convergence process in application of the Powell method to optimization in Scheme 8.21 is shown in Tables 8.8 and 8.9 for the cases of a single and two parameters, respectively. For optimization of a single parameter the convergence is better with the Powell method than with the gradient method as given in Table 8.4. For optimization of two parameters, rate constant $k_{+2}$ approaches the optimal solution, while $k_{+2}''$ deviates con-

Table 8.8  Optimization of one parameter by the Powell method

| Step | Starting value | 1 | 2 | 3 | 4 | 5 | 6 | 7 | 8 | 9 | 10 |
|---|---|---|---|---|---|---|---|---|---|---|---|
| $k_{+2}$ | 0.10 | 9.83 | 12.5 | 10.1 | | | | | | | |
| $J(P)$ | 16.4 | $3.95 \times 10^{-3}$ | 0.560 | $1.27 \times 10^{-3}$ | | | | | | | |
| $k_{+2}$ | 90.0 | 9.00 | 26.2 | 10.7 | 12.6 | 7.69 | 9.94 | | | | |
| $J(P)$ | 12.1 | 0.152 | 5.96 | $6.29 \times 10^{-2}$ | 0.607 | 0.851 | $5.45 \times 10^{-4}$ | | | | |
| $k_{+2}$ | 400. | 4.00 | 102 | 13.8 | 30.3 | 13.9 | 14.7 | 1.47 | 13.4 | 10.9 | 10.2 |
| $J(P)$ | 12.3 | 5.95 | 12.2 | 1.09 | 7.06 | 1.18 | 1.51 | 12.2 | 0.915 | 0.101 | $5.86 \times 10^{-3}$ |

Optimal value : $k_{+2} = 10.0$

Table 8.9  Optimization of two parameters by the Powell method

| Step | Starting value | 1 | 2 | 3 | 4 | 5 | 6 |
|---|---|---|---|---|---|---|---|
| $k_{2+}$ | 50.0 | 50.0 | 0.500 | 0.500 | 10.7 | 10.7 | 11.1 |
| $k_{+2}''$ | 10.0 | 0.100 | 0.100 | 45.3 | 36.0 | 0.36 | $3.60 \times 10^{-3}$ |
| $J(P)$ | 10.3 | 10.3 | 5.61 | 17.4 | 6.57 | 1.60 | $5.86 \times 10^{-2}$ |

Optimal value : $k_{+2} = 10.0$, $k_{+2}'' = 0.100$

siderably from the optimal solution $0.1 \sec^{-1}$ at the fifth and sixth iterations. This is due to a simplified linear search, which generates a new search direction not coincident with the conjugate direction. Alternative linear search or application of the modified Powell method should be considered for improvement.

**Remarks for Parameter Evaluation**

In this section the gradient and Powell methods are described and compared in their application to parameter evaluation in a reaction system. Correctness of the obtained values could be examined here, since the optimal solution (*i.e.,* true parameter values) was known beforehand. This is never possible in actual problems. It thus becomes important in applying optimization methods to actual problems that the degree of agreement be specified between the data from simulation using the evaluated parameters and the experimental data (*i.e.,* time course of substrate concentration, in this example). The specification corresponds to the sensitivity analysis as shown in Fig. 8.30. The gradient method is regarded more efficient in the sense that the sensitivity analysis can be performed in association with the optimization process.

**5. Evaluation of Binding Free Energy in Lysozyme**

An example of application of the Powell method is given here by the evaluation of free energy change of substrate binding in the lysozyme-catalyzed reaction [9]. The reaction mechanism of lysozyme is described in Section 4.1. It is well known that lysozyme has six substrate-binding subsites, A~F. In the formation of 1:1 enzyme-substrate complex, a substrate (GlcNAc)₁ binds to each of the six subsites according to the value of binding free energy of the corresponding subsite. Similarly, the binding of other chitooligosaccharides is dependent on the binding free energy of each subsite. The binding constant and binding free energy of each subsite of lysozyme have ordinarily been evaluated from the experimental data under the assumption of the binding mode of the substrate and additivity of binding free energy of each subsite [10]. The evaluated values of binding free energy still remain uncertain in some cases. We here attempt to accurately evaluate the binding constants and binding free energies and to examine the difference in binding free energy between the native and modified lysozymes.

In the optimization procedure for the parameter evaluation, some values are first taken as an initial guess of the objective parameters. The rate equation is numerically solved with these given parameter values and the cost function is calculated to obtain the degree in deviation of the numerical solution from the experimental data. Then, the initial evaluation is revised so that the cost function decreases its value toward the minimum. These steps are iterated until the evaluated values of the parameters yield the minimum value in cost function, and the values from the final iteration are taken as the optimal parameter values. In practical optimization the iterative revision of evaluated values becomes the primary procedure, requiring computation of the direction and size of revision by the success and failure routine.

The overall binding constants of $(GlcNAc)_1$, $(GlcNAc)_2$ and $(GlcNAc)_3$ experimentally obtained are available for evaluation of the binding free energies of subsites A, B and C. We employ the modified Powell method for the optimization procedure, assuming no specific binding mode for either of the chitooligosaccharides. The binding free energies of the three subsites are set to initially have an equal value of $-1.0$ kcal/mol. The results of evaluation are summarized in Table 8.10. Numerals in parentheses in the table indicate the overall binding constants determined spectrophotometrically. The measurement is based on the fact that the binding of substrate

Table 8.10    Binding free energies and binding constants

|  | | Binding free energy (kcal/mol) | | | Binding constants[a] $(M^{-1})$ | | |
|---|---|---|---|---|---|---|---|
|  | | $\Delta G_A$ | $\Delta G_B$ | $\Delta G_C$ | $(GlcNAc)_1$ | $(GlcNAc)_2$ | $(GlcNAc)_3$ |
| Native | 1) | $-2.41$ | $-3.13$ | $-4.80$ | 3.0 | 4700 | 170000 |
|  | 2) | $-2.42$ | $-3.02$ | $-4.89$ | 39 | 4700 | 170000 |
|  |  |  |  |  | (39 | 4700 | 170000)[b] |
| Kyn–lysozyme | 1) | $-1.52$ | $-2.67$ | $-4.24$ | 12.8 | 819 | 8695 |
|  | 2) | $-1.54$ | $-3.26$ | $-3.62$ | 8.9 | 820 | 8699 |
|  |  |  |  |  | (8.9 | 820 | 8700)[b] |
| NPS–lysozyme | 1) | $-1.48$ | $-2.87$ | $-2.80$ | 1.4 | 120 | 1200 |
|  | 2) | $-1.61$ | $-2.72$ | $-2.81$ | 3.8 | 120 | 1200 |
|  |  |  |  |  | (3.9 | 120 | 1200)[b] |
| NBS–lysozyme | 2) | $+0.34$ | $-3.90$ | $-3.31$ | 30 | 1300 | 770 |
|  |  |  |  |  | (30 | 1300 | 770)[b] |

The values of subsites A, B, and C are evaluated by optimization assuming that 1) substrates bound in all binding modes are measured experimentally; 2) only substrates spanning the C-site are measured experimentally.
[a] Computed binding constants using estimated binding free energies of subsites A, B, and C, and $\Delta G_D = +4.5$, $\Delta G_E = -2.5$, and $\Delta G_F = -1.5$ (in kcal/mol).
[b] Experimentally obtained overall binding constant.

affects the spectroscopic property of chromophores, Trp 62, Trp 63 and Trp 108, locating proximally at subsite C.

For the native lysozyme the values evaluated by the optimization method are not much different from the averages of the experimental data. For the NBS-, Kyn- and NPS-lysozymes, in which only Trp 62 is chemically

**Table 8. 11**  Binding modes of $(GlcNAc)_n$ to native lysozyme

| $(GlcNAc)_n$ | Binding mode (%) |
|---|---|
| $n=6$ | F 0.00007 · EF 0.003 · ⌐DEF 0.000003 · CDEF 0.006 · BCDEF 0.7 · ABCDEF 30.3 · ABCDE 2.0⌐ · ABCD 0.06 · ABC 66.0 · AB 0.03 · A 0.0003 |
| $n=5$ | F 0.0001 · EF 0.005 · ⌐DEF 0.000004 · CDEF 0.009 · BCDEF 1.0 · ABCDE 4.2⌐ · ABCD 0.09 · ABC 94.6 · AB 0.05 · A 0.0004 |
| $n=4$ | F 0.001 · EF 0.005 · ⌐DEF 0.000005 · CDEF 0.01 · BCDE 0.1⌐ · ABCD 0.09 · ABC 99.7 · AB 0.05 · A 0.0004 |
| $n=3$ | F 0.0001 · EF 0.005 · ⌐DEF 0.000005 · CDE 0.0009⌐ · BCD 0.002 · ABC 99.9 · AB 0.05 · A 0.0004 |
| $n=2$ | F 0.005 · EF 0.2 · ⌐DE 0.0002⌐ · CD 0.0008 · BC 97.7 · AB 2.1 · A 0.02 |
| $n=1$ | F 0.4 · E 2.2 · D 0.00004 · C 90.5 · B 4.9 · A 1.9 |

The productive complexes are enclosed by the broken line.

**Table 8. 12**  Binding modes of $(GlcNAc)_n$ to NPS-lysozyme

| $(GlcNAc)_n$ | Binding mode (%) |
|---|---|
| $n=6$ | F 0.01 · EF 0.5 · ⌐DEF 0.0004 · CDEF 0.03 · BCDEF 2.4 · ABCDEF 29.4 · ABCDE 2.8⌐ · ABCD 0.06 · ABC 64.0 · AB 0.8 · A 0.01 |
| $n=5$ | F 0.01 · EF 0.7 · ⌐DEF 0.0006 · CDEF 0.05 · BCDEF 3.4 · ABCDE 4.0⌐ · ABCD 0.08 · ABC 90.6 · AB 1.1 · A 0.02 |
| $n=4$ | F 0.01 · EF 0.7 · ⌐DEF 0.0007 · CDEF 0.05 · BCDE 0.4⌐ · ABCD 0.09 · ABC 97.5 · AB 1.2 · A 0.02 |
| $n=3$ | F 0.02 · EF 0.7 · ⌐DEF 0.0007 · CDE 0.05⌐ · BCD 0.007 · ABC 98.0 · AB 1.2 · A 0.02 |
| $n=2$ | F 0.2 · EF 7.4 · ⌐DE 0.0006⌐ · CD 0.001 · BC 79.9 · AB 12.3 · A 0.2 |
| $n=1$ | F 4.7 · E 22.3 · D 0.0004 · C 36.1 · B 31.4 · A 5.6 |

The productive complexes are enclosed by the broken line.

**Table 8.13**  Binding modes of $(GlcNAc)_n$ to NBS-lysozyme

| $(GlcNAc)_n$ | Binding mode (%) | | | | | | | | | | |
|---|---|---|---|---|---|---|---|---|---|---|---|
| **n=6** | F | EF | DEF | CDEF | BCDEF | ABCDEF | ABCDE | ABCD | ABC | AB | A |
|  | 0.01 | 0.49 | 0.00045 | 0.078 | 33.8 | 19.9 | 1.9 | 0.039 | 43.4 | 0.25 | 0.00058 |
| **n=5** | F | EF | DEF | CDEF | BCDEF | ABCDE | ABCD | ABC | AB | A | |
|  | 0.012 | 0.62 | 0.00056 | 0.097 | 42.3 | 2.4 | 0.05 | 54.2 | 0.31 | 0.0007 | |
| **n=4** | F | EF | DEF | CDEF | BCDE | ABCD | ABC | AB | A | | |
|  | 0.021 | 1.05 | 0.0009 | 0.16 | 6.88 | 0.083 | 91.2 | 0.53 | 0.0012 | | |
| **n=3** | F | EF | DEF | CDE | BCD | ABC | AB | A | | | |
|  | 0.023 | 1.12 | 0.001 | 0.017 | 0.15 | 98.1 | 0.56 | 0.0013 | | | |
| **n=2** | F | EF | DE | CD | BC | AB | A | | | | |
|  | 0.014 | 0.67 | 0.00006 | 0.0002 | 98.9 | 0.33 | 0.0008 | | | | |
| **n=1** | F | E | D | C | B | A | | | | | |
|  | 1.55 | 7.35 | 0.00014 | 25.9 | 65.1 | 0.088 | | | | | |

The productive complexes are enclosed by the broken line.

modified, it is expected that the binding free energy of subsite C might be changed considerably by the modification. The evaluation reveals that the decrease in binding free energy of subsite C is much less than expected; in contrast, that of subsite A is surprisingly large.

Using the optimal values of binding free energy evaluated above, we can delineate the binding mode of substrates with calculation of the relative fraction (in %) of each binding mode. Table 8.11 shows the binding mode with the native lysozyme. The productive complexes are indicated by the enclosure with the broken line. The Kyn- and NPS-lysozymes exhibit very similar binding modes for substrates; the binding mode with the NPS-lysozyme is presented in Table 8.12. It is revealed that the relative fraction of productive complexes increases with the modification. On the other hand, the binding mode with the NBS-lysozyme is distinguished from those with other lysozymes, as shown in Table 8.13.

# References

[1]  Koga, D. and K. Hayashi (1976). Activation process of pepsinogen. *J. Biochem.*, **79**, 549-558.

[2]  Karfunkel, H.R. and F.F. Seeling (1972). Reversal of inhibition of enzymes and the model of a spike oscillator. *J. Theor. Biol.*, **36**, 237-253.

[3]  Kowalik, J. and M.R. Osborne (1968). "Methods for Unconstrained Optimization Problems," American Elsevier, New York.

[ 4 ]   Powell, M.J.D. (1964). An efficient method for finding the minimum of a function of several variables without calculating derivatives. *Computer J.,* **7**, 155-162.

[ 5 ]   Gear, C.W. (1971). "Numerical Initial Value Problems in Ordinary Differential Equations," Prentice-Hall, Englewood Cliffs.

[ 6 ]   Hemker, P.W. (1972). Numerical methods for differerential equations in system simulation and in parameter estimation. *In* "Analysis and Simulation of Biochemical Systems," ed. by H.C. Hemker and B. Hess, pp.59-80, North-Holland, Amsterdam.

[ 7 ]   Powell, M.J.D. (1968). On the calculations of orthogonal vectors. *Computer J.,* **11**, 302-304.

[ 8 ]   Brent, R.P. (1982). "Algorithms for Minimization without Derivatives," Prentice-Hall, Englewood Cliffs.

[ 9 ]   Kuhara, S., M. Ezaki, T. Fukamizo and K. Hayashi (1982). Estimation of the binding free energy changes of substrate binding in lysozyme-catalyzed reactioon. *J. Biochem.,* **92**, 121-127.

[10]    Imoto, T., L. N. Johnson, A. C. T. North, D.C. Phillips and J.A. Rupley (1972). Vertebrate lysozymes. *In* "The Enzymes," 3 rd ed., ed. by P.D. Boyer, Vol.7, pp.665-868, Academic Press, New York.

# CHAPTER 9

# Related Topics in Dynamic Analysis

In the preceding chapters the dynamic analysis of enzyme systems is performed by formulation and numerical solution of differential equations based on the deterministic approach. Since the reactions in living systems take place in various modes and structures, it is readily apparent that the deterministic approach may not be applicable in some cases and that other new approaches are necessary to analyze the underlying problems. Such new approaches are usually constructed on elegant and sophisticated theories and mathematical principles. In this chapter, some topics which require new approaches are briefly introduced in order to outline the recent progress in problems closely related to the dynamic analysis of enzyme systems. We are concerned here with stochastic and control processes in enzyme systems, and large-scale systems as well as thermodynamics in living systems.

## 9.1 Chemical Reaction as Stochastic Process

### 1. Stochastic Process

In the deterministic approach to chemical processes, the variables are assumed to be continuous, representing mean values of chemical quantity such as concentration and number of molecules of chemical species. The assumption on continuity of the variable is not unreasonable, for a huge number of molecules are generally contained in a reaction system; there exist about $10^{18}$ molecules in 100 ml reaction mixture at $10^{-5}$ M concentration.

Chemical reaction starts with the contact or collision between reactants

---

This chapter was written by Katsuya Hayashi and Naoto Sakamoto.

or reactant and catalyzer, which is inherently of a probabilistic or stochastic nature. Therefore, a chemical reaction is an event governed by probability and the variable (concentration or number of molecules) is regarded as a probabilistic one. The process undergoing the successive probabilistic events is called the stochastic process. In this process, starting from an initial condition, the value of the variable at a given time cannot be determined uniquely, but it falls to the inherent statistical fluctuation around a mean value. Thus, the concentration or the number of molecules in the stochastic process is a random variable and the observed concentration represents the mean of distributed values.

To ensure exact analysis, the chemical reaction should be discussed with respect to random variables. However, for most chemical reactions, the deterministic approach results in the same solution as that obtained by the stochastic approach, owing to the small fluctuations of variables. It has been well recognized, nevertheless, that the stochastic approach should be applied to a number of special reactions. This is especially the case when the very early and final stages of reaction are analyzed in detail. Other systems which require the stochastic approach are reaction at extremely low concentrations of reactants, degradation and synthesis of polymeric substances, and reaction for which the detailed mechanism is examined at the level of elementary reaction.

In the stochastic approach, the probability, Prob $\{X(t) = x\} \equiv P_x(t)$, is defined as a basic variable, where $X(t)$ is a random variable which has a value $x$ for the number of molecules of a chemical species. $P_x(t)$ represents the probability that in a single experiment the number of molecules in the system at time $t$ is just $x$, starting from an initial value $x_0$. When similar experiments are repeated under the same conditions, the value of $X(t)$ fluctuates with a certain distribution function. Thus, the number of molecules in the system at a given time has no specific value, in contrast to the situation in the deterministic approach.

Furthermore, in the stochastic process the change in $X(t)$ with time is described as a probabilistic event with respect to transition probability; for example, in the process from $X(t) = x$ to $X(t + \Delta t) = x - 1$, the probability for occurrence of the transition is introduced to a description of the stochastic process. Therefore, the dynamics or the rate equation of the stochastic process could be represented by two primary terms, $P_x(t)$ and transition probability. Since the rate equation generally involves the difference expression on $P_x(t)$, the equation is called difference-differential equation (DDE).

As in the deterministic approach, a DDE is solved by suitable methods employing complicated mathematical techniques. In the following, we introduce the stochastic formulation of unimolecular and bimolecular reactions to outline the stochastic approach to an analysis of reaction systems.

## 2. Unimolecular Reaction

We consider a unimolecular reaction [1] simply represented by

$$A \xrightarrow{k} B.$$

Let the initial number of molecule A be $n_0$ and the number of molecule A at time $t$ be $X(t) = n$. The transition probability of $n \to (n-1)$ in the time interval $(t, t+\Delta t)$ is $kn\Delta t + O(\Delta t)$, and then the probability of no transition, i.e., that of $n \to n$, becomes $(1 - kn\Delta t) + O(\Delta t)$. It is assumed that $\Delta t$ is small enough so that there is at most a single transition during the interval.

The probability that $n$ molecules of A are in the reaction mixture at $t + \Delta t$ may be represented by

$$P_n(t+\Delta t) = k(n+1)\Delta t P_{n+1}(t) + (1 - kn\Delta t) P_n(t) + O(\Delta t). \qquad (1)$$

The first and second terms on the right-hand side indicate the respective probabilities that $(n+1)$ molecules at $t$ reduce to $n$ in time interval $\Delta t$, and that $n$ molecules at $t$ remain unchanged at $t + \Delta t$. Taking limit with $\Delta t \to 0$, we have

$$\lim_{\Delta t \to 0} \left( \frac{P_n(t+\Delta t) - P_n(t)}{\Delta t} \right) = \frac{dP_n(t)}{dt},$$

and then, from equation (1), the difference-differential equation,

$$\frac{dP_n(t)}{dt} = k(n+1)P_{n+1}(t) - kn P_n(t), \qquad (2)$$

is obtained as the rate equation.

Introduction of the generating function,

$$G(z, t) = \sum_{n=0}^{\infty} z^n P_n(t), \quad |z| < 1, \qquad (3)$$

leads to rearrangement of equation (2) in a simple form. We can rewrite equation (2) in terms of the generating function such that

$$
\begin{aligned}
\frac{\partial G(z,t)}{\partial t} &= \sum_{n=0}^{\infty} z^n \frac{\partial P_n(t)}{\partial t} = k \sum_{n=0}^{\infty} z^n [(n+1)P_{n+1}(t) - nP_n(t)] \\
&= k \left[ \frac{\partial}{\partial z} \sum_{n=0}^{\infty} z^{n+1} P_{n+1}(t) - z \frac{\partial}{\partial z} \sum_{n=0}^{\infty} z^n P_n(t) \right] \\
&= k \left\{ \frac{\partial}{\partial z} [G(z,t) - P_0(t)] - z \frac{\partial G(z,t)}{\partial z} \right\} \\
&= k(1-z) \frac{\partial G(z,t)}{\partial z},
\end{aligned}
\qquad (4)
$$

where the relationship,

$$\frac{\partial G(z,t)}{\partial z} = \sum_{n=0}^{\infty} nP_n(t) z^{n-1},$$

is used in the rearrangement of the equation. With definition of new variables,

$$\tau = kt, \quad d\tau = kdt,$$

$$\zeta = -\ln(1-z), \quad d\zeta = \frac{1}{1-z} dz,$$

the partial differential equation (4) may be reduced to

$$\frac{\partial G(z,t)}{\partial \tau} = \frac{\partial G(z,t)}{\partial \zeta}, \tag{5}$$

indicating that the generating function has the form of

$$G(z,t) = f(\tau + \zeta). \tag{6}$$

From an initial condition of $P_n(0) = \delta_{n,n_0}$ and $X(0) = n_0$,

$$G(z,0) = z^{n_0} \tag{7}$$

is obvious. Taking $\tau = 0$ in equation (6), we get the relationship,

$$f(\zeta) = G(z,0) = z^{n_0}, \tag{8}$$

or

$$f(\zeta) = (1 - e^{-\zeta})^{n_0}, \tag{9}$$

where the definition $\zeta = -\ln(1-z)$ is used. Equation (6) thus suggests that the solution of equation (5) may be represented by

$$G(z,t) = (1 - e^{\ln(1-z)-kt})^{n_0} = [1 - (1-z)e^{-kt}]^{n_0}. \tag{10}$$

Binomial expansion is done to rewrite equation (10):

$$G(z,t) = \sum_{n=0}^{n_0} \binom{n_0}{n} (1 - e^{-kt})^{n_0-n} (ze^{-kt})^n, \tag{11}$$

which is compared with equation (3) to give

$$P_n(t) = \binom{n_0}{n} (1 - e^{-kt})^{n_0-n} e^{-nkt}, \tag{12}$$

*i.e.*, the solution of rate equation (2).

The first moment of the generating function $G(z,t)$ corresponds to the averaged number of molecule A at time $t$:

$$\mu_1 = <n>_t = z \frac{\partial G(z,t)}{\partial z}\bigg|_{z=1} = n_0 e^{-kt}. \tag{13}$$

This value is identical to that obtained from the rate equation in the deterministic approach. The variance of $n$ at time $t$ is given by

$$\left.\frac{\partial^2 G(z, t)}{\partial z^2}\right|_{z=1} + \left.\frac{\partial G(z, t)}{\partial z}\right|_{z=1} - \left.\left(\frac{\partial G(z, t)}{\partial z}\right)^2\right|_{z=1} = n_0 e^{-kt}(1 - e^{-kt}). \quad (14)$$

## 3. Bimolecular Reaction

We now consider a bimolecular reaction [2] represented by

$$A + B \xrightarrow{k_1} C + B, \quad B \xrightarrow{k_2} D,$$

where B is an unstable catalyzer, decomposing spontaneously to D, and $k_1$ and $k_2$ are the rate constants. Let $X_1(t) = n_1$ and $X_2(t) = n_2$ be the numbers of molecules A and B, respectively, at time $t$ in the reaction mixture, and $P_{n_1, n_2}(t)$ be the joint probability that at time $t$, $X_1(t) = n_1$ and $X_2(t) = n_2$. $n_{10}$ and $n_{20}$ denote the initial numbers of molecules A and B, respectively. The probabilities that a molecule A reacts in the time interval $[t, t + \Delta t]$ and that a molecule B decomposes are given by $k_1 n_1 n_2 \Delta t$ and $k_2 n_2 \Delta t$, respectively. The rate equation for this system may thus be expressed as

$$\frac{dP_{n_1, n_2}(t)}{dt} = k_1(n_1 + 1) n_2 P_{n_1 + 1, n_2}(t) + k_2(n_2 + 1) P_{n_1, n_2 + 1}(t)$$

$$- [k_1 n_1 n_2 + k_2 n_2] P_{n_1, n_2}(t). \quad (15)$$

The generating function is defined as

$$G(z_1, z_2, t) = \sum_{n_1=0}^{n_{10}} \sum_{n_2=0}^{n_{20}} z_1^{n_1} z_2^{n_2} P_{n_1, n_2}(t), \quad (16)$$

to transform equation (15) to

$$\frac{\partial G}{\partial t} = k_1 z_2(1 - z_1) \frac{\partial^2 G}{\partial z_1 \partial z_2} + k_2(1 - z_2) \frac{\partial G}{\partial z_2}. \quad (17)$$

Employing a similar treatment to the generating function in unimolecular reaction, we obtain

$$G(z_1, z_2, 0) = z_1^{n_{10}} z_2^{n_{20}},$$

and

$$G(z_1, z_2, t) = \sum_{j=0}^{n_{10}} \binom{n_{10}}{j} (z_1 - 1)^j \left[ \frac{k_2 + [k_1 j z_2 + k_2(z_2 - 1)] e^{-(k_2 + j k_1) t}}{k_2 + j k_1} \right]^{n_{20}}. \quad (18)$$

Binomial expansion of the right-hand side of equation (18) leads to

$$P_{n_1, n_2}(t) = \binom{n_{10}}{n_1} \binom{n_{20}}{n_2} \sum_{j=0}^{n_{10} - n_1} \binom{n_{10} - n_1}{j} (-1)^j \left\{ \left( \frac{k_2}{k_2 + (n_1 + j) k_2} \right) \right.$$

$$\left. \times (1 - e^{-[k_2 + (n_1 + j) k_2] t}) \right\}^{n_{20} - n_2} e^{-n_2 [k_2 + (n_1 + j) k_1] t}. \quad (19)$$

The average numbers of molecules A and B at time $t$ are given by

$$\langle n_1 \rangle_t = z_1 \frac{\partial G(z_1, z_2, t)}{\partial z_1}\bigg|_{z_1=1} = n_{10}\left[\frac{k_2 + k_1 e^{-(k_1+k_2)t}}{k_1 + k_2}\right]^{n_{20}}$$

and

$$\langle n_2 \rangle_t = z_2 \frac{\partial G(z_1, z_2, t)}{\partial z_2}\bigg|_{z_2=1} = n_{20}e^{-k_2 t}, \tag{20}$$

respectively, while the number of molecule A at time $t$ is expressed by

$$n_1(t) = n_{10}\exp\left[-\frac{k_2}{k_1}n_{20}(1 - e^{-k_2 t})\right] \tag{21}$$

with the deterministic method. $n_1(t)$ is different from $\langle n_1 \rangle_t$, but both values become equal in the limit with $n_{10}$, $n_{20} \to \infty$.

## 4. Application to Enzymatic Reactions
### Michaelis-Menten-Type Reaction
Application of the stochastic approach to simple enzymatic reactions was first attempted by Darvey and Staff [3]. In the enzymatic reaction, at least four chemical species, free enzyme, substrate, enzyme-substrate complex and product, are involved in the reaction mixture. Consequently, it is necessary in principle to formulate the joint probability of four variables. In a closed reaction system, however, only the numbers of substrate and product may be taken as variables, because, with the initial (or total) concentrations of enzyme and substrate specified, the conservation law makes it possible to eliminate the variables for enzyme species (free enzyme and enzyme-substrate complex). Thus, the rate equation for the enzymatic reaction of Michaelis-Menten type can be expressed with respect to two variables and solved by the same procedure as applied to bimolecular reaction. Detailed analyses of simple enzymatic reactions have been further performed by Heyde and Heyde [4], and Staff [5].
### Complex Enzymatic Reactions
The stochastic rate equation gets complicated with the complexity of enzymatic reaction. In general, it becomes quite a difficult task to solve the rate equation. For such a complex enzymatic reaction, therefore, the steady-state approximation or simplified modeling of enzymatic reaction has been introduced to the analysis [6]. The queue theory also has often been employed to analyze open reaction systems. Smith applied this theory to the open system including a competitive inhibitor [7]. Further, some other complex systems such as the immobilized-enzyme system and drug-inflict-

ed pharmacokinetics [8] have been analyzed successfully by the stochastic approach.

**Biosynthesis of Biopolymers**

The characteristics in biosynthesis of biopolymers like nucleic acids and proteins may stem from a feature wherein the reaction point moves step by step in sequential order along a template of nucleic acid. This process can be formulated by diffusion or queue models of the reaction point in chain elongation. However, the rate equation thus formulated precisely reflecting the actual experimental knowledge cannot be solved due to a lack of suitable mathematical procedures. In fact, most biosynthetic processes of nucleic acids and proteins have been analyzed by queue models under the steady-state assumption [1,9,10].

As stated above, it is obvious that application of the stochastic approach is indispensable in constructing the exact rate equation of enzymatic reaction. Nevertheless, the deterministic approach is still widely used for the dynamic analysis of enzymatic reactions because the stochastic approach requires skillful but sometimes cumbersome mathematical treatment. In practice, the deterministic approach generally does not lead to the wrong conclusion, so that it is preferable in most cases. However, it should still be noted that there are some specific cases in which only the stochastic approach is effective for the dynamic analysis of enzyme systems.

## 9.2  Control Processes in Enzyme Systems

### 1.  Control Modes of Enzyme Systems *in vivo*

The main role of enzyme systems in living organisms is to produce constituent components and energy from nutrients. The enzyme system attains such goals by operating under precise regulations in accord with the requirements of the entire organism. The control modes functioning in enzyme systems can be basically classified into three categories :

(1) Homeostatic or constant-value control which keeps the concentration of product or intermediates at a desired level against the external perturbation disturbing the system and causing the deviation.

(2) Adaptive control in response to a transient requirement of the system, for instance, the control for a suitable supply of energy to meet rapid muscle-work.

(3) Control supporting the scheduled variation of the system, such as circadian rhythm, development, differentiation, and aging.

At present, the molecular components and basic mechanisms for realizing such control modes are well understood. However, it is very obvious that a structural knowledge of molecular devices does not necessarily lead to the complete elucidation of control mechanisms with the desired functions. There arise associated problems to which information science should be applied in order to allow the molecular devices to accomplish the required control. The solution to such problems may be obtained only through the analysis of control systems by appropriate control theories. Analytical treatment of homeostatic and adaptive controls using control theories are in progress and a considerable amount of knowledge has already been accumulated, while control problems of scheduled variations in living systems are entirely unsolved because we have not been able to find even a basic strategy to solve them.

The analysis of enzyme systems through the control theory may consist of several steps, such as (1) finding the dynamic behavior of the system, (2) evaluation of controllability of the system against external perturbation, (3) determination of system response to the control variable, and (4) clarification of the mechanisms of production, storage, translation and operation of information relevant to the control. Some topics related to these steps are presented in the preceding chapters.

## 2. Elementary Subsystems Participating in Control of Enzyme Systems

### Feedback System

Feedback is the most basic subsystem common to all regulatory aspects of living and engineering systems. The output or state variables of the system are used for generating the control variable which in general negatively affects the control element. In input control, the control variable regulates the flow of input, and in parameter control it acts on the system element. As previously described, in the enzymatic feedback system, the product directly inhibits the regulatory enzyme. Most regulatory enzymes are found to be allosteric (see Chapters 4 and 5).

### Coupled Reaction

In coupled reaction, the progress of one reaction is indispensable for another reaction. Thus, one reaction is passively controlled by another. The dynamics of coupled reaction is described in Chapters 5 and 6.

### Cyclic Reaction

When some intermediates or coenzymes in a coupled reaction are used in a cyclic way or reversibly, the reaction becomes cyclic, so that the rate of reaction depends strongly on the concentration of the intermediate or coenzyme. The intermediate or coenzyme is thus regarded as a control variable. On the other hand, there are several cyclic reactions in living organisms in which overall rate of the reaction is governed by a rate-limiting step. This is also regarded as a control device in the enzymatic reaction system.

## Reversible Reaction

Most enzymatic reaction systems consist of a sequence of reversible enzymatic reactions. In such a system, the abnormal accumulation of intermediate metabolites may be released automatically and partially by mass action. This type of autocontrol is thought to play a role in homeostatic control. In open system, the distribution of reactants and product to several reaction steps is not simply determined by the equilibrium constants evaluated in closed system. This fact may be related to the autocontrol of sequentially linked enzymatic reactions (see Chapter 6).

## Compartment and Membrane

The existence of compartments and membranes substantially affects the rate of reaction accompanying the transport of reactants. Change in the permeability of membrane serves as a control variable to the inner enzyme system. This is a typical input-control for the enzyme system, whereas other regulatory modes belong to parameter control in which the activity of the enzyme is controlled directly by a control variable.

## Pool

The pools of metabolites are closely associated with the regulation of enzyme systems. For example, in storage pools like glycogen and glycerides, substances take different forms from the substrates for enzymatic reactions in aqueous medium. The transfer of a substance from and to the pool is usually of zeroth-order reaction. The addition of sugar residue to a glycogen molecule and release of glucose-1-phosphate from it occur only at the non-reducing terminals which are kept at a constant number during the reactions. In the soluble pool, the transfer of a metabolite is of zeroth-order when the pool size is sufficiently large. These imply that the supply of a metabolite from its pool to the enzymatic reaction is always kept constant, contributing to the homeostatic control.

## 3. Techniques for Solution of Control Problems

The control problems related with the enzyme system may be classified into two categories: one is concerned with determination of the control mechanism in the enzymatic reaction system *in vivo*, and the other with the synthesis of control systems for bioreactors. The mathematical formulation and solution of the control problems essentially require a knowledge of control theories at various levels specific to the feature of the problems. Here we introduce briefly the fundamental control theories.

## Linear System

The theory of automatic control is the most classical one, by which the behavior of a linear system with single input and single output can be analyzed quantitatively. The main mathematical tool for the theory is the

Laplace transform and the behavior of the linear system is discussed using the transfer function from the state equation. Accordingly, as the practical necessity arises for solving more complicated problems, general theories have been developed to generate basic procedures. In these control theories, the dynamics of the system in homogeneous condition is represented by linear ordinary differential equations and the desired controllability is formulated using the cost function. For the heterogeneous (inhomogeneous) system, partial differential equations are formulated to represent the system.

**Optimal Regulator**

Various mathematical procedures are used according to the feature of the optimal control problems. The most fundamental procedure among them may be the variational method, the outline of which is introduced in the following. Let $x(t)$ be the state variable of dimension $n$ (concentrations of $n$ chemical species) and $u(t)$ be the control variable of dimension $r$. The state equation of linear constant-parameter system is written as

$$\frac{dx(t)}{dt} = Ax(t) + Bu(t), \tag{22}$$

where $A$ and $B$ are $n \times n$ and $r \times n$ constant matrices, respectively. When we consider the homeostatic control, the cost function may be represented by

$$I = \int_0^t (x_d'(t) R_1 x_d(t) + u'(t) R_2 u(t)) \, dt$$

$$= \int_0^t F(x_d(t), u(t)) \, dt, \quad t \in [0, t_1], \tag{23}$$

where prime indicates the transpose of column vector, and $R_1$ and $R_2$ are the weighting constant matrices. The deviation $x_d$ is defined by

$$x_d(t) = x(t) - \bar{x},$$

where $\bar{x}$ is a desired constant-value and $x_d(t)$ is the deviation of state variable from the desired value. The problem is then to obtain the optimal control variable $u^*(t)$ which minimizes the cost function (23) under the constraint of the state equation (22).

If there exists no constraint, the necessary condition under which the cost function takes the minimum value is stated as

$$\left.\begin{aligned} \frac{\partial F}{\partial x} - \frac{d}{dt}\left(\frac{\partial F}{\partial \dot{x}}\right) &= 0 \\ \frac{\partial F}{\partial u} - \frac{d}{dt}\left(\frac{\partial F}{\partial \dot{u}}\right) &= 0. \end{aligned}\right\} \tag{24}$$

These equations are called Euler equations for functional $F$ derived by the variational method. The solution of partial differential equations (24) results in the optimal state $x^*(t)$ and control $u^*(t)$.

In enzyme systems, $x(t)$ and $u(t)$ should satisfy the differential equation (22) as the constraint condition. If we use the expression,

$$\frac{dx(t)}{dt} - Ax(t) - Bu(t) = g(x, \dot{x}, u) = 0,$$

and the Hamiltonian defined by

$$H = F - \lambda(t) g(x, \dot{x}, u),$$

the Euler equation for the Hamiltonian gives the necessary condition for the optimal control variable $u^*(t)$. $\lambda(t)$ is a vector called Lagrange multiplier. Thus, the solution of differential equations,

$$\left. \begin{aligned} \frac{\partial H}{\partial x} - \frac{d}{dt}\left(\frac{\partial H}{\partial \dot{x}}\right) &= 0 \\ \frac{\partial H}{\partial u} - \frac{d}{dt}\left(\frac{\partial H}{\partial \dot{u}}\right) &= 0 \\ \frac{\partial H}{\partial \lambda} - \frac{d}{dt}\left(\frac{\partial H}{\partial \dot{\lambda}}\right) &= 0, \end{aligned} \right\} \tag{25}$$

leads to the optimal control $u^*(t)$ :

$$u^*(t) = R_1^{-1} B\lambda(t). \tag{26}$$

Substituting equation (26) into equation (22) and again using equation (25), we obtain the simultaneous differential equations:

$$\left. \begin{aligned} \frac{dx(t)}{dt} &= Ax(t) + B'\lambda(t) \\ \frac{d\lambda(t)}{dt} &= A'\lambda(t) - x(t). \end{aligned} \right\} \tag{27}$$

The solution of equation (27) has the form of

$$\lambda(t) = Px^*(t), \tag{28}$$

where $P$ is a constant matrix and the solution of Riccati matrix equation. From equations (26) and (28), the optimal control is represented as

$$u^*(t) = R_1^{-1} BPx^*(t). \tag{29}$$

Equation (29) indicates that the optimal control is determined by the state variable through the feedback loop with proportional coefficent of $R_1^{-1}BP$. Thus, it is obvious that the optimal control in homeostasis is realized by feedback. The structure of the optimal feedback system has already been described in Chapter 6.

## Maximum Principle

Let $x(t)$ and $u(t)$ be the $n$-dimensional state vector and $r$-dimensional control vector, respectively :

$$x(t) = (x^1(t), x^2(t), \cdots\cdots, x^n(t)) \left.\right\}$$
$$u(t) = (u^1(t), u^2(t), \cdots\cdots, u^r(t)) . \left.\right\} \tag{30}$$

When the cost function $I$ is defined by

$$I = \int_0^t f^0(x(t), u(t)) dt = x^0(t), \tag{31}$$

the state differential equation of control system may be represented by

$$\frac{dx^i(t)}{dt} = f^i(x(t), u(t)), \quad i = 0, 1, 2, \cdots, n. \tag{32}$$

Now we define the adjoint variable,

$$\psi(t) = (\psi_0(t), \psi_1(t), \cdots, \psi_n(t)), \tag{33}$$

which satisfies the adjoint equation :

$$\frac{d\psi_i(t)}{dt} = -\sum_{j=0}^{n} \frac{\partial f^j(x(t), u(t))}{\partial x^i} \psi_j(t), \quad i = 0, 1, \cdots, n. \tag{34}$$

When the Hamiltonian is defined by

$$H(\psi, x, u) = \sum_{i=0}^{n} \psi_i f^i(x, u), \tag{35}$$

equations (32) and (34) are written as

$$\frac{dx^i(t)}{dt} = \frac{\partial H}{\partial \psi_i} = f^i(x(t), u(t)) \left.\right\}$$
$$\frac{d\psi_i(t)}{dt} = \frac{\partial H}{\partial x^i} = -\sum_{j=0}^{n} \frac{\partial f^j(x(t), u(t))}{\partial x_i} \psi_j. \left.\right\} \tag{36}$$

The Hamiltonian takes the maximum value with the optimal control $u^*(t)$ as

$$H(\psi, x, u^*) = \max_v H(\psi, x, v)$$
$$\frac{\partial H}{\partial u^*} = 0, \tag{37}$$

where $v$ is an admissible control. The necessary condition for the optimal control represented by equation (37) is called Pontryagin's maximum principle. Thus, for obtaining optimal control, equation (36) should be solved with respect to the state and adjoint variables and the control variable $u$ satisfying equation (37) should be sought.

## Nonlinear System

As described in the preceding chapters, all the enzyme systems are nonlinear and hence their control problems are formulated by nonlinear equations. Furthermore, the cellular conditions for enzyme systems are heterogeneous and sometimes an element in an enzyme system generates a time delay. The state equations for such systems are formulated by nonlinear partial differential equations or by difference-differential (functional) equations. Nonlinear differential equations cannot be solved analytically. That is, the control problems of enzyme systems cannot be solved by direct application of analytical procedures such as the variational method and the maximum principle. Thus, some numerical techniques are indispensable for obtaining the solution of control problems of enzyme systems.

**Gradient Methods**

Numerical solution of the control problem to find the optimal control variable is essentially based upon the iterative computation starting from the initially assumed values of the control variable until final attainment of the optimal values. In the iterative process, the value of control variable is successively revised toward the optimal value. Therefore, the most important part of the iteration is to establish the method by which the previous value of the control variable is correctly and efficiently improved. For this purpose, gradient methods are widely employed. We introduce the direct gradient method in the following (see also Section 8.2).

We consider the problem of determining the optimal control variable which minimizes the cost function,

$$I = \int_0^{t_1} f^0(x(t), u(t))\, dt, \quad t \in [0, t_1],\tag{38}$$

under constraint of the state equation,

$$\frac{dx^i(t)}{dt} = f^i(x(t), u(t)), \quad i = 1, 2, \cdots\cdots, n.\tag{39}$$

In the direct gradient method, equation (39) is first solved using the initial value, $u^{(0)}(t)$, of the control variable to obtain the solution $x^{(0)}(t)$, and then the value of cost function is calculated with $x^{(0)}(t)$ and $u^{(0)}(t)$. Next, $u^{(0)}(t)$ is successively revised toward the optimal control $u^*(t)$ by

$$u^{(i)}(t) = u^{(i-1)}(t) + \delta u^{(i)}(t), \quad i = 1, 2, \cdots, n,\tag{40}$$

where $\delta u^{(i)}(t)$ is the value of the revision. When $u^{(n+1)}(t) = u^{(n)}(t)$, $u^{(n)}(t)$ is regarded to be the optimal control $u^*(t)$. In this iteration, most important step is to find the suitable value of $\delta u^{(i)}(t)$. In gradient methods, this value is evaluated from the gradient of functional $f^0(x(t), u(t))$ in the cost function with respect to the control variable. Thus, we can represent $\delta u(t)$ by

$$\delta u(t) = \alpha g(t)$$
$$g(t) = \beta \frac{\partial f^0(x(t), u(t))}{\partial u},$$
(41)

where $\alpha$ and $\beta$ are the step width of revision and proportional constant, respectively. With equation (41), equation (40) is rewritten as

$$u^{(i)}(t) = u^{(i-1)}(t) + \alpha_i g_i(t).$$
(42)

The step width $\alpha$ at each step is determined by the condition,

$$\frac{\partial f^0(x(t), u(t))}{\partial \alpha} = 0.$$
(43)

Practically, $\alpha$ is estimated by a suitable direct-search-method such as the golden division method.

**Numerical Solution through Maximum Principle**

The optimal control problems take various forms according to the feature of the enzyme systems. The direct gradient method does not necessarily provide an effective procedure for a special problem. In practice, several other gradient methods are employed for such problems. We mention here the method in which a nonlinear optimal control problem is formulated by the maximum principle. For a nonlinear enzyme system, equation (36) becomes nonlinear state and adjoint equations, and the analytical method is not available at present. Equation (36) therefore should be solved by numerical technique. Because the initial conditions are unknown for the adjoint equation, this equation is inversely solved from the final conditions. The same principle of gradient methods as described above is used for the numerical solution of equation (36) to obtain the optimal control. Equation (36) is solved with an initial estimate of control variable and the Hamiltonian is evaluated. The control variables estimated are successively revised using the gradient of Hamiltonian with respect to the control variable to reach the optimal value, which yields the maximum value of Hamiltonian.

## 9.3 Analysis of Large-Scale Systems

The dynamic analysis of enzyme systems at any level from the stochastic approach to the microscopic or macroscopic treatment of deterministic processes must, of necessity, to extend to the problems of analysis of large-scale systems. The simulation of molecular dynamics of metabolic pathways in a cell might have to deal with rate equations describing the behavior of more than one thousand chemical species. Even the macroscopic treatment of input-output relations in cells and organs requires too many variables and parameters to represent the system in the appropriate form.

In fact, we recognize at present major discrepancies between the results of the respective analyses of stochastic, microscopic and macroscopic aspects of biological processes. The analysis of large-scale systems could lead to the integration of these analyses to formulate and explain precisely the biological phenomena at the molecular level. In this section we introduce effective methods for modeling and analyzing large-scale systems, which are described in a survey of Mahmound [11]. These methods solve the problems arising from the increase in number of variables for analysis at each level, as well as the inherent difference in structures and processes relevant to the respective analyses.

## 1. Hierarchical Structure in a Large-Scale System

A large-scale system consists of a number of interdependent and resource-sharing subsystems which are governed by a set of interrelated goals and constraints to attain their particular functions for organization of the

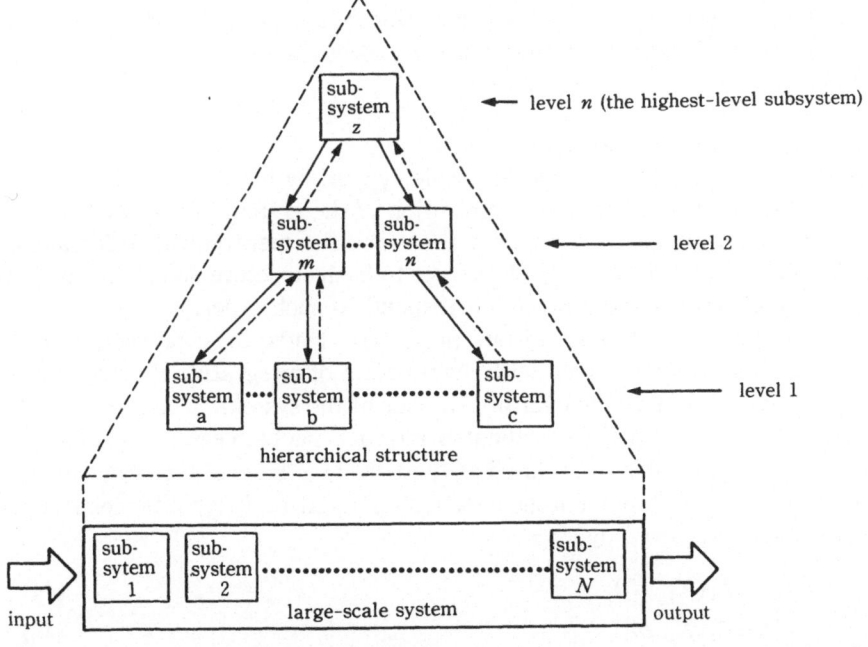

**Fig. 9. 1**  Hierarchical model of large-scale system. $\longrightarrow$ : command; $\cdots\rightarrow$ : response.

structure and behavior in the whole system. The principle in modeling such a large-scale system is based on an assumption that there exists a hierarchical structure in the functions of constituent subsystems and their interactions. Every subsystem is represented by a structurally valid model. These models are arranged to reflect the hierarchical structure in the subsystems in order to construct a structurally valid model for the large-scale system.

The hierarchical structures have certain basic features as indicated in Fig. 9.1 [11]. First, the interacting subsystems in a large-scale system are related with each other, forming a multi-level relationship through the nature of their interactions. Secondly, the integrated control of over-all behavior of the system is attained by information processing through the hierarchical arrangement of the subsystems. The information is only permitted to flow vertically between any two adjacent levels. Furthermore, to meet the overall objectives of the system, an assessment mechanism operates to judge the performance of the whole complex of subsystems. The response of the local units (*i.e.,* subsystems at a lower level) to the command of the supremal units (*i.e.,* subsystems at a higher level) is evaluated by the supremal units in comparison with the prescribed objectives. Based on the resulting deviation, the supremal units provide the local units with new commands, and the process is repeated until the desired objectives are satisfied for the system.

## 2.  Analysis of Hierarchical Model

The dynamic analysis of a large-scale system by hierarchical model gives rise to the first problem in decomposition of the system into subsystems. In large-scale systems of enzymatic reactions the hierarchical order among the subsystems is inherent with respect to both structure and function. The decomposition would naturally correspond to such order.

Now that the decomposition procedure leads to construction of a hierarchical model, we deal with the problem of integrating the behavior of every subsystem into the overall behavior of the system. The procedure for dynamic analysis of the coordinated interactions between subsystems in a hierarchical model is formulated as follows [11]. Each of the $N$ subsystems in a system has input-output relations as well as inequality constraints, which are expressed by

$$Z_i = T_i(X_i, M_i) \tag{44}$$

$$Y_i = S_i(X_i, M_i) \tag{45}$$

$$X_i = \sum_{i=1}^{N} c_{ij} Z_j \tag{46}$$

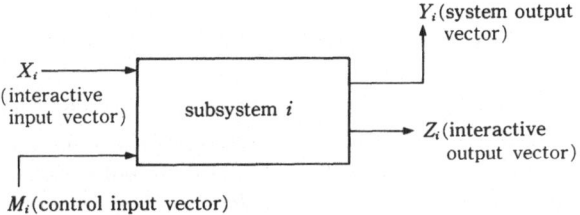

**Fig. 9. 2**  Inputs and outputs of a subsystem.

$$G_i(M_i, X_i, Y_i) \geq 0, \quad i=1,\cdots\cdots, N, \tag{47}$$

for the $i$th subsystem. As shown in Fig. 9.2, the vector $Y_i$ represents the system output, the vector $Z_i$ stands for the output serving as the input to other subsystems, the vector $M_i$ is the control input, and the vector $X_i$ accounts for the input from other subsystems. $G$, $T$ and $S$ represent operators, and $c$ denotes coefficients. We define the overall objective function of the system as the sum of objective functions of every subsystem:

$$F = \sum_{i=1}^{N} f_i(M_i, X_i), \tag{48}$$

where $F$ and $f_i$ indicate the overall and local objective functions, respectively. The problem is thus reduced to determine the optimal controls $M_i$ ($i= 1,2,\cdots, M$) which minimize (or maximize) $F$ subject to the constraints (44) $\sim$(47) (refer to the preceding section). The solution is derived from optimization procedure of a functional by defining an appropriate Lagrangian functional in the form:

$$L = \sum_{i=1}^{N}\Big[ f_i(M_i, X_i) + \lambda_i{}^t\{T_i(M_i, X_i) - Z_i\} + U_i{}^t G_i(M_i, X_i, S_i)$$
$$+ P_i{}^t\Big\{\sum_{j=1}^{N} c_{ij}Z_j - X_j\Big\}\Big], \tag{49}$$

where $P_i$ and $\lambda_i$ are Lagrangian multipliers for the equality constraints (44) and (45), and $U_i$ are the nonnegative Kuhn-Tacker multipliers for the inequality constraint (47).

The analysis of large-scale systems has commonly treated only the replicatively or predictively valid models. The dynamic analysis of biological processes has to deal with the structurally valid models to reveal the relationships between the structures and functions of real systems. Achievement of the dynamic analysis of large-scale systems of enzymatic reactions is most desirable, employing the procedures introduced in this section.

## 9.4   Thermodynamics in Biological Processes

Living systems function and maintain their homeostasis through metabolic processes which are constantly active and appropriately regulated. This is possible due to the highly inhomogeneous distribution of substances (*i.e.,* formation of structure) in a system. In a biological organism numerous chemical reactions, several thousands even in a single-cell organism like a bacterium, proceed to generate such a distribution in an integrated manner by complex control mechanisms. Organisms have acquired their functions and structures through the process of evolution and these are retained, transferred and expressed by genetic information.

One of the approaches in dynamic analysis aims at the understanding of these biological phenomena from the functions of the organized network of biochemical reactions. In this book we have mainly analyzed the dynamic behavior of every chemical species in the biochemical networks. The analysis derives the molecular mechanisms in the networks of matter-energy and information in biological processes from the remarkable accomplishments in biochemistry and molecular biology, leading to a complete molecular representation of biological phenomena.

Advancement in these fields is also associated with an integration of biology with chemistry and physics to establish a thermodynamical approach to the biological phenomena. The thermodynamic laws, which are general physical laws concerned with state changes in a system due to the flows of matter and energy, have been investigated for their adaptation to the nonlinear interactions and nonequilibrium processes specific to the biological phenomena.

Both molecular and thermodynamical approaches are essential to the understanding of mechanisms in biological phenomena. The findings from the dynamic analysis should always be examined as to whether they are consistent with the thermodynamic laws governing the biological phenomena. In this section we introduce some basic concepts necessary for thermodynamics applicable to the biological processes [12].

### 1.   Limitation of Equilibrium Thermodynamics

Thermodynamics is mostly concerned with determination of the direction of change in states of a system and the nature and stability of the state established. The first and second laws of thermodynamics lead to the answers for the isolated and closed systems. The first law regarding the energy-conservation principle states that increase in energy of a thermodynamic system is equal to the difference between heat from its surroundings and work against the exterior force, which is expressed by the differential form,

$$dE = dQ - pdV,  \tag{50}$$

where $E$, $Q$ and $pdV$ denote the internal energy of the system, heat from its surroundings and work associated with the thermodynamically reversible change in volume against the static pressure $p$, respectively.

The second law, which prescribes the direction of change in natural phenomena, is represented by the Clausius inequality,

$$dS \geq \frac{dQ}{T},  \tag{51}$$

employing the concept of entropy first introduced by Clausius. Entropy is an extensive variable, and the entropy change $dS$ is decomposed into two terms as follows:

$$dS = d_i S + d_e S,  \tag{52}$$

where $d_i S$ indicates entropy production due to the change inside the system, while $d_e S$ denotes entropy flux due to the interaction with its surroundings. In a closed system, which has an energetic exchange only by transfer of heat, but not a material exchange between the system and its surroundings, we have

$$d_e S = \frac{dQ}{T},  \tag{53}$$

with which the inequality (51) is rewritten as

$$d_i S \geq 0.  \tag{54}$$

The equality in (54) corresponds to the reversible change inside the system, while the inequality corresponds to the change due to irreversible processes such as diffusion, heat conduction and chemical reactions. The second law thus prescribes the direction of change in natural phenomena by stating that the entropy production never becomes negative in association with the state change inside the closed system.

The second law can specify the direction of chemical reactions in biological organisms as well. Substitution of $dQ$ from equation (50) into the inequality (51) yields the relationship,

$$dE - TdS + pdV < 0,  \tag{55}$$

where the equality is ignored because we deal with an irreversible process of chemical reaction. From the definition of Gibbs free energy $G$ as

$$G = E - TS + pV,  \tag{56}$$

the left-hand side in inequality (55) is equal to the change $dG$ in the progress of biochemical reaction under constant temperature and pressure, and then we have

$$dG < 0. \tag{57}$$

This implies that the biochemical reaction (enzymatic reaction) proceeds in the direction of decrease in Gibbs free energy under constant temperature and pressure.

We now have the laws established in physics that the entropy production is nonnegative in a physico-chemical system and that the system invariably reaches an equilibrium state with time, at which the entropy production cannot occur. Moreover, some equilibrium states are stable, making it possible for the systems to maintain various structures (*i.e.,* equilibrium structures) such as crystalline structure and gas in a completely random state. On the other hand, the biological phenomena cannot proceed in isolated or closed systems. Nor can the biological structures be described in terms of the equilibrium structures. Then, we cannot remove the contradiction between biological organization and the second law of thermodynamics as long as we employ the methods of equilibrium thermodynamics for the phenomenological approach to living systems.

## 2. Dissipative Structures [12]

Biological organisms always exchange matter and energy with their surroundings, functioning far from equilibrium. The thermodynamical treatment of such nonequilibrium and open systems bears on the problems as to definition of the boundary between the system and exterior, and state variables like entropy at states far from equilibrium. In order to treat the open nonequilibrium systems with the extension of classic equilibrium thermodynamics, Prigogine *et al.* have developed nonequilibrium thermodynamics, especially the theory of dissipative structures, in which the local equilibrium is assumed and the local state variables are related with each other in the same functions as those in an equilibrium state.

Similarly in a closed system, the entropy production $d_iS$ can never be negative, while the entropy flux $d_eS$ does not have a definite sign. It follows that the system may evolve to attain a state of lower entropy than the initial state:

$$\Delta S = \int_{\text{path}} dS < 0. \tag{58}$$

This state can be maintained indefinitely if the system is allowed to remain in a steady state such that

$$dS = 0, \quad \text{or} \quad d_e S = -d_i S < 0 . \tag{59}$$

This could never arise from the viewpoint of equilibrium thermodynamics. Thus, it is possible in principle that the system maintains an ordered structure by gaining an appropriate amount of negative entropy flow, which must occur solely under nonequilibrium conditions. We therefore regard nonequilibrium as a source of order. Such a generated order is called dissipative structure, and results from the nonequilibrium exchange of matter and energy with the surroundings.

Subsequently, Prigogine *et al.* have derived general (or universal) evolution criterion on the entropy production in a nonequilibrium state. The local entropy production $\sigma$ is expressed in general as the sum of contribution from every thermodynamic process:

$$\sigma = \sum_i J_i X_i , \tag{60}$$

where $J_i$ and $X_i$ represent the generalized flow and generalized force, respectively. In the chemical reaction $J_i$ corresponds to reaction velocity and $X_i$ is related to chemical affinity. In the diffusion process $J_i$ corresponds to diffusion flow and $X_i$ is related to chemical potential.

The total entropy production $P$ in the whole system is obtained from the volume integration:

$$P = \int \sigma dV = \int \sum_i J_i X_i dV , \tag{61}$$

where $\int dV$ denotes the volume integration over the whole system. Its temporal change is represented by

$$\begin{aligned}
\frac{dP}{dt} &= \int dV \sum_i J_i \frac{dX_i}{dt} + \int dV \sum_i X_i \frac{dJ_i}{dt} \\
&= \frac{d_X P}{dt} + \frac{d_J P}{dt}
\end{aligned} \tag{62}$$

This equation and the second law of thermodynamics lead to the general evolution criterion in the form of

$$\frac{d_X P}{dt} \leq 0, \tag{63}$$

in which the equality corresponds to steady states of the system. The criterion implies that the term due to the variation in force $X_i$ decreases with time in the entropy production in the system. On the other hand, the general characterization of $d_J P/dt$ is not attained.

In the linear range (*i.e.*, the region close to equilibrium), in which Onsager's reciprocity relations are valid between the generalized flows and forces, the general evolution criterion leads to the theorem of minimum entropy-production:

$$\frac{dP}{dt} \leq 0. \tag{64}$$

It follows that entropy production becomes minimum in a steady nonequilibrium state.

We now define the excess entropy production $\delta_x P$ by

$$\delta_x P = \int dV \sum_i J_i \delta X_i, \tag{65}$$

where $\delta X_i$ represents the deviation from the steady-state value. Arguments on the general evolution criterion lead to the stability condition for a steady nonequilibrium state:

$$\delta_x P \geq 0. \tag{66}$$

The necessary condition for formation of dissipative structure then becomes

$$\delta_x P < 0, \tag{67}$$

since the destabilization of equilibrium structure gives rise to dissipative structure.

As described above, Prigogine *et al.* have derived the general evolution criterion to prove theoretically the possibility of formation of dissipative structures in far-from-equilibrium states. The development of nonequilibrium thermodynamics thus provides us with a phenomenological approach to clarification of the mechanisms in biological processes, demonstrating that the second law of thermodynamics ($d_i S \geq 0$) is consistent with the decrease in total entropy of a system ($dS < 0$). The mechanisms in generating and maintaining organizations in biological systems are studied to derive the universal physical laws capable of describing the biological phenomena macroscopically. Rapid progress in this direction is expected in parallel with the analytical approach to molecular structures and mechanisms in individual organisms.

## References

[ 1 ]  McQuarrie, D. A. (1967). Stochastic approach to chemical kinetics. *J . Appl. Prob.*, **4**, 413-478.
[ 2 ]  Oppenheim, I., K. E. Shuler and G. H. Weiss (1977). "Stochastic Processes

in Chemical Physics," MIT Press, Cambridge.

[ 3 ]  Darvey, I.G. and P.J. Staff (1967). The application of the theory of Markov processes to the reversible one substrate-one intermediate-one product enzymic mechanism. *J. Theor. Biol.,* **14**, 157-172.

[ 4 ]  Hyde, C.C. and E. Hyde (1969). A stochastic approach to a one substrate-one product enzyme reaction in the initial velocity phase. *ibid.,* **25**, 159-172.

[ 5 ]  Staff, P.J. (1970). A stochastic development of the reversible Michaelis-Menten mechanism. *ibid.,* **27**, 227-232.

[ 6 ]  Hasstedt, S. (1978). A stochastic model for a closed biochemical system at equilibrium. *ibid.,* **70**, 199-212.

[ 7 ]  Smith, W.L. (1971). Stochastic models for an enzyme reaction in an open linear system. *Bull. Math. Biophys.,* **33**, 97-115.

[ 8 ]  Feldman, U. and B. Schneider (1976). A general approach to multicompartment analysis and models for the pharmacodynamics. *In* "Mathematical Models in Medicine," ed. by J. Berger, W. Bühler, R. Repqes and P. Tautu, pp. 243-279, Springer-Verlag, Berlin.

[ 9 ]  Goel, N.S. and V. Richter-Dyn (1974). "Stochastic Models in Biology," Academic Press, New York.

[10]  McDonald, C.T., J. H. Gibbs and A. C. Pipkin (1968). Kinetics of bio-polymerization on nucleic acid templates. *Biopolymers,* **6**, 1-25.

[11]  Mahmound, M.S. (1977). Multilevel system control and applications: a survey. *IEEE Trans. Systems, Man and Cyber.,* **SMC-7**, 125-143.

[12]  Nicolis, G. and I. Prigogine (1977). "Self-Organization in Nonequilibrium Systems," Wiley-Interscience, New York.

GENERAL

*Enzyme Control*

1.  Hemker, H.C. and B. Hess, ed. (1972). "Analysis and Simulation of Biochemical Systems," North-Holland, Amsterdam.

2.  Johnes, R.W. (1973). "Principle of Biological Regulation," Academic Press, New York.

3.  van den Driessche, P. (1974). "Mathematical Problems in Biology," Springer-Verlag, Berlin.

4.  Banks, H.T. (1975). "Modeling and Control in the Biomedical Science," Springer-Verlag, Berlin.

5.  Thomas, D. and J.-P. Kernevez (1976). "Analysis and Control of Immobilized Enzyme Systems," North-Holland, Amsterdam.

6.  Savageau, M.A. (1976). "Biochemical Systems Analysis," Addison-Wesley, Reading.

*Control Theory*

1.  Bellman, R. (1967). "Introduction to the Mathematical Theory of Control Processes," Vol.1, Academic Press, New York.

2.  Bellman, R. (1971). "Introduction to the Mathematical Theory of Control Processes," Vol.2, Academic Press, New York.

3. Luenberger, D.G. (1969). "Optimization by Vector Space Method," John–Wiley & Sons, New York.
4. Kwakernaak, H. and R. Sivan (1972). "Linear Optimal Control Systems," Wiley-Interscience, New York.
5. Hasdorff, L. (1976). "Gradient Optimization and Nonlinear Control," Wiley-Interscience, New York.

*Large-Scale System*
1. Mesarovic, M. D., D. Macko and Y. Takahara (1970). "Theory of Hierarchical Multilevel Systems," Academic Press, New York.
2. Singh, M. G. (1980). "Dynamical Hierarchical Control," North-Holland, Amsterdam.

*Nonequilibrium Thermodynamics*
1. Glausdorff, P. and I. Prigogine (1971). "Thermodynamic Theory of Structure, Stability and Fluctuations," Wiley-Interscience, New York.
2. Nicolis, G. and I. Prigogine (1977). "Self-Organization in Nonequilibrium Systems," Wiley-Interscience, New York.

# Index